Constructed Wetlands as a Sustainable Technology for Wastewater Treatment: Current Trends and Future Potential

Constructed Wetlands as a Sustainable Technology for Wastewater Treatment: Current Trends and Future Potential

Guest Editor

Zizhang Guo

MDPI

Basel • Beijing • Wuhan • Barcelona • Belgrade • Novi Sad • Cluj • Manchester

Guest Editor
Zizhang Guo
School of Environmental
Science and Engineering
Shandong University
Qingdao
China

Editorial Office
MDPI AG
Grosspeteranlage 5
4052 Basel, Switzerland

This is a reprint of the Special Issue, published open access by the journal *Water* (ISSN 2073-4441), freely accessible at: https://www.mdpi.com/journal/water/special_issues/A9K1A31VIK.

For citation purposes, cite each article independently as indicated on the article page online and as indicated below:

Lastname, A.A.; Lastname, B.B. Article Title. *Journal Name* **Year**, *Volume Number*, Page Range.

ISBN 978-3-7258-6220-7 (Hbk)
ISBN 978-3-7258-6221-4 (PDF)
https://doi.org/10.3390/books978-3-7258-6221-4

Contents

water

MDPI

Editorial

Constructed Wetlands as a Sustainable Technology for Wastewater Treatment: Current Trends and Future Potential

Zizhang Guo

Shandong Key Laboratory of Water Pollution Control and Resource Reuse, School of Environmental Science and Engineering, Shandong University, Qingdao 266237, China; guozizhang@sdu.edu.cn

1. Introduction to the Special Issue

The world is currently facing a dual challenge of water scarcity and pollution, exacerbated by the rapid development of the social economy and continuous improvements in living standards, which have heightened public concern regarding and the demand for water quality and safety [1]. Globally, approximately 359.4×10^9 m^3/year of wastewater is generated, with nearly half being discharged into the environment without adequate treatment [2]. Although conventional wastewater treatment plants (WWTPs) have been widely adopted for water pollution control in many countries, they are often associated with high energy and chemical consumption, limited efficiency in removing emerging contaminants and pathogens, significant greenhouse gas (GHG) emissions [3,4], and a fundamental disconnection from the natural ecosystems they aim to protect. Therefore, there is an urgent need for solutions that are both environmentally sustainable and economically viable.

Constructed wetlands (CWs), also referred to as treatment wetlands, emulate the synergistic interactions among plants, substrates, and microorganisms to efficiently remove pollutants in a controlled environment [5,6]. Recognized as environmentally sustainable, cost-effective, and efficient nature-based solutions, CWs have become increasingly important for wastewater treatment [7]. Similar to natural wetlands, CWs provide a range of ecological and societal benefits, such as water storage, purification, resource recovery, and carbon sequestration [8]. They contribute to urban water management by enhancing rainwater retention and infiltration, supporting climate change mitigation, restoring biodiversity, and serving as venues for environmental education, recreation, and ecotourism [9,10].

The purpose of this Special Issue is to address contemporary challenges in CWs, such as carbon emission reduction and the treatment of emerging contaminants. The collected contributions offer valuable insights derived from diverse research studies, with a focus on advancing innovative applications and enhancing the mechanistic understanding of CWs for effective pollution control and sustainable development.

2. Main Contributions of This Special Issue

This Special Issue comprises seven research articles and four review articles that focus on advancing the design of constructed wetlands, enhancing treatment performance, elucidating the removal and transformation mechanisms of emerging contaminants, and investigating geochemical processes within the context of carbon neutrality.

A comprehensive understanding and optimization of pollutant removal mechanisms in CWs are essential to improving treatment efficiency and ensuring long-term system stability. Nitrogen removal in CWs is primarily driven by microbially mediated nitrification and denitrification processes [contribution 1], whereas phosphorus dynamics are

 https://doi.org/10.3390/w17223295

largely regulated by the synergistic interactions between plant uptake and substrate adsorption. Long-term monitoring of estuarine CWs has demonstrated average removal efficiencies of 36.2% for total nitrogen (TN), 26.7% for total phosphorus (TP), and 30.7% for the permanganate index (COD_{Mn}). However, prolonged operation may lead to pollutant accumulation and substrate saturation, ultimately resulting in reduced treatment performance [contribution 2]. These findings enhance the mechanistic understanding of pollutant removal in CWs and provide critical data for optimizing system performance and sustaining long-term operational effectiveness.

Synergistic purification mediated by plants and microorganisms plays a pivotal role in enhancing the efficiency of pollutant removal in CWs. Root exudates released by submerged macrophytes have been shown to vary in response to environmental factors such as light intensity and nutrient availability, thereby influencing the composition and structure of microbial communities in both planktonic and biofilm phases [contribution 3]. These shifts in microbial community dynamics directly affect biofilm development and the rates of pollutant degradation. In *Vallisneria*-based wetland systems, optimal TN removal was observed at an air-to-water ratio of 15:1, which coincided with a marked enrichment of aerobic denitrifying bacteria within the biofilm [contribution 4]. Collectively, these findings highlight the essential contribution of plant–microbe interactions to the functional stability and resilience of wetland ecosystems, underscoring the potential of plant-mediated microbial regulation as a sustainable, nature-based strategy for water purification.

Design parameters and site-specific conditions play a critical role in determining the treatment efficiency and ecological functionality of CWs. The effectiveness of ecological buffer zones in intercepting non-point source pollution is jointly influenced by buffer width, vegetation composition, and slope gradient, with nitrogen and phosphorus removal rates reaching up to 90% under optimal configurations [contribution 5]. A study conducted in high-altitude regions of Ecuador demonstrated that surface flow CWs exhibit superior performance compared to vertical subsurface flow CWs in removing organic matter and microbial contaminants, which can be attributed to longer hydraulic retention times (HRTs) and the better adaptation of native plant species [contribution 6]. In contrast, wetlands dominated by low-density *Nelumbo nucifera* may contribute to increased nitrogen accumulation. To enhance denitrification, this study recommends expanding stands of *Phragmites australis* [contribution 7].

Research on CWs has progressively advanced beyond the conventional objective of pollutant removal, expanding to encompass broader evaluations of ecological impacts and system sustainability [contributions 8 and 9]. CW systems employed for swine wastewater treatment have demonstrated significant efficacy in removing suspended solids (SS) and nutrients. However, their operation necessitates strict control of organic loading rates to mitigate clogging risks and minimize emissions of greenhouse gases, particularly nitrous oxide (N_2O) and methane (CH_4) [contribution 10]. In response to these environmental challenges, innovative approaches, such as the integration of microbial fuel cells (MFCs) into CWs, are being investigated to enhance treatment performance while concurrently reducing carbon footprints [contribution 11]. Furthermore, CWs are increasingly acknowledged for their potential in mitigating non-point source pollution originating from agricultural runoff and livestock effluents, especially under adverse conditions such as high altitudes or in cold climates. Nonetheless, these systems continue to face critical challenges associated with fluctuating pollutant loads, climatic variability, and long-term operational reliability, underscoring the need for future designs that balance treatment efficiency with environmental sustainability.

The contributions presented in this Special Issue provide novel insights into the optimization of design, mechanistic understanding, and sustainable application of CWs under evolving environmental challenges. With ongoing research and practical implementation, CWs are expected to assume an increasingly significant role in pollution control, carbon mitigation, and urban ecological restoration, thereby offering substantial support for the development of water ecosystems that reflect the harmonious coexistence between humans and nature.

Conflicts of Interest: The authors declare no conflict of interest.

List of Contributions:

1. Dong, J.; Kuang, S. Bibliometric Analysis of Nitrogen Removal in Constructed Wetlands: Current Trends and Future Research Directions. *Water* **2024**, *16*, 1453. https://doi.org/10.3390/w16101453
2. Li, H.; Xu, Q.; Jiang, S.; Liu, Y.; Wang, R.; Xu, Y.; Feng, J.; Shen, J.; Wang, X. Decline in Water Treatment Efficiency of an Estuarine Constructed Wetland over Its Operating Years. *Water* **2025**, *17*, 352. https://doi.org/10.3390/w17030352
3. Shi, Y.; Zhang, X.; Zhao, M.; Zheng, X.; Gu, J.; Wang, Z.; Fan, C.; Gu, W. The Status of Research on the Root Exudates of Submerged Plants and Their Effects on Aquatic Organisms. *Water* **2024**, *16*, 1920. https://doi.org/10.3390/w16131920
4. Mao, H.; Lu, S.; Huang, G.; Jia, X.; Bao, C.; Yan, X.; Zheng, X.; Zhao, M.; Jin, Z. Impact of Gas-to-Water Ratio on Treatment Efficiency of Submerged-Macrophyte Constructed Wetland Systems. *Water* **2024**, *16*, 1790. https://doi.org/10.3390/w16131790
5. Wang, D.; Gao, X.; Wu, S.; Zhao, M.; Zheng, X.; Wang, Z.; Zhang, Y.; Fan, C. A Comprehensive Review on Ecological Buffer Zone for Pollutants Removal. *Water* **2024**, *16*, 2172. https://doi.org/10.3390/w16152172
6. Jerves-Cobo, R.; Maldonado, E.; Hidalgo-Cordero, J.F.; García-Herazo, H.; Mora-Serrano, D. Comparative Assessment of Wastewater Treatment Technologies for Pollutant Removal in High-Altitude Andean Sites. *Water* **2025**, *17*, 1800. https://doi.org/10.3390/w17121800
7. Liu, B.; Gao, Y.; Zhou, J.; Wang, Y.; He, J. Jinluo Low-Density Lotus Pond Wetland Water Purification Practice Experiment—A Case of Limited Efficacy. *Water* **2025**, *17*, 1985. https://doi.org/10.3390/w17131985
8. Xing, M.; Yan, D.; Hai, M.; Zhang, Y.; Zhang, Z.; Li, F. Arsenic Contamination in Sludge and Sediment and Relationship with Microbial Resistance Genes: Interactions and Remediation. *Water* **2024**, *16*, 3633. https://doi.org/10.3390/w16243633
9. Yan, J.; Yang, P.; Wang, B.; Wu, S.; Zhao, M.; Zheng, X.; Wang, Z.; Zhang, Y.; Fan, C. Green Roof Systems for Rainwater and Sewage Treatment. *Water* **2024**, *16*, 2090. https://doi.org/10.3390/w16152090
10. Gómez-Garrido, M.; Turbí, M.A.T.; El Bied, O.; Cano, Á.F. Impact of an Integral Management System with Constructed Wetlands in Pig Slurry Traceability and GHG/NH_3 Emissions. *Water* **2024**, *16*, 2351. https://doi.org/10.3390/w16162351
11. Qi, R.; Dong, J.; Kang, Y.; Xie, H.; Wu, H.; Hu, Z.; Guo, Z. Migration and Transformation of Greenhouse Gases in Constructed Wetlands: A Bibliometric Analysis and Trend Forecast. *Water* **2025**, *17*, 412. https://doi.org/10.3390/w17030412

References

1. Jones, E.R.; Bierkens, M.F.P.; van Vliet, M.T.H. Current and future global water scarcity intensifies when accounting for surface water quality. *Nat. Clim. Chang.* **2024**, *14*, 629–635. [CrossRef]
2. Wu, H.; Wang, R.; Yan, P.; Wu, S.; Chen, Z.; Zhao, Y.; Cheng, C.; Hu, Z.; Zhuang, L.; Guo, Z.; et al. Constructed wetlands for pollution control. *Nat. Rev. Earth Environ.* **2023**, *4*, 218–234. [CrossRef]
3. Zhu, L.; Wu, Q.; Gao, S.; Lin, Y.; Wu, Y. Short-chain PFAS predominate in large-scale lithium battery industrial parks, Eastern China: Source apportionment and downstream impact implications. *J. Hazard. Mater.* **2025**, *499*, 139964. [CrossRef] [PubMed]
4. He, M.; Xu, Z.; Hou, D.; Gao, B.; Cao, X.; Ok, Y.S.; Rinklebe, J.; Bolan, N.S.; Tsang, D.C.W. Waste-derived biochar for water pollution control and sustainable development. *Nat. Rev. Earth Environ.* **2022**, *3*, 444–460. [CrossRef]

5. Yue, J.; Hu, X.; Xie, H.; Sun, B.; Hu, Z.; Zhang, J. Organic matter-driven hydroxyl radical generation enhances emerging pollutant removal in plant rhizosphere of constructed wetlands: Coupled direct and indirect effects. *Water Res.* **2025**, *287*, 124336. [CrossRef] [PubMed]
6. Zhang, F.; Shen, C.; Zhao, Y.; Kong, L.; Ju, F.; Ji, B.; Man, Y.B.; Shan, S.; Li, Y.; Zhang, J. Biofilm-enhanced PFAS removal in constructed wetlands: Sewage sludge biochar drives adsorption and microbial synergy. *Chem. Eng. J.* **2025**, *522*, 167257. [CrossRef]
7. Lu, J.; Yu, P.; Zhang, J.; Guo, Z.; Li, Y.; Wang, S.; Hu, Z. Biotic/abiotic transformation mechanisms of phenanthrene in iron-rich constructed wetland under redox fluctuation. *Water Res.* **2024**, *261*, 122033. [CrossRef] [PubMed]
8. Zhao, Y.; Ji, B.; Liu, R.; Ren, B.; Wei, T. Constructed treatment wetland: Glance of development and future perspectives. *Water Cycle* **2020**, *1*, 104–112. [CrossRef]
9. Zheng, Z.; Duan, X.; Lu, S. The application research of rainwater wetland based on the Sponge City. *Sci. Total Environ.* **2021**, *771*, 144475. [CrossRef] [PubMed]
10. Stefanakis, A.I. The Role of Constructed Wetlands as Green Infrastructure for Sustainable Urban Water Management. *Sustainability* **2019**, *11*, 6981. [CrossRef]

water

MDPI

Article

Bibliometric Analysis of Nitrogen Removal in Constructed Wetlands: Current Trends and Future Research Directions

Jiahao Dong and Shaoping Kuang *

College of Environment and Safety Engineering, Qingdao University of Science and Technology, Qingdao 266042, China; dongjiahao624@126.com
* Correspondence: kuangshaoping@126.com

Abstract: Nitrogen pollution in water environments has reached critical levels globally, primarily stemming from agricultural runoff, industrial discharges, and untreated sewage. The excessive presence of nitrogen compounds poses a significant threat to water quality, leading to adverse impacts on ecosystems and human health. Reaching a breakthrough in the technology of constructed wetlands (CWs) for mitigating nitrogen pollution is hindered by existing knowledge gaps regarding the mechanisms involved in the removal process. Reaching this understanding, we offer a comprehensive summary of current advancements and theories in this research field. Initially, bibliometric techniques were employed to identify yearly patterns in publications and areas of research focus. Subsequently, the chosen documents underwent statistical analysis using VOSviewer_1.6.20 to determine countries' annual productivity, significant publication years, influential authors, keyword clustering analysis, and more. Finally, a comprehensive overview is provided on the elimination of nitrogen through CWs, encompassing insights into microbial communities and structure types. This analysis aims to uncover potential strategies for optimizing the rate of nitrogen removal. Furthermore, this study elucidates the current research trend concerning the nitrogen removal performance of CWs and identifies challenges and future research directions in this field.

Keywords: constructed wetlands; nitrogen removal; bibliometrics; VOSviewer; current trend

1. Introduction

Nitrogen stands as a linchpin element in aquatic ecosystems, driving essential ecological processes that shape the dynamics of these environments. The cycling of nitrogen, encompassing ammonium, nitrate, and nitrite, influences primary production, nutrient availability, and trophic interactions [1]. Ascertaining the significance of nitrogen in aquatic ecosystems is fundamental for advancing our comprehension of the intricate interplay of biogeochemical cycles.

However, the problem of nitrogen pollution in aquatic ecosystems is worsening due to human activities, primarily stemming from agricultural, urban, and industrial sources [2,3]. Agricultural runoff, laden with excess nitrogen from fertilizers and animal waste, represents a major contributor. Urban stormwater runoff, carrying dissolved inorganic nitrogen from impervious surfaces and atmospheric deposition, also significantly impacts water quality [4]. Industrial discharges containing nitrogen-based compounds further exacerbate the issue, introducing elevated nitrogen concentrations into aquatic environments. The ecological impacts of nitrogen pollution in water bodies are multifaceted and profound. Excessive nitrogen inputs often result in eutrophication and disrupt the nitrogen-fixing capabilities of certain symbiotic organisms, affecting nutrient cycling and overall ecosystem productivity [5]. Furthermore, nitrogen pollution can disrupt nitrogen cycling carried out by microbes. Nitrification and denitrification, crucial processes in nitrogen transformation, may experience imbalances, affecting the availability of nitrogen in the ecosystem [6,7]. This disturbance in nitrogen cycling has cascading effects on nutrient dynamics and can ultimately influence the overall productivity of aquatic ecosystems. Recovering and recycling nitrogen is encouraged in this context due to concerns about the stability of aquatic ecosystems [8].

There exist numerous effective technologies for the elimination of nitrogen. However, many of these options are associated with elevated expenses and energy requirements, intricate operational procedures, and excessive maintenance needs. Examples include activated sludge [9], photobioreactors [10], and membrane bioreactors [11]. Consequently, constructed wetlands (CWs) incorporating sorption filter media offer a comprehensive and cost-effective treatment solution by effectively removing various contaminants through precipitation, microbial activity, and plant absorption processes. CWs represent a proactive and innovative approach to ecological conservation and water management. In the realm of wastewater treatment, CWs play a crucial role in purifying contaminated water [12]. The intricate network of plants and microorganisms within these systems acts as a natural filter, breaking down pollutants and enhancing water quality [13]. This sustainable approach to water treatment not only helps mitigate pollution but also contributes to the preservation of aquatic ecosystems.

Currently, the primary method utilized in CWs for nitrogen removal is the process of phytoremediation. Wetland plants, particularly emergent and submerged species, play a crucial role in absorbing and assimilating nitrogen compounds, such as nitrate and ammonium, through their roots [14]. This biological uptake, facilitated by the specific metabolic activities of wetland vegetation, significantly contributes to reducing nitrogen levels in water. Additionally, the substrate composition of CWs, often designed with materials such as gravel, sand, and organic matter, fosters conditions conducive to denitrification [15]. Denitrifying bacteria found within the sediments of wetlands transform nitrate into nitrogen gas, thereby effectively eliminating nitrogen from the aquatic ecosystem [6,16]. The efficiency of nitrogen removal in CWs is influenced by several factors, including the hydraulic residence time (HRT), plant species selection, and overall design considerations [17]. Optimizing these parameters enhances the performance of CWs in reducing nitrogen concentrations in treated water. Therefore, there is a need for long-term monitoring to assess the sustained efficiency of CWs and the optimization of design parameters for varying environmental conditions. Continued research and technological advancements are crucial to refining the performance of CWs and ensuring their role as effective tools in the ongoing efforts to combat nitrogen pollution in water bodies. Thus, the current situation and research frontiers in nitrogen removal technology for constructed wetlands need to be examined, considering advancements in theoretical knowledge and new domains.

Bibliometrics is employed for analyzing the research status and development trend of a specific field of study and can be utilized to recognize and establish connections between crucial elements within a certain subject [18,19]. For example, Colares et al. conducted a bibliometric analysis to investigate the key factors influencing the performance and feasibility of floating treatment wetlands (FTWs) in terms of design and operational conditions. Through bibliometric mapping, the authors observed correlations between HRT, water depth, and phosphorus removal efficiency in these systems [20]. In summary, bibliometrics serves as a valuable tool for understanding the dynamics of nitrogen removal research in CWs. Its applications contribute to the identification of trends, assessment of research impact, recognition of key contributors, and the overall advancement of knowledge in this specialized environmental domain [21].

Considering the aforementioned circumstances, a textual corpus was established by conducting a search via the Web of Science. It cannot be denied that there is currently insufficient discussion regarding controlled operational parameters for nitrogen removal using CWs, both in laboratory and field studies. Additionally, there is a dearth of comprehensive mechanistic studies pertaining to the elimination process. The primary aim of this study is to identify key knowledge gaps that need filling, based on the current understanding of conclusions regarding the effectiveness of CWs in nitrogen management. This study aims to (1) summarize bibliometric data on nitrogen removal in CWs including publication title, author information, affiliations, and field of study; (2) analyze research keywords within the text corpus to predict trends for better domain support; (3) monitor research progress on nitrogen removal by CWs; and (4) evaluate the potential nitrogen removal capacity of CWs and identify challenges and future research directions in the nitrogen removal of CWs.

2. Materials and Methods

2.1. Bibliometric Data Sources

In order to assess the current status of research, a bibliometric analysis was conducted using data from the Web of Science platform. A total of 4557 scholarly articles on the subject of "constructed wetlands" and "nitrogen" were retrieved for examination. The search results included various details about each document such as author(s), title, source (journal title), language, document type, author keywords, addresses, cited reference count, times cited, publisher information, page count, ISSN, and subject category. Complete records were downloaded for further investigation. Contributions from different countries and institutions were estimated based on the location affiliation of at least one author mentioned in each published paper [22]. The analysis focused on the examination of articles published in the period between 2008 and 2024 that revolve around the topics of "constructed wetlands" and "nitrogen". Various factors were taken into consideration, including document type, language of publication, characteristics of publication output (such as authors involved, the average number of authors per article, cited references count, average number of references per article, and page count), patterns of publications (percentage distribution of articles across different categories, impact factor, and subject category classification for journals and their respective positions within those categories), as well as research interests related to wetlands determined through author keyword analysis and analyses based on title words and Key Words Plus.

2.2. Bibliometric Methodology

Bibliometric mapping was conducted using VOSviewer_1.6.20 software, following the methodology of De Souza et al. and considering all research records and document types throughout the entire period [23]. The 2008–2023 period was investigated because the most advanced research on nitrogen removal using CWs was performed during this period. Hence, the complete period was regarded as a reflection of the progress in research within this field from its inception to the current era. The maps generated in this study were network visualizations, with labels and circles representing items based on their importance. Lines indicate links between items, and the distance reflects the strength of their connection [24]. Previous studies have also demonstrated the high efficiency of VOSviewer in bibliometric analysis [19]. Bibliometric mapping data were used for literature research to better understand the highlighted topics and their connections and investigate recent scientific advances related to the study subject.

3. Results and Discussion

3.1. Research Area Analysis

By utilizing the search keywords "constructed wetlands" and "nitrogen", we acquired data from various sources on the Web of Science platform, including articles, reviews, and more. The research direction is primarily divided into ten distinct fields, as can be observed from Figure 1. These categories are environmental sciences (70.4%), environmental engineering (33.3%), water resources (22.1%), ecology (16.5%), chemical engineering (8.9%), biotechnology applied microbiology (8.4%), agricultural engineering (6.4%), energy fuels (6.1%), green sustainable science technology (3.8%), and marine freshwater biology (2.6%). It is demonstrated that the use of CW for nitrogen removal is an important water treatment and environmental protection technology. This could be because the nitrogen removal capacity of CWs directly contributes to water treatment objectives. By mitigating nitrogen pollution, these wetlands enhance the quality of water bodies, reducing the risk of eutrophication and maintaining ecological balance [25]. The natural processes within the wetland act as a biological filter, promoting the purification of water. Additionally, the synergy between CWs, nitrogen removal, water treatment, and environmental protection technology collectively contributes to enhanced environmental sustainability [26]. Therefore, CWs, as a sustainable environmental protection technology, play a crucial role in removing nitrogen from water.

Figure 1. The publications related to the search terms "constructed wetlands" and "nitrogen", obtained from the Web of Science, are categorized.

3.2. Publication Years and Authors Analysis

The keywords "constructed wetlands" and "nitrogen" were used to search for long-term publications. The nitrogen removal capacity of CWs was rarely studied before 2009, with subsequent fluctuations observed from 2009 to 2014 (Figure 2a). The number of articles published per year exhibited a marked increase in 2014–2021. The number of articles published per year increased from 99 in 2008 to 492 in 2021, spanning the period from 2008 to 2021, and the difference in cumulative trends was evident between 2008 and 2015 and between 2015 and 2021. Therefore, the curve fitting shows a high growth rate for articles published annually from 2014 to 2021. However, there is a downward trend in the number of articles published annually from 2021 to 2023. This can be attributed to the gradual maturation of conventional CW nitrogen removal technology and the absence of novel enhanced techniques and methodologies, resulting in stagnation in the advancement of this field. Additionally, the publication rate of relevant articles has witnessed a decline in the last two years as an outcome. Since the 1990s, global scientific exchange has facilitated the widespread adoption of CW technology as a globally preferred approach for addressing diverse wastewater-related issues [27]. Therefore, the nitrogen removal technology of CWs has attracted increasing attention and exploration from scientists, further promoting the growth in the number of related articles published each year.

A total of 11,217 authors were identified from the analysis of 4557 articles. As shown in Figure 2b, the author with the highest publication volume has published a total of 125 articles (2.7%) on nitrogen removal using CWs. The number of articles published by authors ranked second, third, and fourth is, respectively, 66 (1.5%), 65 (1.4%), and 61 (1.3%). The remaining authors ranked in the top fifteen have an average publication volume of around 50 articles. The researchers were likely experts in various academic areas beyond CWs, thanks to the interconnectedness of physical, chemical, and biological processes in complex research [13]. This necessitated a diverse and multidisciplinary knowledge base, requiring collaboration among researchers from different disciplines. Furthermore, the diverse interests of researchers may lead to a decentralization of their publications, potentially resulting in a lower number of articles on CW research [28]. This could help explain the variation in the number of publications among different researchers. We predict that an increased publication rate, as discussed above, will lead to more researchers being involved in studies on CWs. The analysis may be biased for authors who consistently use the same name or different names in their published works [29]. The increasing number of articles on CWs suggests a promising future with more papers and researchers.

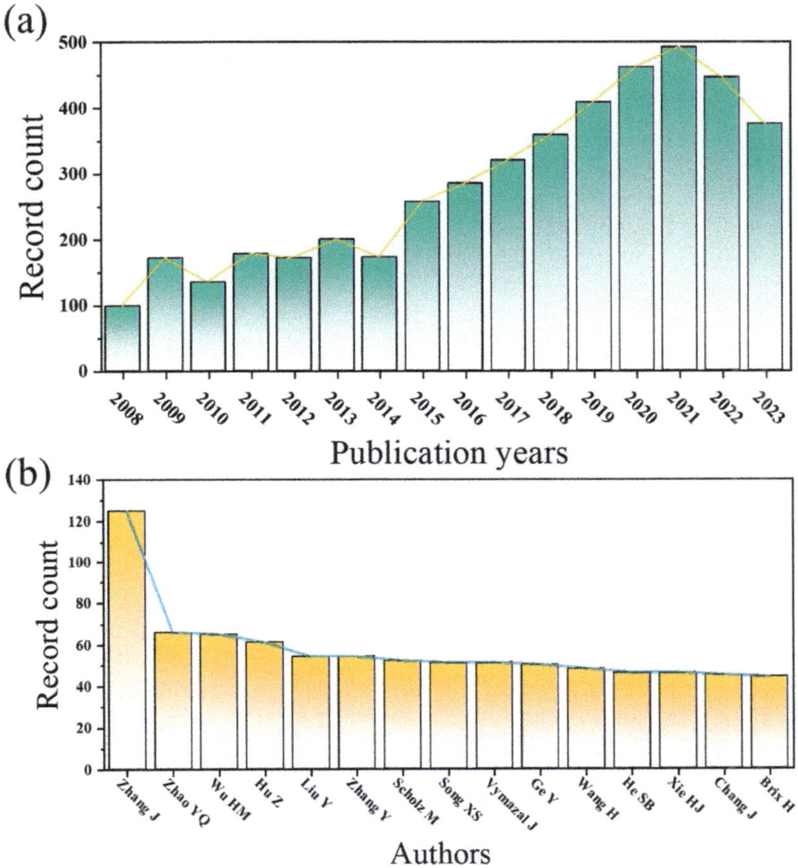

Figure 2. The years (**a**) and authors (**b**) of publications on the topic of "constructed wetlands" and "nitrogen", obtained from the Web of Science, were analyzed.

3.3. Publication Regions and Languages Analysis

The 4557 articles primarily originate from 10 different countries (Figure 3a), with 2125 being independent publications from a single country (53.4%). Over the past 16 years, the nitrogen removal capacity of CWs has been extensively researched by more than one-third of the world's countries. Notably, China, America, India, and Spain emerged as the dominant regions in terms of generating publications on CWs for nitrogen removal. Among these regions, it is noteworthy that developing countries exhibited the highest total number of relevant publications. One potential explanation is that CWs offer a cost-effective method for wastewater treatment and are particularly favored by developing nations [30]. The publication trend results for the ten most productive countries, namely China, USA, India, Spain, Australia, Canada, England, Brazil, Italy, and Denmark, revealed the predominant contribution of China in nitrogen removal capacity research conducted on CWs since 2008. The overall increase in the total number of articles published in these countries from 2014 to 2021 can be attributed to the rapid global development of nitrogen removal research in CWs. Meanwhile, slight fluctuations observed in the total article count in each country are likely due to publication delays resulting from the extensive processes involved in correlational research [31]. The rapid urbanization, industrialization, and accelerated economic development in China position it as a promising leader in the field of research papers on CWs in the coming years [32,33]. These factors have resulted in significant environmental challenges, rendering traditional energy-intensive sanitation systems inadequate and ineffective in meeting the country's demands. Consequently, there has been a promotion of wetland-related research.

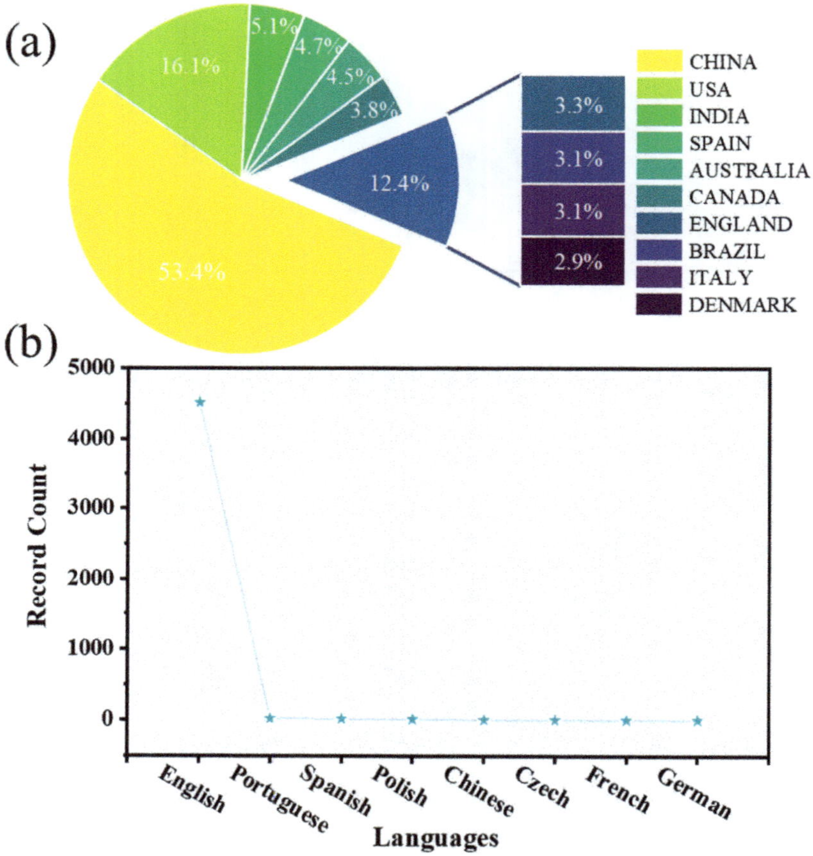

Figure 3. The regions (**a**) and languages (**b**) of publications pertaining to the search terms "constructed wetlands" and "nitrogen", as derived from data obtained through Web of Science.

The articles were published in eight languages (Figure 3b), with English being the predominant language for articles on nitrogen removal capacity in constructed wetlands, accounting for 99% of all publications. This can be attributed to the fact that SCI is an American-based database, and most ISI-listed journals are published in English [29]. The English language is also extensively utilized in international conferences and communications and continues to maintain its dominance across various academic disciplines [34]. However, during the period from 2008 to 2023, a total of 43 non-English articles pertaining to the nitrogen removal capacity of CWs were identified in the Web of Science database. These articles encompassed various languages, including Portuguese (16; 0.35%), Spanish (14; 0.31%), Polish (8; 0.17%), Chinese (2; 0.04%), Czech (1; 0.02%), French (1; 0.02%) and German (1; 0.02%).

3.4. Publication Keywords and Main Items Analysis

The analysis of keyword records involved 4557 articles from the Web of Science database. The utilization of bibliometric analysis through keywords, as demonstrated by Garfield [35], has proven successful in identifying future directions for scientific research. In recent years, the application of keyword-based bibliometric analysis has emerged as an effective approach to analyzing research trends and advancements [29]. It should be noted that highly important items may overlap in the bibliometric map. To improve clarity, we selected only the most relevant terms (with 60% or more relevance) using VOSviewer and generated a new visualization, as shown in Figure 4.

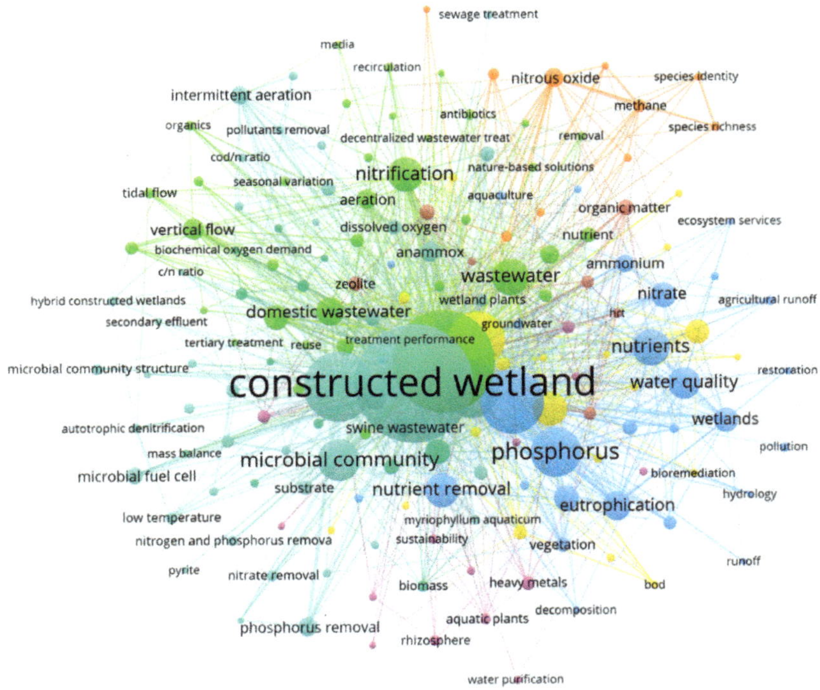

Figure 4. A bibliometric network map was generated in VOSviewer using the search terms "constructed wetlands" and "nitrogen", based on data obtained from the Web of Science.

The accuracy of the results was enhanced by selecting keywords that have a frequency of occurrence exceeding 15 times. Findings revealed that out of the total 8841 keywords, only 2.1% were utilized more than 15 times. The prevalence of low-frequency usage (less than 15 times) may be attributed to research discontinuity or a diverse research focus. Furthermore, the presence of various synonymous terms, spelling variations, and abbreviations might deviate from standardized or widely accepted conventions among researchers [36]. The keywords "constructed wetland", "wastewater" and "nutrient removal" had higher frequencies of use compared to others, denoted as 819, 433, and 130, respectively. This is because CWs play a crucial role in wastewater treatment, particularly in the context of nutrient removal. Wastewater, which contains various pollutants, including nutrients like nitrogen and phosphorus, can have detrimental effects on the environment if discharged untreated [37]. CWs act as natural filtration systems where aquatic plants and microorganisms help break down and absorb contaminants from the wastewater. Moreover, the intricate relationship between CWs, wastewater, and nutrient removal is facilitated by a range of biochemical and physical processes such as benthic bioturbation, substrate adsorption, and microbiological deterioration that collectively contribute to the effective purification of wastewater [38]. The wetland ecosystem harbors a diverse array of plants that serve as a conducive habitat for microorganisms actively engaged in the decomposition of organic matter. Moreover, the wetland substrate effectively facilitates nutrient removal through processes such as adsorption, precipitation, and microbial transformations [39]. This integrated approach results in the effective reduction in nutrient concentrations in the wastewater, making it environmentally safer before discharge. Therefore, the high frequency of usage is attributed to the correlation between these keywords.

The nitrogen removal mechanism in CWs was further investigated, elucidating the pivotal role of microorganism-mediated nitrification and denitrification in the chemoautotrophic process (Figure 5). The keyword nitrification was used 139 times in nitrogen removal in CWs. CWs are efficient in removing nitrogen through a process that involves both nitrification and denitrification. Nitrification is the conversion of ammonia (NH_4^-) to nitrate (NO_3^-), while denitrification is the reduction of nitrate to nitrogen gas (N_2), completing the nitrogen removal cycle [40]. Nitrification is the aerobic process where ammonia (NH_3) is oxidized to nitrite (NO_2^-) and then further to NO_3^-. This process is typically executed by two distinct groups of bacteria, namely ammonia-oxidizing bacteria (AOB) and nitrite-oxidizing bacteria (NOB) [41]. Denitrification is the anaerobic process where NO_3^- is reduced to N_2 or other gaseous nitrogen compounds. This process is carried out by denitrifying bacteria. The relationship between CWs and these nitrogen transformations lies in the controlled conditions provided by the wetland environment. CWs create a suitable habitat for both aerobic and anaerobic microorganisms, allowing for a sequential occurrence of nitrification and denitrification processes [42]. The subsurface flow constructed wetland facilitates the percolation of wastewater through the wetland matrix, thereby supplying oxygen to support surface-level nitrification [43]. As water moves deeper into the wetland, anaerobic conditions prevail, promoting denitrification. The intricate network of plant roots further enhances nitrogen removal by providing surfaces for microbial attachment and facilitating the transport of oxygen into the wetland matrix.

Therefore, in CWs designed for nitrogen removal, managers often optimize conditions to encourage both processes. This may involve controlling the water flow, maintaining suitable oxygen levels, and providing organic carbon sources to stimulate denitrification [44]. Balancing these factors within CWs of diverse structures ensures efficient removal of nitrogen from wastewater while minimizing the release of nitrogen compounds that could have negative environmental impacts. However, the structure of CWs varies due to the different design methods employed, resulting in significant disparities in achieved efficiency when balancing these factors. This integrated approach plays a crucial role in mitigating nitrogen pollution in wastewater and promoting a more sustainable water treatment process.

3.5. Current Research Hotspot Analysis

Based on a timeline of 4557 existing studies, we analyzed the latest research trends in nitrogen removal in CWs (Figure 6). Recently, the primary research focus has been directed toward microbial-enhanced nitrogen removal technology in CWs. The keyword was extensively discussed 276 times between 2020 and 2023.

Firstly, microorganisms such as *Nitrocellulose*, *Bacillu*, *Pseudomonas*, *Micrococcus*, and others play a crucial role in the biotransformation of nutrients within CWs. They facilitate processes like decomposition and enzymatic breakdown to convert complex nutrients such as dissolved organic matter into simpler forms [45]. This microbial activity not only helps in nutrient cycling but also promotes water quality by reducing the accumulation of nutrients. Secondly, microorganisms are crucial for nitrogen cycling in CWs (Figure 7). These microbes actively participate in nitrification, where ammonia is converted into nitrite and nitrate, and denitrification, where nitrate is transformed back into nitrogen gas [6]. This nitrogen removal process is vital for preventing water contamination and maintaining a balanced nitrogen cycle within the CWs. Additionally, microorganisms contribute to the breakdown of organic matter present in the CWs. The decomposition of organic material releases nitrogen in various forms, and microorganisms assist in converting these nitrogen compounds into more stable forms or facilitating their removal from the system [46]. In addition, microorganisms support the establishment of a diverse and stable ecosystem within CWs. They form symbiotic relationships with plants, promoting nutrient uptake and enhancing plant growth [47]. This synergy between microorganisms and vegetation contributes to the overall resilience and sustainability of the wetland ecosystem.

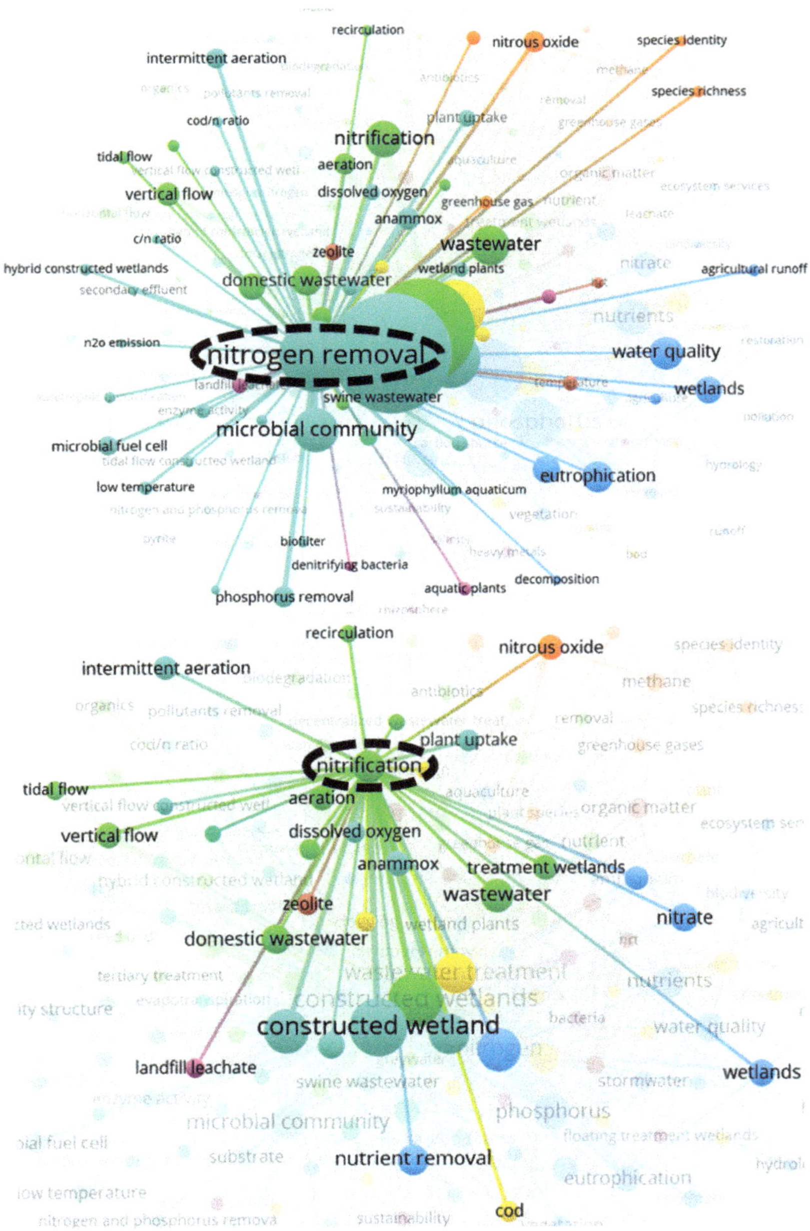

Figure 5. Highlighted terms associated with nitrogen removal were selected based on data obtained from the Web of Science.

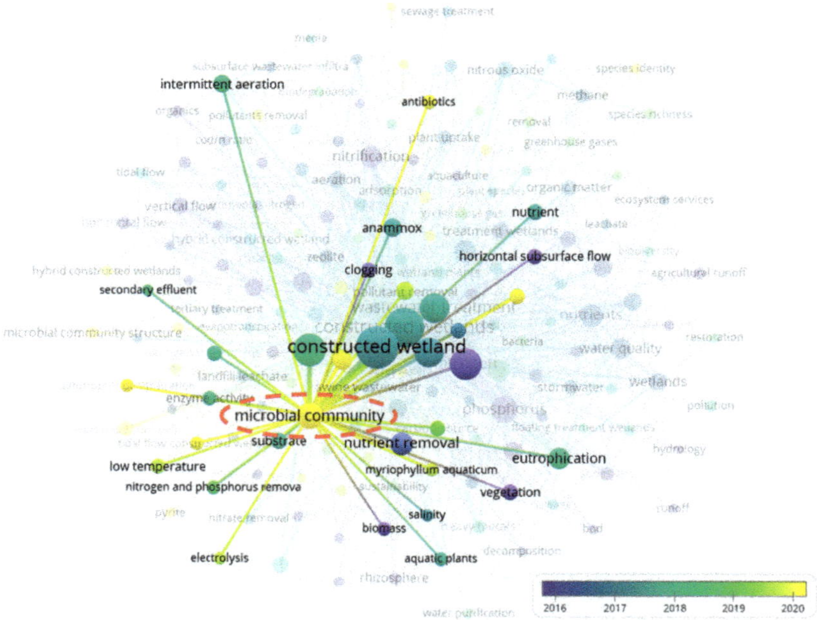

Figure 6. The highlighted terms related to "constructed wetlands" and "nitrogen" were selected based on timeline analysis using data from the Web of Science.

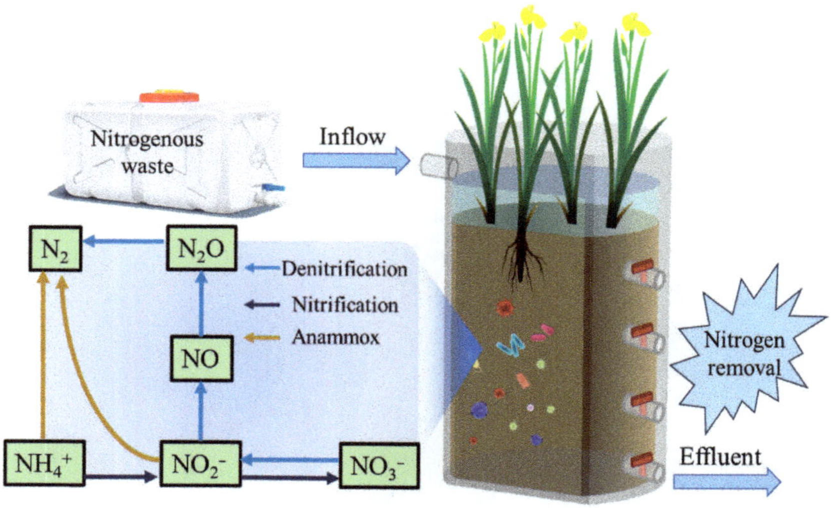

Figure 7. Conceptual model for the mechanisms of microbial nitrogen removal in CWs.

This provides evidence that the significance of microorganisms in CWs lies in their ability to mediate the transformation of nitrogen compounds, playing a pivotal role in the overall nitrogen removal process. Through nitrification and denitrification, these microorganisms significantly contribute to the overall effectiveness and ecological balance of these engineered ecosystems.

4. Knowledge Gaps and Future Studies

Limited studies have provided comprehensive insights into the genetic mechanisms underlying microbial nitrogen removal in CWs, while a significant gap still exists in linking theoretical frameworks to practical applications. Therefore, further investigations should focus on elucidating the migration and fate of nitrogen within CWs, with particular emphasis on understanding the biological effects involved, in order to unravel the intricate response mechanisms at play. Furthermore, researchers should remain vigilant regarding the residual nitrogen by-products that may persist on plant surfaces or be absorbed by plants for prolonged periods due to varying water flow impacts. The disruption of roots and leaves, decomposition of plants, or disturbances caused by aquatic animals could reintroduce these nitrogen compounds and their by-products into the CWs environment, which is critical for ensuring the sustainability of CWs. Based on this review, future research on nitrogen removal in CWs should prioritize the following areas:

(1). Efficient nitrogen removal materials should be further explored to continuously optimize the substrate composition in CWs, taking into account both economic viability and environmental sustainability.
(2). Researchers should investigate the use of genetic and ecological interventions for enhancing microbial populations' ability to perform nitrogen conversion processes, such as nitrification, denitrification, ammonification, and anaerobic ammonia oxidation, at higher rates and with greater efficiency.
(3). The advancement of bioscience control technologies in CWs, encompassing aquatic organisms, plants, biofilms, and microbial techniques, still encounters a substantial knowledge gap. Future research is anticipated to prioritize the investigation of the interplay between living organisms and nitrogen capture.
(4). The integration of advanced sensors and real-time monitoring systems is essential for advancing microbial engineering in CWs. It enables continuous surveillance of microbial activity, environmental conditions, and nitrogen levels, facilitating the precise control and optimization of nitrogen removal performance. Smart data-driven systems can revolutionize this field by improving adaptability to changing nitrogen loading conditions.

The complex nitrogen removal process in CWs should be comprehensively analyzed through model development and software fitting when sufficient data are available. It will be necessary for future research to establish emission limit standards and toxicological limit standards for nitrogen, aiming to provide better guidance for environmental management practices.

5. Conclusions

We conducted a comprehensive review encompassing 4557 publications sourced from the Web of Science database, elucidating the removal of nitrogen using CWs. This study encompasses pertinent countries and organizations, publication years, authors, and sources, as well as clustering and co-occurrence analysis of keywords. Furthermore, we examined the parameters and mechanisms influencing nitrogen removal in CWs. Overall, our survey provides an extensive knowledge base for potential collaborators or researchers in this field. Additionally, our research underscores the urgent need to enhance legislative and policy frameworks to mitigate nitrogen pollution at its source by addressing future threats. Extensive in-depth research is imperative to focus on developing controlled conditions for CWs that are cleaner and offer potential methods for efficient nitrogen removal.

Author Contributions: Conceptualization, J.D.; methodology, J.D.; software, J.D.; validation, Jiahao Don; investigation, J.D.; data curation, J.D.; writing—original draft preparation, J.D.; writing—review and editing, S.K.; supervision, S.K.; project administration, S.K.; funding acquisition, S.K. All authors have read and agreed to the published version of the manuscript.

Funding: This work was supported by the Natural Science Foundation of Shandong Province (No. ZR202102280181), the National Natural Science Foundation of China (No. 52200196 and 52270158), and the Major Innovation Project of Shandong Province (No. 2021CXGC011206).

Data Availability Statement: The data presented in this study are available on request from the corresponding author.

Conflicts of Interest: The authors declare that they have no known competing financial interests or personal relationships that could have appeared to influence the work reported in this paper. The raw data supporting the conclusions of this article will be made available by the authors on request.

References

1. Vymazal, J. The use of hybrid constructed wetlands for wastewater treatment with special attention to nitrogen removal: A review of a recent development. *Water Res.* **2013**, *47*, 4795–4811. [CrossRef] [PubMed]
2. Li, J.; Zheng, B.; Chen, X.; Li, Z.; Xia, Q.; Wang, H.; Yang, Y.; Zhou, Y.; Yang, H. The use of constructed wetland for mitigating nitrogen and phosphorus from agricultural runoff: A review. *Water* **2021**, *13*, 476. [CrossRef]
3. Keffala, C.; Ghrabi, A. Nitrogen and bacterial removal in constructed wetlands treating domestic waste water. *Desalination* **2005**, *185*, 383–389. [CrossRef]
4. Troitsky, B.; Zhu, D.Z.; Loewen, M.; van Duin, B.; Mahmood, K. Nutrient processes and modeling in urban stormwater ponds and constructed wetlands. *Can. Water Resour. J.* **2019**, *44*, 230–247. [CrossRef]
5. Hu, X.; Yue, J.; Yao, D.; Zhang, X.; Li, Y.; Hu, Z.; Liang, S.; Wu, H.; Xie, H.; Zhang, J. Plant development alters the nitrogen cycle in subsurface flow constructed wetlands: Implications to the strategies for intensified treatment performance. *Water Res.* **2023**, *246*, 120750. [CrossRef] [PubMed]
6. Tang, S.; Liao, Y.; Xu, Y.; Dang, Z.; Zhu, X.; Ji, G. Microbial coupling mechanisms of nitrogen removal in constructed wetlands: A review. *Bioresour. Technol.* **2020**, *314*, 123759. [CrossRef] [PubMed]
7. Fu, G.; Wu, J.; Han, J.; Zhao, L.; Chan, G.; Leong, K. Effects of substrate type on denitrification efficiency and microbial community structure in constructed wetlands. *Bioresour. Technol.* **2020**, *307*, 123222. [CrossRef] [PubMed]
8. Spiller, M.; Moretti, M.; De Paepe, J.; Vlaeminck, S.E. Environmental and economic sustainability of the nitrogen recovery paradigm: Evidence from a structured literature review. *Resour. Conserv. Recycl.* **2022**, *184*, 106406. [CrossRef]
9. Xiang, S.; Liu, Y.; Zhang, G.; Ruan, R.; Wang, Y.; Wu, X.; Zheng, H.; Zhang, Q.; Cao, L. New progress of ammonia recovery during ammonia nitrogen removal from various wastewaters. *World J. Microbiol. Biotechnol.* **2020**, *36*, 144. [CrossRef]
10. Luo, Y.; Le-Clech, P.; Henderson, R.K. Assessing the performance of membrane photobioreactors (MPBR) for polishing effluents containing different types of nitrogen. *Algal Res.* **2020**, *50*, 102013. [CrossRef]
11. Mao, X.; Myavagh, P.H.; Lotfikatouli, S.; Hsiao, B.S.; Walker, H.W. Membrane bioreactors for nitrogen removal from wastewater: A review. *J. Environ. Eng.* **2020**, *146*, 03120002. [CrossRef]
12. Kang, Y.; Zhang, J.; Xie, H.; Guo, Z.; Ngo, H.H.; Guo, W.; Liang, S. Enhanced nutrient removal and mechanisms study in benthic fauna added surface-flow constructed wetlands: The role of Tubifex tubifex. *Bioresour. Technol.* **2017**, *224*, 157–165. [CrossRef] [PubMed]
13. Vymazal, J.; Zhao, Y.; Mander, Ü. Recent research challenges in constructed wetlands for wastewater treatment: A review. *Ecol. Eng.* **2021**, *169*, 106318. [CrossRef]
14. Yao, D.; Dai, N.; Hu, X.; Cheng, C.; Xie, H.; Hu, Z.; Liang, S.; Zhang, J. New insights into the effects of wetland plants on nitrogen removal pathways in constructed wetlands with low C/N ratio wastewater: Contribution of partial denitrification-anammox. *Water Res.* **2023**, *243*, 120277. [CrossRef]
15. Xu, R.; Zhang, Y.; Liu, R.; Cao, Y.; Wang, G.; Ji, L.; Xu, Y. Effects of different substrates on nitrogen and phosphorus removal in horizontal subsurface flow constructed wetlands. *Environ. Sci. Pollut. Res.* **2019**, *26*, 16229–16238. [CrossRef] [PubMed]
16. Li, X.; Guo, Q.; Wang, Y.; Xu, J.; Wei, Q.; Chen, L.; Liao, L. Enhancing nitrogen and phosphorus removal by applying effective microorganisms to constructed wetlands. *Water* **2020**, *12*, 2443. [CrossRef]
17. Qi, R.; Zhang, L.; Yang, F.; Yan, C.-Z. Effect of hydraulic residence time on removal efficiency of pollutants in subsurface flow constructed wetlands and analysis of denitrification mechanism. *Huan Jing Ke Xue Huanjing Kexue* **2021**, *42*, 4296–4303. [PubMed]
18. Tan, H.; Li, J.; He, M.; Li, J.; Zhi, D.; Qin, F.; Zhang, C. Global evolution of research on green energy and environmental technologies: A bibliometric study. *J. Environ. Manag.* **2021**, *297*, 113382. [CrossRef] [PubMed]
19. Dell'Osbel, N.; Colares, G.S.; de Oliveira, G.A.; de Souza, M.P.; Barbosa, C.V.; Machado, Ê.L. Bibliometric analysis of phosphorous removal through constructed wetlands. *Water Air Soil Pollut.* **2020**, *231*, 117. [CrossRef]
20. Colares, G.S.; Dell'Osbel, N.; Wiesel, P.G.; Oliveira, G.A.; Lemos, P.H.Z.; da Silva, F.P.; Lutterbeck, C.A.; Kist, L.T.; Machado, Ê.L. Floating treatment wetlands: A review and bibliometric analysis. *Sci. Total Environ.* **2020**, *714*, 136776. [CrossRef]
21. Alsharif, A.H.; Salleh, N.; Baharun, R. Bibliometric analysis. *J. Theor. Appl. Inf. Technol.* **2020**, *98*, 2948–2962.
22. Zhi, W.; Ji, G. Constructed wetlands, 1991–2011: A review of research development, current trends, and future directions. *Sci. Total Environ.* **2012**, *441*, 19–27. [CrossRef] [PubMed]
23. de Souza, M.P.; Hoeltz, M.; Gressler, P.D.; Benitez, L.B.; Schneider, R.C. Potential of microalgal bioproducts: General perspectives and main challenges. *Waste Biomass Valoriz.* **2019**, *10*, 2139–2156. [CrossRef]

24. Van Eck, N.; Waltman, L. Manual for VOSviwer version 1.6. 10. In *CWTS Meaningful Metrics*; Universiteit Leiden: Leiden, The Netherlands, 2019.
25. Vymazal, J. Removal of nutrients in various types of constructed wetlands. *Sci. Total Environ.* **2007**, *380*, 48–65. [CrossRef]
26. Nyer, S.C.; Volkenborn, N.; Aller, R.C.; Graffam, M.; Zhu, Q.; Price, R.E. Nitrogen transformations in constructed wetlands: A closer look at plant-soil interactions using chemical imaging. *Sci. Total Environ.* **2022**, *816*, 151560. [CrossRef]
27. Vymazal, J. Constructed wetlands for wastewater treatment: Five decades of experience. *Environ. Sci. Technol.* **2011**, *45*, 61–69. [CrossRef]
28. Farooq, R. A review of knowledge management research in the past three decades: A bibliometric analysis. *VINE J. Inf. Knowl. Manag. Syst.* **2024**, *54*, 339–378. [CrossRef]
29. Chiu, W.-T.; Ho, Y.-S. Bibliometric analysis of tsunami research. *Scientometrics* **2007**, *73*, 3–17. [CrossRef]
30. Bastian, R.; Hammer, D. The use of constructed wetlands for wastewater treatment and recycling. In *Constructed Wetlands for Water Quality Improvement*; CRC Press: Boca Raton, FL, USA, 2020; pp. 59–68.
31. Vymazal, J. The use of constructed wetlands for nitrogen removal from agricultural drainage: A review. *Sci. Agric. Bohem.* **2017**, *48*, 82–91. [CrossRef]
32. Wei, D.; Singh, R.P.; Li, Y.; Fu, D. Nitrogen removal efficiency of surface flow constructed wetland for treating slightly polluted river water. *Environ. Sci. Pollut. Res.* **2020**, *27*, 24902–24913. [CrossRef]
33. Zhang, H.; Tang, W.; Wang, W.; Yin, W.; Liu, H.; Ma, X.; Zhou, Y.; Lei, P.; Wei, D.; Zhang, L. A review on China's constructed wetlands in recent three decades: Application and practice. *J. Environ. Sci.* **2021**, *104*, 53–68. [CrossRef] [PubMed]
34. Zhang, G.; Xie, S.; Ho, Y.-S. A bibliometric analysis of world volatile organic compounds research trends. *Scientometrics* **2010**, *83*, 477–492. [CrossRef]
35. Garfield, E. KeyWords Plus-ISI's breakthrough retrieval method. 1. Expanding your searching power on current-contents on diskette. *Curr. Contents* **1990**, *32*, 5–9.
36. Li, J.; Zhang, Y.; Wang, X.; Ho, Y.-S. Bibliometric analysis of atmospheric simulation trends in meteorology and atmospheric science journals. *Croat. Chem. Acta* **2009**, *82*, 695–705. [CrossRef]
37. Lin, Y.-F.; Jing, S.-R.; Lee, D.-Y.; Wang, T.-W. Nutrient removal from aquaculture wastewater using a constructed wetlands system. *Aquaculture* **2002**, *209*, 169–184. [CrossRef]
38. Vymazal, J. Removal of nutrients in constructed wetlands for wastewater treatment through plant harvesting–Biomass and load matter the most. *Ecol. Eng.* **2020**, *155*, 105962. [CrossRef]
39. Wang, H.X.; Xu, J.L.; Sheng, L.X.; Liu, X.J. A review of research on substrate materials for constructed wetlands. In *Materials Science Forum*; Trans Tech Publications Ltd.: Zurich, Switzerland, 2018; pp. 917–929.
40. Tao, W.; Wen, J.; Han, Y.; Huchzermeier, M.P. Nitrogen removal in constructed wetlands using nitritation/anammox and nitrification/denitrification: Effects of influent nitrogen concentration. *Water Environ. Res.* **2012**, *84*, 2099–2105. [CrossRef] [PubMed]
41. Fu, G.; Yu, T.; Ning, K.; Guo, Z.; Wong, M.-H. Effects of nitrogen removal microbes and partial nitrification-denitrification in the integrated vertical-flow constructed wetland. *Ecol. Eng.* **2016**, *95*, 83–89. [CrossRef]
42. Platzer, C. Design recommendations for subsurface flow constructed wetlands for nitrification and denitrification. *Water Sci. Technol.* **1999**, *40*, 257–263. [CrossRef]
43. Butterworth, E.; Richards, A.; Jones, M.; Dotro, G.; Jefferson, B. Assessing the potential for tertiary nitrification in sub-surface flow constructed wetlands. *Environ. Technol. Rev.* **2016**, *5*, 68–77. [CrossRef]
44. Wu, T.; Yang, S.-S.; Zhong, L.; Pang, J.-W.; Zhang, L.; Xia, X.-F.; Yang, F.; Xie, G.-J.; Liu, B.-F.; Ren, N.-Q. Simultaneous nitrification, denitrification and phosphorus removal: What have we done so far and how do we need to do in the future? *Sci. Total Environ.* **2023**, *856*, 158977. [CrossRef] [PubMed]
45. Wang, J.; Long, Y.; Yu, G.; Wang, G.; Zhou, Z.; Li, P.; Zhang, Y.; Yang, K.; Wang, S. A review on microorganisms in constructed wetlands for typical pollutant removal: Species, function, and diversity. *Front. Microbiol.* **2022**, *13*, 845725. [CrossRef] [PubMed]
46. Xu, W.; Yang, B.; Wang, H.; Wang, S.; Jiao, K.; Zhang, C.; Li, F.; Wang, H. Improving the removal efficiency of nitrogen and organics in vertical-flow constructed wetlands: The correlation of substrate, aeration and microbial activity. *Environ. Sci. Pollut. Res.* **2023**, *30*, 21683–21693. [CrossRef] [PubMed]
47. Hu, X.; Xie, J.; Xie, H.; Huo, J.; Wu, H.; Hu, Z.; Xue, K.; Song, M.; Liang, S.; Zhang, J. Towards a better and more complete understanding of microbial nitrogen transformation processes in the rhizosphere of subsurface flow constructed wetlands: Effect of plant root activities. *Chem. Eng. J.* **2023**, *463*, 142455. [CrossRef]

water

MDPI

Article

Decline in Water Treatment Efficiency of an Estuarine Constructed Wetland over Its Operating Years

Huaqing Li [1], Qian Xu [2,3], Shiyi Jiang [2,4,*], Yanping Liu [2,3,5], Ronghui Wang [2,3], Yong Xu [2,3], Jimeng Feng [2,3,5,*], Jian Shen [2,3,5] and Xinze Wang [2,3,5]

1 The Fifth Engineering Co., Ltd. of China Railway 10th Bureau Group, Suzhou 215000, China; ztsjwgsr@163.com
2 Shanghai Jiao Tong University Yunnan Dali Research Institute, Dali 671000, China; sjtuyanjiuyuan@163.com (Q.X.); yplex1230@163.com (Y.L.); 15125285426@163.com (R.W.); 18508783598@163.com (Y.X.); sjlnts@sjtu.edu.cn (J.S.); xinzewang@sjtu.edu.cn (X.W.)
3 National Observation and Research Station of Erhai Lake Ecosystem in Yunnan, Dali 671000, China
4 Yunnan Zhaohong Environmental Engineering Co., Ltd., Dali 671000, China
5 School of Environmental Science and Engineering, Shanghai Jiao Tong University, Shanghai 200240, China
* Correspondence: jsy_18213452479@163.com (S.J.); fjm520@sjtu.edu.cn (J.F.)

Abstract: Estuarine constructed wetlands (ECWs) play a role as ecological barriers in the control of external pollution in lakes. Usually, ECWs show reduced water treatment efficiency after many years of operation compared to their initial performance. However, it is unclear how the water purification efficiency of an ECW changes over time. After over a decade of tracking analysis on an ECW, this study found that it indeed played a significant role in achieving water quality improvement effects. The average removal rates for total nitrogen (TN) and total phosphorus (TP) and the permanganate index (COD_{Mn}) were 36.2%, 26.7%, and 30.7%, respectively, with annual reductions of 1.6 t/a, 20.8 t/a, and 44.6 t/a. The surface hydraulic load is a critical indicator for the design and operational management of ECWs. The reduction loads of TP, TN, and COD_{Mn} increased with the rise in surface hydraulic load, indicating that this ECW project had certain advantages in treating large-volume water bodies. However, when strict COD_{Mn} treatment is needed, the surface hydraulic load should be reduced. During the high-efficiency period (2010–2015), the treatment effects on TN and TP were more than twice those during the degradation period (2016–2021), and the effect on COD_{Mn} was about 1.5 times greater. With increased operation years, the TN removal rate declined most rapidly due to pollutant accumulation and sediment release.

Keywords: estuarine constructed wetland; surface hydraulic load; removal rate

1. Introduction

Most pollutants from sources such as domestic sewage, industrial wastewater, and agricultural tailwater flow into lakes through rivers or surface runoff, causing a series of problems such as lake eutrophication and aquatic organism toxicity [1,2]. Therefore, controlling external pollution is particularly important for lake protection and management. Establishing constructed wetlands at the entrance of lakes is one of the key measures to reduce the pollution load entering the lake. While numerous studies have examined the impact of various factors on pollutant removal in wetlands, there is limited evidence of the long-term variations in pollutant removal efficiency, especially over extended periods. Initially, wetlands are expected to exhibit a more stable depuration capacity shortly after establishment due to the clean wetland sediment, reasonable aquatic plant configuration, and good hydraulic management [3,4]. However, this stability is threatened by substrate saturation, wetland clogging, decreased pollutant uptake by senescent plants, and substantial deposition and release of pollutants during long-term operation [5]. Therefore, it is essential to estimate the ability of wetlands to purify polluted wastewater under various internal and external environmental conditions overlong time scales to address the challenges posed by water quality standards [6].

Estuarine constructed wetlands (ECWs) play a role as ecological barriers in the control of external pollution in lakes [7]. Although the treatment effect is not as high as that of sewage treatment plants, they can effectively intercept pollutants entering the lake, restore the functions of aquatic and terrestrial ecosystems, maintain the stability of lake water quality, and enhance the ecological landscape of lake bays [8,9]. Although ECWs, as an open wetland type, do not face the issues of clogging and substrate adsorption saturation that are relevant for many subsurface flow wetlands, they still encounter numerous challenges. Anthropogenic changes and alterations in hydrological conditions pose a risk of degradation to estuarine wetlands [10]. Due to the deposition of pollutants, the death of aquatic plants, or other irregular management measures, constructed wetlands often exhibit a poorer water treatment efficiency after a certain number of years of operation compared to their initial performance [11]. However, due to the lack of continuous tracking and evaluation survey data, it is generally unclear how the water purification efficiency of ECWs changes with an increasing operational duration. In addition, changes in upstream water quality and load have a significant impact on the treatment efficiency of ECWs. Although various countries and regions have design specifications for the load of constructed wetlands, the actual intensity of the received pollution load may not be consistent with expectations, such as encountering an impactful water volume and pollution load during the main flood season. Government management departments have also noticed the decline in ECW efficiency, but existing knowledge makes it difficult to determine when and what measures to take to address it.

This study tracked the water quality and pollution load of an ECW over more than 10 years following its initial construction, analyzed the interannual and intra-annual variations in the wetland's operational status, investigated the reasons for the decline in water purification efficiency, and proposed some improvement suggestions, providing valuable references for the operation and management of similar estuarine wetlands.

2. Materials and Methods

2.1. Overview of Study Area

The ECW involved in this study is located in the Erhai Lake basin of Dali, Yunnan Province, China and is named the Luoshi River Estuarine Constructed Wetland. Its latitude and longitude are 25.950657° N, 100.101517° E, and its altitude is about 1966 m. The Erhai Lake, covering an area of 252 km^2 and with a catchment area of 2565 km^2, stands as a significant freshwater lake in China and the second-largest plateau lake in Yunnan Province [12]. Erhai Lake not only provides water for local people and tourist resources but also delivers many other ecosystem services to human beings, e.g., biodiversity maintenance. The northern part of the Erhai Lake is primarily a surface runoff generation area, accounting for approximately 60% of the total runoff that flows into the lake [13]. Additionally, there is a large rural population and extensive farmland in this region, placing significant pressure on pollution prevention and control. Luoshi River is one of the three main rivers in the northern part of Erhai Lake, accounting for about 15% of the water volume entering the lake.

Since the first large-scale blue–green algal bloom broke out in Erhai Lake in 2003, the water quality of Erhai Lake has failed to meet the Class II standard for surface water

(TN ≤ 0.5 mg/L, TP ≤ 0.025 mg/L) required by the Chinese government. Considering that the nitrogen and phosphorus levels in the rivers flowing into the lake are significantly higher than those in Erhai Lake, in order to improve the water quality entering the lake, in 2009, the government constructed the ECW at the entrance of Luoshi River, serving as an important water quality purification functional area before entering the lake. The effectiveness of its water quality purification has a significant impact on the overall water quality of Erhai Lake. The Luoshi River ECW covers a total area of 48.467 hm^2, with a water area of 44.467 hm^2. The designed treatment capacities vary in different seasons, namely the flood season, normal water level period, and drought period, which are 1.12 million m^3/d, 300,000 m^3/d, and 170,000 m^3/d, respectively. After the completion of the ECW, the area has transformed from a single farmland ecosystem that was seasonally flooded to an ecological environment suitable for the habitats and reproduction of more aquatic plants and animals, promoting biodiversity and optimizing the internal biological community structure of the Erhai Lake shore zone, ensuring ecosystem stability. The vegetation in the ECW is mostly artificially planted aquatic plants, including *Typha orientalis* C. Presl, *Nelumbo* sp., *Phragmites australis* (Cav.) Trin. ex Steud, *Ceratophyllum demersum* L., *Nymphaea* L., *Potamogeton maackianus*, *Myriophyllum spicatum*, and *Zizania caduciflora*. As a unique ecosystem, the completed Luoshi River ECW plays a bridging role between land and lake, playing a key role in intercepting pollution inputs within the catchment area and restoring the ecological function of the water–land interface. It is an important ecological purification functional area for improving the water quality of the lake, and its water purification effect has a significant impact on the overall water quality of Erhai Lake. The ECW at the estuary has been in operation for more than 10 years and is gradually facing the problem of aging.

2.2. Sampling Point Setting and Sampling Time

After conducting an on-site investigation and considering the characteristics of the river basin, a total of 8 monitoring and sampling points (Figure 1) were set up along the Luoshi River, labeled S1–S8. Among them, S7 is located upstream of the ECW section of the Luoshi River, while S8 is situated downstream of the ECW. From June 2010 to December 2021, water samples were regularly collected at these points at a monitoring frequency of twice a month. The sampling of wetland sediments was conducted in August 2022.

2.3. Analysis Indicators and Methods

Water quality monitoring indicators include total nitrogen (TN), total phosphorus (TP), and the permanganate index (COD_{Mn}), among which TN was measured according to HJ 636-2012 [14], TP was measured according to GB 11893-89 [15], and COD_{Mn} was measured according to GB 11892-89 [16]. Sediment monitoring indicators include the TN, TP, and organic matter (OM), among which TN was measured according to HJ 717-2014 [17], TP was measured according to HJ 632-2011 [18], and OM was measured according to NY/T 1121.6-2006 [19].

Sampling was conducted in accordance with HJ 494-2009 [20]. At least one standard sample was measured for each batch of samples. To ensure the precision of the test results, each indicator was measured three times and the average value was taken. To ensure method performance, at least two blank experiments were conducted for each batch of samples.

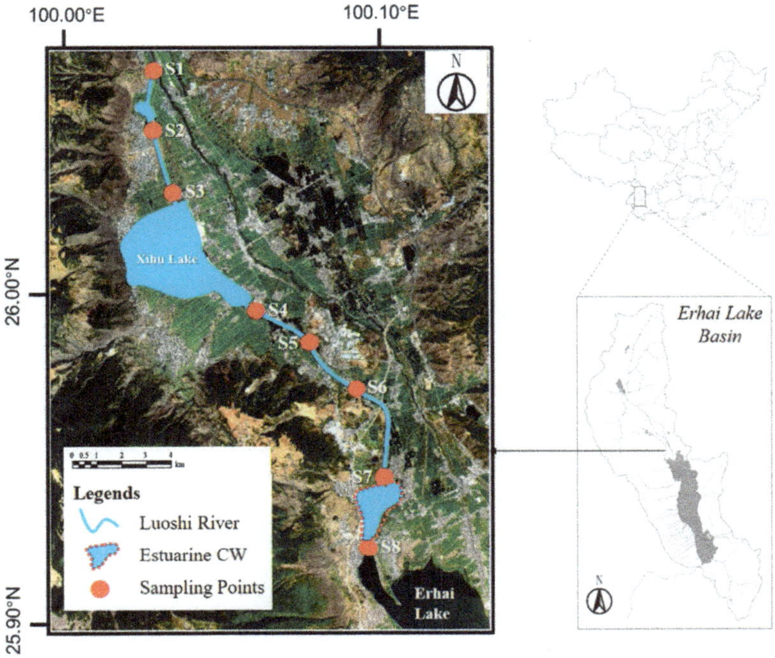

Figure 1. Sampling points and the studied ECW location.

2.4. Data Processing and Plotting

Data processing was conducted using Microsoft Excel 2019. Origin Pro 2021 was utilized for plotting and data analysis. The Simple Fit App in Origin Pro 2021 was used for linear fitting analysis. The values of the correlation coefficients, slopes, intercepts, and their standard errors were determined by the linear regression equation using the same program. The Kolmogorov–Smirnov normality test was used to analyze the reliability of the data.

3. Results and Discussion

3.1. Water Quality Changes Along the Luoshi River

Between June 2010 and December 2021, the average concentrations of TN, TP, and CODMn at eight sampling points on the Luoshi River ranged from 1.26 to 2.97 mg/L, 0.065 to 0.130 mg/L, and 3.72 to 6.06 mg/L, respectively. As shown in Figure 2, the water quality upstream of the Luoshi River (S1) was relatively good. Where the river flows through S2 and S3, several villages and some farmland are distributed around it, leading to an increase in TP, TN, and COD concentrations. However, after S3, the river enters a small lake, the Xihu Lake Wetland, where TP and TN concentrations improved due to sedimentation and biological utilization, while COD increased due to algal proliferation and algal-derived organic matter. Subsequently, as a large amount of mountain runoff and major towns flowed into the lake, both the water volume and pollutant concentrations rose again. Fortunately, after purification in the ECW, the concentrations of nitrogen and phosphorus that ultimately flowed into the Erhai Lake were better than those upstream, and the concentration of COD_{Mn} was similar to that upstream. Considering the gradual changes in water quantity from upstream to downstream, the difficulty of purifying water quality also increased accordingly. Therefore, ECW indeed played a significant role in achieving water quality improvement effects [8,21,22].

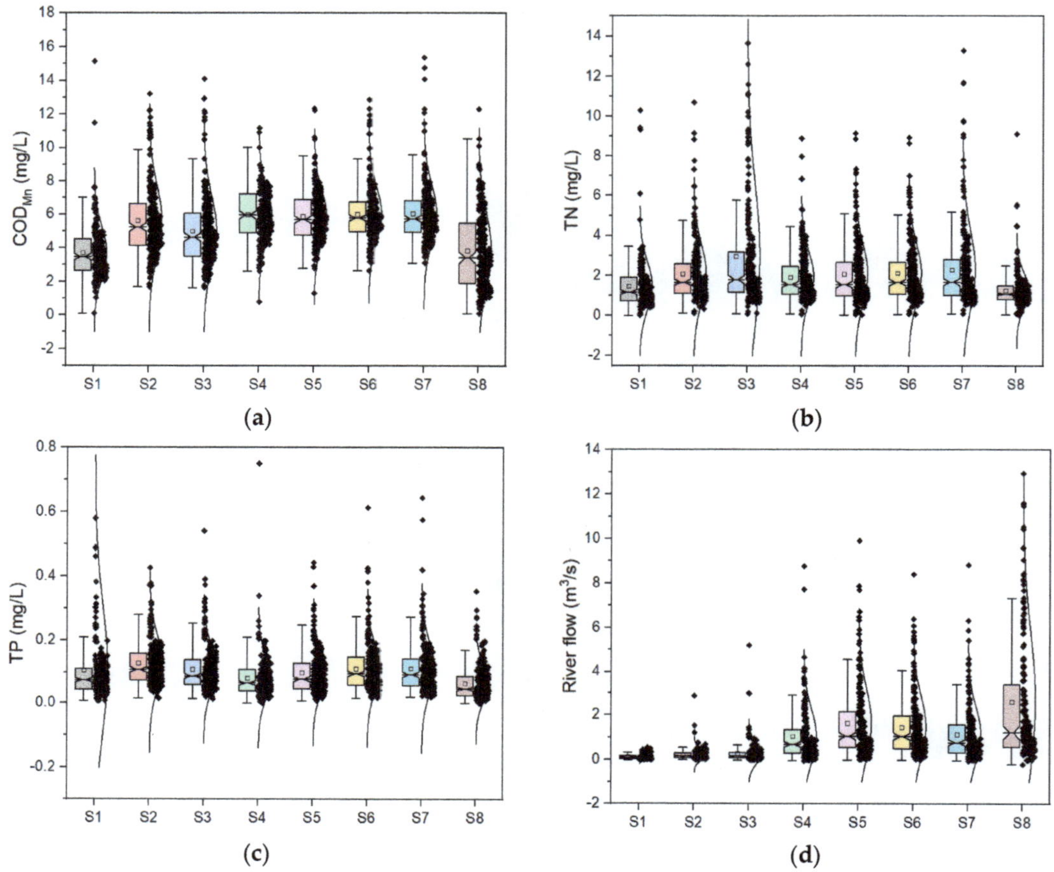

Figure 2. Changes in water quality and river flow along the river course ((**a**) COD_{Mn}; (**b**) TN; (**c**) TP; (**d**) river flow; positions S1–S8 refer to Figure 1).

3.2. Overall Reduction Effect of Pollution Load in ECW

The water quality indicators for TP, TN, and COD_{Mn} in the ECW inlet ranged from 0.020 to 0.644 mg/L, 0.27 to 13.28 mg/L, and 3.45 to 15.37 mg/L, respectively. The effluent concentrations of TP, TN, and COD_{Mn} were nd-0.274 mg/L, nd-9.12 mg/L, and nd-12.31 mg/L, respectively. The average removal rates of TP, TN, and COD_{Mn} were 36.2%, 26.7%, and 30.7%, respectively. As the inlet flow rate changed, the ECW wetland received different pollution load intensities. Based on the inlet and outlet water quality and flow rate, we calculated the reduction loads of TP, TN, and COD_{Mn} for the ECW (Figure 3). The reduction load of TP ranged from −0.0221 to 0.0554 g/m^2/d, with an average of 0.0042 g/m^2/d. The reduction load of TN ranged from −0.2718 to 1.0616 g/m^2/d, with an average of 0.0543 g/m^2/d. The reduction load of COD_{Mn} was −0.7982–0.8034 g/m^2/d, with an average of 0.1162 g/m^2/d. When converted, the cumulative reductions in TP, TN, and COD_{Mn} during the operating period from 2010 to 2021 were 17.6 t, 229 t, and 491 t, respectively, with annual reductions of 1.6 t/a, 20.8 t/a, and 44.6 t/a.

3.3. Impact of Hydraulic Load on ECW Treatment Effect

The surface hydraulic load is a key indicator for the design and operational management of constructed wetlands [23,24]. Detailed regulations on surface hydraulic load, TP, TN, and COD_{Mn} reduction load are stipulated in the design specifications of China's environmental protection industry. This wetland project is a special estuarine wetland that requires the treatment of all river water. The quality of upstream water is difficult to control, making it challenging to manage the reduction load. However, the control of the surface hydraulic load can be adjusted through the operation and management of river gates. Therefore, we analyzed the impact of the ECW surface hydraulic load on the TP, TN, and COD_{Mn} reduction load and removal rates, aiming to guide and optimize the subsequent operation and management of the ECW. According to Figure 4, TP, TN, and COD_{Mn} reduction loads increased with the increase in surface hydraulic load, indicating that this ECW project had certain advantages in treating large-flow water bodies. However, only the TP reduction load showed a stronger linear relationship with the increase in surface hydraulic load, while the linear relationship between TN and COD_{Mn} reduction loads and surface hydraulic load was not as strong, suggesting that ECW has a stronger advantage in TP removal. The average removal rate of TP reaching 36.2% also proved this point. We know that TP settlement characteristics may be characteristic of a large-scale ECW [25,26]. When the rainy season arrived, rivers carried a large amount of sediment and soil particles into the ECW. A higher hydraulic load indicated a larger water volume, which may carry more particulate matter. The phosphorus adsorbed by these particles could settle well in the ECW within 1–3 days, leading to a decrease in TP. As the surface hydraulic load increased, the removal rates of TP, TN, and COD_{Mn} concentrations decreased, with the most significant decrease in the COD_{Mn} removal rate. This is mainly due to the fact that the biodegradation process for COD_{Mn} removal depends on the action time of microorganisms [27,28]. Therefore, when there are high requirements for COD_{Mn} treatment, we must appropriately control the surface hydraulic load to keep it at a lower level.

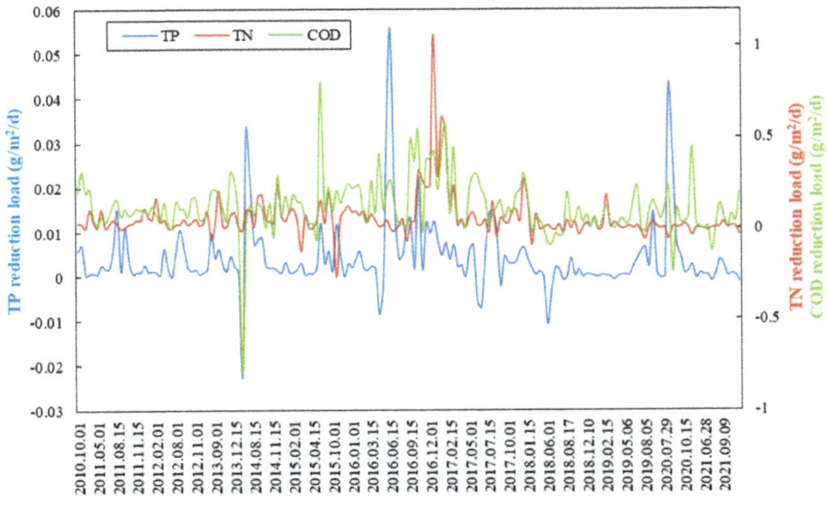

Figure 3. Reduction effect of pollution load in ECW.

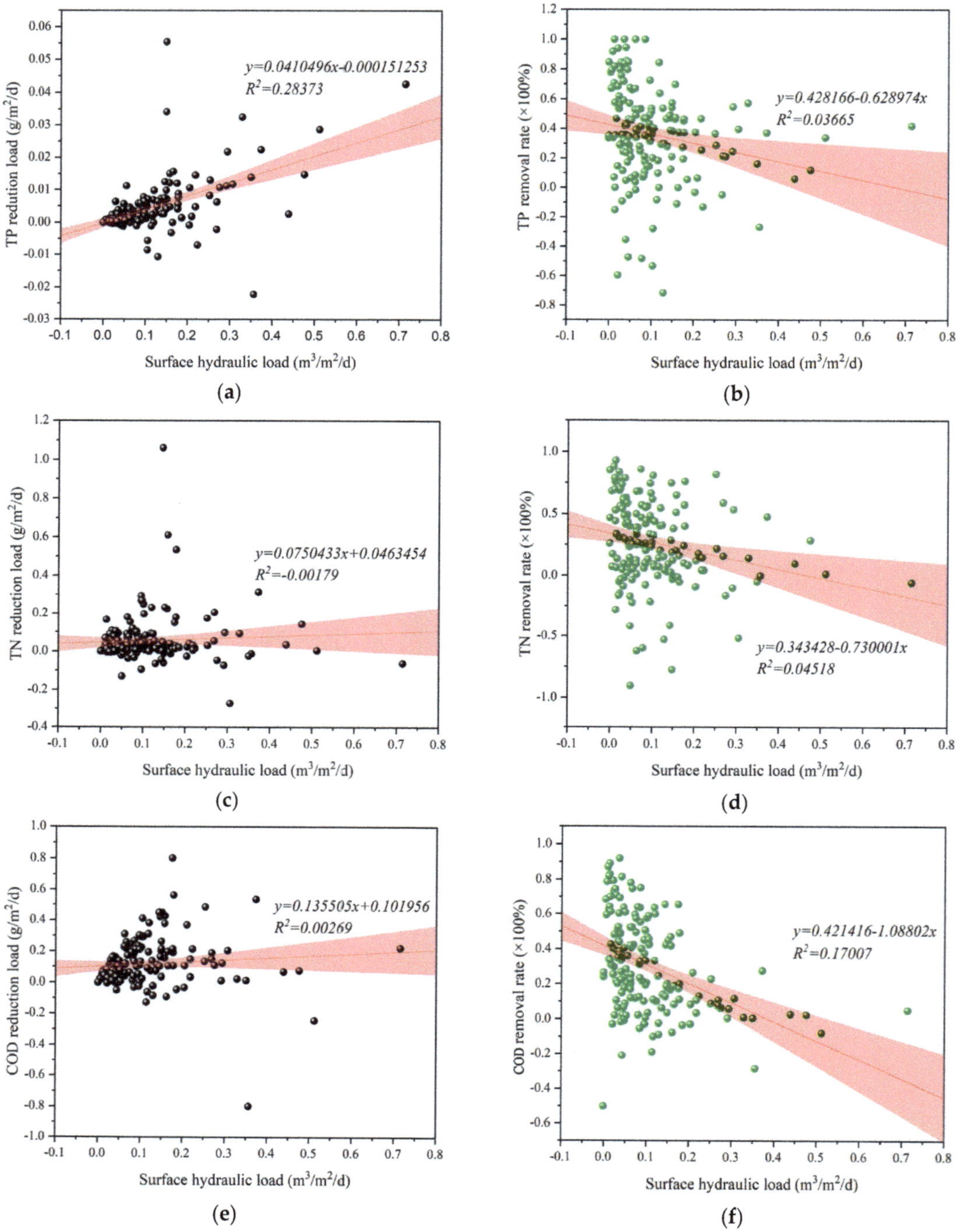

Figure 4. Linear fitting of surface hydraulic load with the reduction load and removal rate of TP, TN, and COD_{Mn} ((**a**) TP reduction load; (**b**) TP removal rate; (**c**) TN reduction load; (**d**) TN removal rate; (**e**) COD_{Mn} reduction load; (**f**) COD_{Mn} removal rate).

3.4. Impact of Operational Duration on ECW Treatment Effect

The Kolmogorov–Smirnov normality test results showed that at the 0.05 level, the data (TN, TP, and COD removal rate) were significantly drawn from a normally distributed population. Based on the tracking and monitoring data from the entire ECW operation period from 2010 to 2021 (Figure 5), we found that the removal rates of TP, TN, and COD_{Mn} significantly decreased, even exhibiting a clear linear relationship. TN had the poorest removal rate (26.9%), which removal rate decreased at the fastest rate as the operation years increased. According to the sediment survey conducted in 2022 in this study, the contents of TN, TP, and organic matter (OM) in ECW sediments were 4362 ± 418 mg/kg, 1080 ± 370 mg/kg, and 72.9 ± 11.0 g/kg, respectively. Compared with the data recorded in 2014 [29], TN increased by 2.37 times, while TP and OM did not change significantly. This further indicated that the deterioration in TN removal rate in the ECW was related to pollutant accumulation and sediment release processes. As the designed water depth of the ECW is less than 2 m, the river maintained a good water quality throughout the year with sufficient dissolved oxygen. However, it may be difficult to remove nitrate and nitrogen from the sediment through the denitrification process. TN removal is indeed a challenge. If further assurance of nitrogen removal functionality is required, it is necessary to add enhanced nitrogen removal treatment technology units in the later stages [4]. The removal rate of TP had also been significantly decreasing. This may be due to sediment disturbance, resuspension, and reactive phosphorus release in the sediments. On the other hand, the ECW is connected to Erhai Lake, and in recent years, the northern part of Erhai Lake has been a high-incidence area for cyanobacteria blooms [30,31]. The water rich in cyanobacteria in the northern part of Erhai Lake could flow into the ECW, leading to an increase in TP in ECW effluent. Unlike TN and TP, the removal rate of COD_{Mn} did not significantly decrease with the extension of ECW operation years, which was very different from other types of constructed wetland. For most constructed wetlands, the continuous decline in COD_{Mn} removal rate due to aquatic plant decay and OM accumulation is a common phenomenon [32,33]. This indicates that as the ECW is a relatively open form of water area, OM can be quickly discharged with the water flow, and no significant accumulation of OM occurs in the sediments.

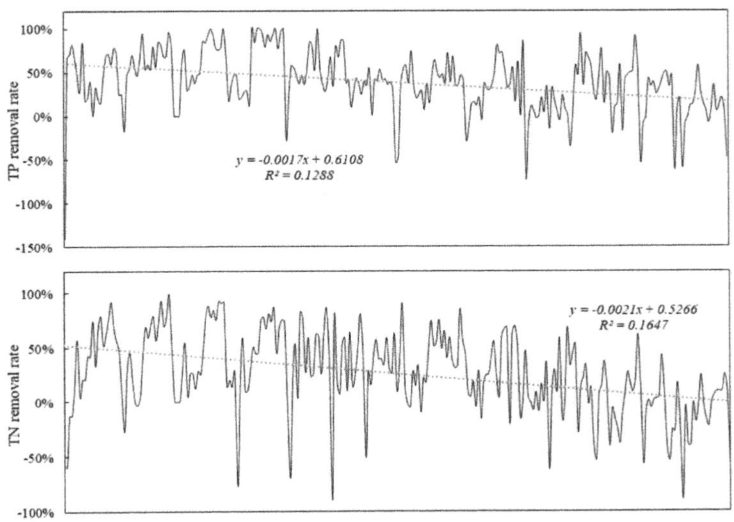

Figure 5. *Cont.*

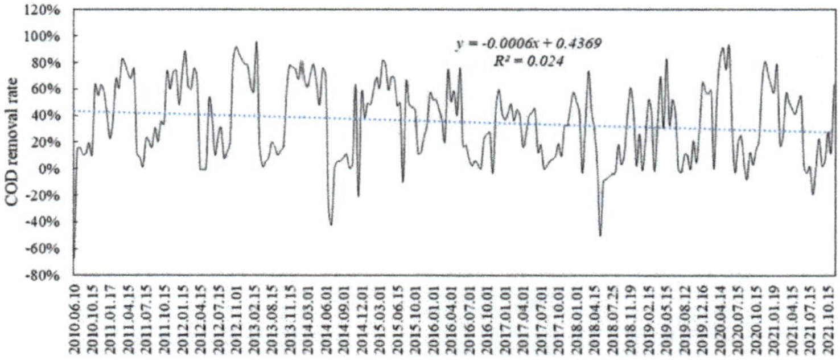

Figure 5. Trends in TP, TN, and COD removal rates over the years in ECW.

3.5. Seasonal Differences in ECW Treatment Effect

The ECW system exhibited significant variations in pollutant purification efficiency across different seasons. To further investigate the differences in pollutant removal efficiency during different periods of wetland operation, a comparison was conducted between the high-efficiency period (2010–2015) and the degradation period (2016–2021) of wetland purification based on seasonal changes (Figure 6). The removal rate of pollutants by wetlands was influenced by seasonal changes, generally showing a trend of winter ≈ spring > autumn > summer. Where the ECW is located, the entire Erhai Lake basin fall within a region with a mild climate, where temperatures rarely drop below 0 °C, even during winter, the average temperature in winter remains above 10 °C, and the water flow is relatively gentle. This extended the hydraulic retention time of wetlands, ensuring a good pollutant removal rate. The low removal rate in summer could be due to the fact that rainfall in the Erhai Lake basin was mainly concentrated in summer. Due to increased non-point source pollution caused by rainfall, doubled water treatment volume in wetlands, and shorter hydraulic retention time, the internal matrix effect of the wetland system was weakened. Especially under high inlet water concentration conditions, the ecological benefits of wetlands were hindered, and the pollution load was not effectively reduced. During the high-efficiency period, the treatment effect on TN and TP was twice or more that of the degradation period, and the treatment effect on COD_{Mn} was about 1.5 times that of the degradation period.

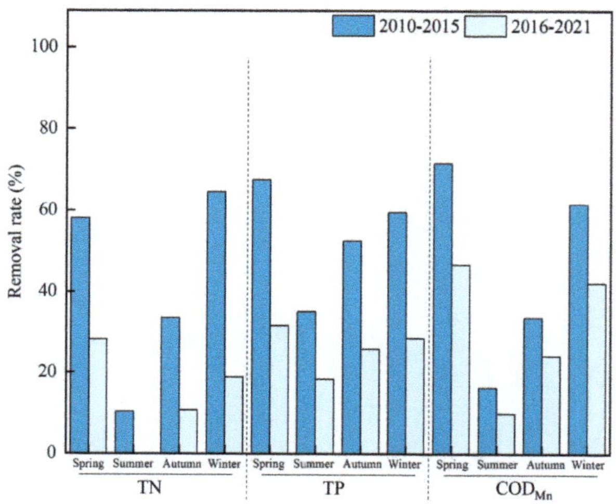

Figure 6. Seasonal differences in ECW treatment effect.

4. Conclusions

Through over a decade of tracking analysis on an ECW, this study found that it indeed played a significant role in achieving water quality improvement effects. The average removal rates of TP, TN, and COD_{Mn} were 36.2%, 26.7%, and 30.7%, respectively, with annual reductions of 1.6 t/a, 20.8 t/a, and 44.6 t/a. The surface hydraulic load is a key indicator for the design and operational management of an ECW. The reduction loads of TP, TN, and COD_{Mn} increased with the increase in surface hydraulic load, indicating that this ECW project had certain advantages in treating large-flow water bodies. However, only the TP reduction load showed a stronger linear relationship with the increase in surface hydraulic load. Nevertheless, when there are high requirements for COD_{Mn} treatment, we must appropriately control the surface hydraulic load to keep it at a lower level. As the operation years increased, the TN removal rate decreased at the fastest rate due to pollutant accumulation and sediment release. Considering the accumulation of TN in the sediment and its inability to be removed through denitrification, measures such as sediment dredging are necessary to ensure the removal rate of TN. As this is a relatively open form of water area, ECW OM can be quickly discharged with the water flow, and no significant accumulation of OM occurs in the sediments. The removal rate of pollutants by wetlands was influenced by seasonal changes, generally showing a trend of winter $\approx$ spring > autumn > summer. During the high-efficiency period (2010–2015), the treatment effect on TN and TP was twice or more that of the degradation period (2016–2021), and the treatment effect on COD_{Mn} was about 1.5 times that of the degradation period. After five years of ECW construction, the decline in treatment efficiency made ECW management more challenging, indicating that ECW construction is not a one-time solution. This provides the government with insights into long-term ECW management approaches.

Author Contributions: Conceptualization, J.S. and H.L.; methodology, Q.X.; software, S.J.; validation, Y.L. and R.W.; investigation, Y.X.; data curation, H.L.; writing—original draft preparation, S.J.; writing—review and editing, J.F.; visualization, S.J.; supervision, X.W.; project administration, J.F.; funding acquisition, J.F. All authors have read and agreed to the published version of the manuscript.

Funding: This research was funded by the National Key Research and Development Program of China (2024YFD1700400; 2021YFD1700400), the Yunnan Fundamental Research Project (202301AT070001), and the Yunnan Province key R&D Research Plan Project (202003AD150015; 202303AC100016; 202303AC100017).

Data Availability Statement: Dataset available on request from the authors.

Conflicts of Interest: Author Huaqing Li was employed by the company The Fifth Engineering Co., Ltd. of China Railway 10th Bureau Group, author Shiyi Jiang was employed by the company Yunnan Zhaohong Environmental Engineering Co., Ltd. The remaining authors declare that the research was conducted in the absence of any commercial or financial relationships that could be construed as a potential conflict of interest.

References

1. Chen, Q.; Huang, M.; Tang, X. Eutrophication assessment of seasonal urban lakes in China Yangtze River Basin using Landsat 8-derived Forel-Ule index: A six-year (2013–2018) observation. *Sci. Total Environ.* **2020**, *745*, 135392. [CrossRef] [PubMed]
2. Zhu, L.; Zhang, H.; Li, Y.; Sun, W.; Zhu, Y.; Song, C.; Wang, L.; Du, G. Evaluation of the optimization effect of a combined constructed wetland with micro-polluted water sources. *J. Water Process Eng.* **2022**, *49*, 103139. [CrossRef]
3. Wu, H.; Zhang, J.; Ngo, H.H.; Guo, W.; Hu, Z.; Liang, S.; Fan, J.; Liu, H. A review on the sustainability of constructed wetlands for wastewater treatment: Design and operation. *Bioresour. Technol.* **2015**, *175*, 594–601. [CrossRef]
4. Vymazal, J. The use of hybrid constructed wetlands for wastewater treatment with special attention to nitrogen removal: A review of a recent development. *Water Res.* **2013**, *47*, 4795–4811. [CrossRef]
5. Salimi, S.; Almuktar, S.A.A.N.; Scholz, M. Impact of climate change on wetland ecosystems: A critical review of experimental wetlands. *J. Environ. Manag.* **2021**, *286*, 112160. [CrossRef]

6. Zhao, L.; Fu, D.; Wu, X.; Liu, C.; Yuan, X.; Wang, S.; Duan, C. Opposite response of constructed wetland performance in nitrogen and phosphorus removal to short and long terms of operation. *J. Environ. Manag.* **2024**, *351*, 120002. [CrossRef] [PubMed]
7. Lemley, D.A.; Lakane, C.P.; Taljaard, S.; Adams, J.B. Inorganic nutrient removal efficiency of a constructed wetland before discharging into an urban eutrophic estuary. *Mar. Pollut. Bull.* **2022**, *179*, 113727. [CrossRef]
8. Hsieh, C.-Y.; Yang, L.; Kuo, W.-C.; Zen, Y.-P. Efficiencies of freshwater and estuarine constructed wetlands for phenolic endocrine disruptor removal in Taiwan. *Sci. Total Environ.* **2013**, *463*, 182–191. [CrossRef]
9. Zhang, H.; Cui, B.; Ou, B.; Lei, X. Application of a biotic index to assess natural and constructed riparian wetlands in an estuary. *Ecol. Eng.* **2012**, *44*, 303–313. [CrossRef]
10. Wu, W.; Deng, S.; Chen, C.; Su, H. Unraveling Multi-Scale dynamics of estuarine wetland vegetation using the multi-resolution analysis wavelet transform and the Landsat time-series. *Ecol. Indic.* **2024**, *158*, 111342. [CrossRef]
11. Wu, H.; Wang, R.; Yan, P.; Wu, S.; Chen, Z.; Zhao, Y.; Cheng, C.; Hu, Z.; Zhuang, L.; Guo, Z. Constructed wetlands for pollution control. *Nat. Rev. Earth Environ.* **2023**, *4*, 218–234. [CrossRef]
12. Peng, J.; Jin, C.; Wu, Y.; Hou, Z.; Gao, S.; Chu, Z.; Zheng, B. Modeling non-point source nutrient loads with different cropping systems in an agricultural lake watershed in southwestern China: From field to watershed scale. *Mathematics* **2022**, *10*, 4047. [CrossRef]
13. Li, C.; Shen, J.; Feng, J.; Chi, L.; Wang, X. Variations of phosphorus in sediments and suspended particulate matter of a typical mesotrophic plateau lake and their contribution to eutrophication. *Sci. Rep.* **2024**, *14*, 26551. [CrossRef] [PubMed]
14. *HJ 636-2012*; Water Quality-Determination of Total Nitrogen-Alkalinepotassium Persulfate Digestion UV Spectrophotometric Method. National Environmental Protection Standards of the People's Republic of China; Ministry of Ecology and Environment the People's Republic of China: Beijing, China, 2012.
15. *GB/T 11893-1989*; Water Quality—Determination of Total Phosphorus—Ammonium Molybdate Spectrophotometric Method. National Standard of the People's Republic of China; Ministry of Ecology and Environment the People's Republic of China: Beijing, China, 1989.
16. *GB/T 11892-1989*; Water Quality—Determination of Permanganate Index. National Standard of the People's Republic of China; Ministry of Ecology and Environment the People's Republic of China: Beijing, China, 1989.
17. *HJ 717-2014*; Soil Quality—Determination of Total Nitrogen—Modified Kjeldahl Method. National Environmental Protection Standards of the People's Republic of China; Ministry of Ecology and Environment the People's Republic of China: Beijing, China, 2014.
18. *HJ 632-2011*; Soil—Determination of Total Phosphorus by Alkali Fusion-Mo-Sb Anti Spectrophotometric Method. National Environmental Protection Standards of the People's Republic of China; Ministry of Ecology and Environment the People's Republic of China: Beijing, China, 2011.
19. *NY/T 1121.6-2006*; Soil Testing Part 6: Method for Determination of Soil Organic Matter. Agricultural Industry Standard of the People's Republic of China; Ministry of Agriculture: Beijing, China, 2006.
20. *HJ 494-2009*; Water Quality—Guidance on Sampling Techniques. National Environmental Protection Standards of the People's Republic of China; Ministry of Ecology and Environment the People's Republic of China: Beijing, China, 2009.
21. Chon, K.; Chang, J.-S.; Lee, E.; Lee, J.; Ryu, J., Cho, J. Abundance of denitrifying genes coding for nitrate (narG), nitrite (nirS), and nitrous oxide (nosZ) reductases in estuarine versus wastewater effluent-fed constructed wetlands. *Ecol. Eng.* **2011**, *37*, 64–69. [CrossRef]
22. Shen, J.; Wang, X.Z.; Zhang., Z.; Hu, X.F. Distribution characteristics of five endocrine disrupting chemicals in Luoshi River. *Res. Environ. Sci.* **2012**, *5*, 495–500.
23. Allen, D.J.; Farrell, M.; Huang, J.; Reynolds, C.; Rupasinghe, M.; Mosley, L.M. Long-term water quality response to increased hydraulic loadings in a field-scale free water surface constructed wetland treating domestic effluent. *J. Environ. Manag.* **2022**, *311*, 114858. [CrossRef] [PubMed]
24. Çakir, R.; Gidirislioglu, A.; Çebi, U. A study on the effects of different hydraulic loading rates (HLR) on pollutant removal efficiency of subsurface horizontal-flow constructed wetlands used for treatment of domestic wastewaters. *J. Environ. Manag.* **2015**, *164*, 121–128. [CrossRef]
25. Johannesson, K.M.; Andersson, J.L.; Tonderski, K.S. Efficiency of a constructed wetland for retention of sediment-associated phosphorus. *Hydrobiologia* **2011**, *674*, 179–190. [CrossRef]
26. Laakso, J.; Uusitalo, R.; Leppänen, J.; Yli-Halla, M. Sediment from agricultural constructed wetland immobilizes soil phosphorus. *J. Environ. Qual.* **2017**, *46*, 356–363. [CrossRef] [PubMed]
27. Saeed, T.; Sun, G. A comparative study on the removal of nutrients and organic matter in wetland reactors employing organic media. *Chem. Eng. J.* **2011**, *171*, 439–447. [CrossRef]
28. Sgroi, M.; Pelissari, C.; Roccaro, P.; Sezerino, P.H.; García, J.; Vagliasindi, F.G.; Ávila, C. Removal of organic carbon, nitrogen, emerging contaminants and fluorescing organic matter in different constructed wetland configurations. *Chem. Eng. J.* **2018**, *332*, 619–627. [CrossRef]

29. Wang, S.; Liu, Y.; Zhang, C.; Hou, L.; Wang, Y. Distribution of nitrogen, phosphorus, and organic matter in wetland sediments at the river mouths entering the Erhai Lake basin, and pollution risk assessment. *Lake Sci.* **2017**, *29*, 69–77.
30. Zhang, Y.; Shen, J.; Feng, J.-M.; Li, X.-Y.; Liu, H.-J.; Wang, X.-Z. Composition, distribution, and source of organic carbon in surface sediments of Erhai Lake, China. *Sci. Total Environ.* **2023**, *858*, 159983. [CrossRef] [PubMed]
31. Zhao, Y.-Q.; Shen, J.; Feng, J.-M.; Wang, X.-Z. Relative contributions of different sources to DOM in Erhai Lake as revealed by PLS-PM. *Chemosphere* **2022**, *299*, 134377. [CrossRef] [PubMed]
32. Vymazal, J.; Kröpfelová, L. Removal of organics in constructed wetlands with horizontal sub-surface flow: A review of the field experience. *Sci. Total Environ.* **2009**, *407*, 3911–3922. [CrossRef]
33. Díaz, F.J.; Anthony, T.; Dahlgren, R.A. Agricultural pollutant removal by constructed wetlands: Implications for water management and design. *Agric. Water Manag.* **2012**, *104*, 171–183. [CrossRef]

water

MDPI

Review

The Status of Research on the Root Exudates of Submerged Plants and Their Effects on Aquatic Organisms

Yahan Shi [1], Xu Zhang [2,*], Min Zhao [1,3,4], Xiangyong Zheng [1,3,4], Jianya Gu [5], Zhiquan Wang [1,3,4], Chunzhen Fan [3,4,*] and Wenwen Gu [5]

1 College of Life and Environmental Science, Wenzhou University, Wenzhou 325000, China; shiyahan1015@163.com (Y.S.); zmzjcn@sina.com (M.Z.); x.zheng@wzu.edu.cn (X.Z.); wangzhiquan@wzu.edu.cn (Z.W.)

2 School of Intelligent Construction, Zhejiang College of Security Technology, Wenzhou 325016, China

3 National and Local Joint Engineering Research Center of Ecological Treatment Technology for Urban Water Pollution, Wenzhou University, Wenzhou 325000, China

4 Institute for Eco-Environmental Research of Sanyang Wetland, Wenzhou University, Wenzhou 325014, China

5 Jiangsu B&P Testing Technology Co., Ltd., Nantong 226000, China; gjy210228@163.com (J.G.); guww988@163.com (W.G.)

* Correspondence: zick079@163.com (X.Z.); fanchunzhen@wzu.edu.cn (C.F.)

Abstract: The ecological restoration of submerged plants is one of the most widely used technologies in the remediation of eutrophic water bodies. This technology mainly removes nitrogen, phosphorus, and other nutrients in water through the absorption effects of plant roots, stems, and leaves and the biotransformation of microorganisms attached to their surfaces. Root exudates can directly affect root-attached microorganisms and other aquatic organisms, thus significantly influencing water remediation by submerged plants. At present, there are few reviews on the root exudates of submerged plants and their effects on aquatic organisms. In this study, the composition, collection, and methods of detecting the root exudates of submerged plants are reviewed. Factors affecting the release of root exudates from submerged plants are analyzed, including abiotic (light, temperature, and nutritional status) and biotic factors (rhizosphere microorganisms). The positive or negative effects of root exudates on phytoplankton, zooplankton, and microorganisms are also discussed. The results show that plant species, growth stages, and environmental factors (light, temperature, and nutritional status) are crucial factors affecting root exudates. In addition, submerged plants can significantly influence phytoplankton, zooplankton, and microorganisms by releasing allelochemicals or other root exudates. Based on the results of this study, the influencing mechanisms of root exudates on ecological restoration processes by submerged plants are clarified. This review provides important guiding significance for applying submerged macrophytes in water restoration.

Keywords: submerged plants; root exudates; allelochemicals; algae inhibition; water restoration

1. Introduction

Root exudates comprise a variety of substances released from plant roots into the growth medium during a plant's growth process [1]. They are adaptive mechanisms developed during plants' long-term evolution and growth that are affected by environmental factors and can change the rhizosphere environment. Research on this topic began at the end of the 18th century. Plenk et al. found that root exudates promoted or inhibited the growth of neighboring plants [2]. Since 1904, after Hiltner proposed the concept of the rhizosphere, studies on root exudates have gradually been carried out. In the 1950s, their role was initially revealed, but more macro studies were conducted due to technological limitations. Since the 1970s, studies on root exudates have flourished with technological improvements. Recent studies showed that root exudates are crucial for maintaining the vitality of the rhizosphere ecosystem. They are also an important part of material migration and regulation in the rhizosphere microecosystem. The systematic analysis of the response of root exudates to environmental factors is a research hotspot in the field of ecological restoration [3].

Lakes account for only a small fraction of the Earth's surface water resources but provide essential material, energy, and information exchange with terrestrial ecosystems. However, lake eutrophication has become a global challenge, with unnaturally high nutrient concentrations destroying about 40% of lakes and reservoirs worldwide [4]. This leads to the excessive growth of algae in water bodies, ultimately causing reductions in dissolved oxygen (DO) levels, declines in water quality, and the death of fish and other aquatic organisms. Technologies for controlling water eutrophication mainly include chemical flocculation [5], microbial dosing [6], and aquatic plant remediation [7]. Among them, remediation technology using aquatic plants is widely used due to its low cost, lack of secondary pollution, and simple operation. It can remove pollutants through the absorption effect of aquatic plants and the biotransformation effect of rhizosphere microorganisms to purify water bodies. Submerged plants play a significant role in remediation processes. However, the role and mechanism of plant root exudates in water remediation are still unclear. Some studies have revealed that the root exudates of submerged plants can provide carbon sources for microorganisms and promote nitrogen removal. Root exudates also play an important role in substance exchange and information transmission, which has important ecological significance [3]. The release of chemical substances from plant roots is part of its normal physiological metabolism, but environmental stress can affect the composition and content of these exudates. These changes can directly reflect the growth and metabolism of plants [8]. Most studies on organic acid exudates have focused on terrestrial plants and rice. Few studies have been conducted on the organic acid exudates of submerged plants. In 2008, Long et al. observed that the phosphorus level in the rhizosphere of seagrass (*T. testudinum*) increased linearly with the concentration of organic acids, which increased with an enhancement in seagrass productivity. Seagrass is an important source of organic acids, which are present at significant levels in the rhizosphere. Although the allelopathy between submerged macrophytes and algae has been frequently reported and the effect of algae exudates on submerged macrophytes has been widely studied, studies on the effect of submerged macrophyte exudates on algae have been insufficient [9]. Xu et al. found that the culture water of mature *Ottelia acuminata* significantly promoted the growth of *Microcystis aeruginosa M. aeruginosa*. In contrast, the culture water of seedlings had no significant effect [10]. Girum Tamire et al. demonstrated that *Potamogeton schweinfurthii* had a significant allelopathic inhibitory effect on cyanobacteria (especially *Microcystis* and *Dolichospermum* spp.). This finding is important for ecological research, but further studies are needed to determine whether exudates are produced by conventional metabolic processes [11]. Wang et al. screened a special bacterial plant-growth-promoting rhizobium (PGPR) from the rhizosphere of *Vallisneria natans* (*V. natans*) under low and high organic matter loads in sediment. It survived in the plant roots and could directly or indirectly promote plant growth. This PGPR used the root exudates of *V. natans* as the sole carbon source, showing high competitiveness for rhizosphere nutrition. This screening method provided a new approach to the artificial restoration of submerged plants [12].

Thus far, many studies have examined the root exudates of submerged plants, but the corresponding reviews have been insufficient. This study reviews the development history, classification, collection and detection methods, and influencing factors of root exudates, as well as the effects of the root exudates on aquatic organisms. The aim of this study is to provide valuable suggestions for the research, development, and application of submerged plants in water restoration.

2. Submerged Plant Exudates

2.1. Definition and Classification

Root exudates refer to various substances secreted or released from different parts of the plant root system to its growth medium during plant growth. The main components of root exudates include organic matter composed of carbonaceous compounds, inorganic ions, H^+, and water [13]. Root exudates can be divided into four categories: (1) exudates, which mainly include low-molecular-weight organic compounds released through cell diffusion, such as sugars, amino acids, and organic acids; (2) secretion, which includes the metabolites actively released by cells in the metabolic process, including phenolic compounds, polysaccharides, and protons; (3) mucilage secreted by root cap cells, epidermal cells without secondary walls, and root hair cells; (4) decomposition and abscission, which are the root cell tissue and its decomposition products [14]. Root exudates can be divided into high- and low-molecular-weight organic compounds according to their molecular weight. High-molecular-weight compounds mainly include polysaccharides, proteins, and enzymes, while low-molecular-weight organic compounds include amino acids, organic acids, sugars, phenols, and secondary metabolites [15]. According to the nature of action, root exudates can be divided into two types: common and specific. Common exudates are common to most plants, while specific exudates are unique to specific plants under specific conditions [16]. Common types of root exudates of submerged plants are shown in Table 1.

Table 1. Common types of root exudates of submerged plants [14,17,18].

Class	Representative Compounds	Major Functions
Saccharide	Glucose, fructose, galactose, rhamnose, ribose, raffinose, xylose, sucrose, lactose, maltose, and arabinose	Promoting rhizosphere microbial growth, regulating soil properties, and affecting rhizosphere microbial community structures
Organic acids	Oxalic acid, tartaric acid, pyruvic acid, malic acid, malonic acid, lactic acid, catalpol, succinic acid, fumaric acid, formic acid, acetic acid, propionic acid, butyric acid, valeric acid, and salicylic acid	Changing the soil's pH value, activating soil nutrients, and improving nutrient absorption by plants
Amino acid	Aspartic acid, threonine, serine, glutamic acid, glycine, alanine, valine, methionine, isoleucine, leucine, tyrosine, phenylalanine, γ-aminobutyric acid, lysine, histidine, arginine, aspartic acid, threonine, serine, glutamic acid, glycine, alanine, valine, methionine, isoleucine, leucine, tyrosine, phenylalanine, γ-aminobutyric acid, lysine, histidine, arginine, and proline	Promoting plant growth and development, improving plant stress resistance, and regulating the soil's microbial community
Long-chain fatty acid	Stearic acid, palmitic acid, oleic acid, and linoleic acid	Promoting plant defense against foliar pathogens, enhancing plant resilience, regulating plant–microbial interactions, and acting as a nutrient source for microorganisms
Steroid	Cholesterol and stigmasterol	Acting as nutrient sources for microorganisms and enhancing the growth potential and stress resistance of plants
Growth hormone	Biotin, vitamin, choline, inositol, and phytohormone	Promoting cell growth, differentiation, division, and biosynthesis
Proteins and enzymes	Amylase, DNA enzyme, phosphatase, polygalacturonase, protease, RNA enzyme, invertase, urease, xylanase, PR protein, etc.	Promoting the absorption and conversion of nutrients and catalyzing the degradation of organic pollutants
Other compounds	Flavonoids, nucleosides, glycosides, and polysaccharides	Genetic information transfer, energy storage and conversion, signal transduction, and storage and transport of substances

2.2. Production Pathway and Mechanism

A total of 28–59% of plant photosynthetic products are transferred to the underground part, 4–70% of which are released into the soil through root exudates. There are two main mechanisms for the release of root exudates, involving metabolic and non-metabolic pathways. However, a unified conclusion about the specific mechanism has not been reached [19]. Figure 1 shows some pathways and proteins that transport certain organic compounds and special metabolites around the cytoplasm and export them to the rhizosphere. Vesicles that sprout from the endoplasmic reticulum and Golgi are loaded with specialized metabolites, guided to the cytoplasm or plasma membrane, and fused with it, releasing content to vacuoles or the extracellular space. The circular symbol indicates the general transporter that loads the compound into the vesicle, and the transport process is completed by the general transporter, involving membrane-bound transporters such as ABC, MATE, MFS, and ALMT families. Although secondary metabolites are not directly involved in plant growth, they are essential for plant disease and stress resistance. Non-metabolic pathways involve the decomposition of root epidermal senescent cells and the release of substances from dead cells, which are not regulated by metabolism. Because the process of root exudates is very complex, both simple and specific root exudates are secreted. Therefore, the evaluation of root exudates should consider various factors, such as the plant environment. Although there are different opinions on the mechanism of root exudates, the consensus is that root exudates help to alleviate stress under environmental stress. Releasing root exudates is an active physiological process, with the energy derived from cell metabolism. Thus, root exudates may result from plant interactions with stressful environments, especially under specific selection pressures. In other words, root exudates are an active adaptation mechanism of plants under environmental stress, and the production of specific exudates is the essence and evolution of plant adaptation to environmental stress [20].

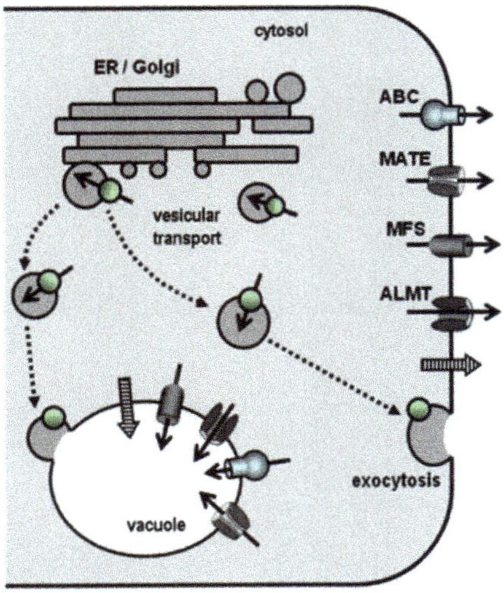

Figure 1. Schematic diagram of root cells. The round symbol depicts a generic transporter loading compounds into the vesicles. The membrane-bound transport proteins known to facilitate the transport of compounds across membranes include the ATP binding cassette family (ABC), the multidrug and toxic compound extrusion family (MATE), the major facilitator superfamily (MFS), and the aluminum-activated malate transporter family (ALMT). The striped arrow indicates the possible diffusion pathway of highly hydrophobic compounds across the lipid bilayers. The other arrows show the direction of substrate movement. Reproduced with permission from [20]. Copyright 2012, Oxford University Press.

Recent studies have shown that plant secretions are released into the environment through various processes. They are leached from decomposed plant residues and the roots or leaves of living plants. These processes are related to the beneficial effects of crop rotation or co-cultivation of certain submerged plants. The exchange of natural products between important plants may explain these ambiguous phenomena [21]. Moreover, the absorbed natural products in some recipient plants are modified, while they simply accumulate in other recipient plants. These modifications include hydroxylation, methylation, and glycosylation processes. In the past, it was thought that these reactions were part of a deliberate detoxification mechanism known as the "green liver concept". However, since the manner and extent of these modifications vary greatly between different plant species, general and universal mechanisms such as the "green liver concept" can be ruled out [22]. The study by Laura Lewerenz et al. was the first to bring to life the phenomenon of "lateral natural product transfer". Figure 2 shows that harmaline is translocated via the xylem into the leaves. Subsequently, the constituents of the xylem are further distributed within the leaf blade, driven by transpiration and root pressure. In the further distribution of the alkaloids within the leaf blade, increasing harmaline is oxidized, resulting in a continuous increase in the ratio of harmine to harmaline [23]. In addition, studies by Tahani Hijazin et al. from the same team also confirmed that various alkaloids are effectively absorbed from the soil, which is strongly influenced by the rhizosphere pH due to their alkaline nature. However, intense caffeine intake is not affected by the various pH values. pH significantly affects the uptake of alkaloids, and the highest uptake appears to be achieved at a specific pH. Thus, the absorption of various alkaloids and their dependence on pH may vary, and the extent of alkaloid absorption cannot be predicted [24].

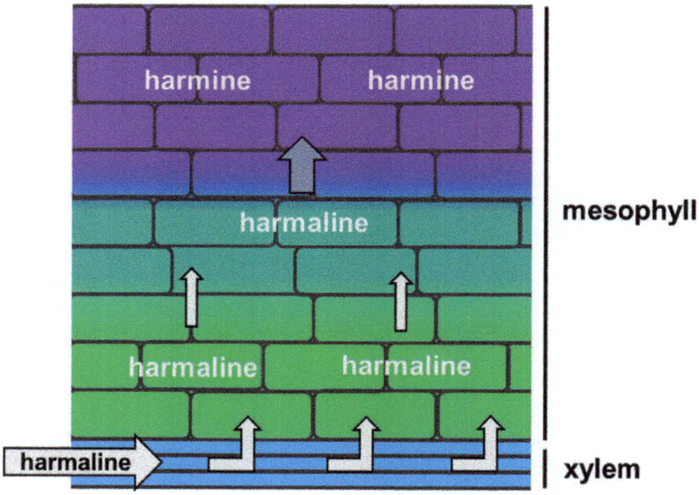

Figure 2. Scheme of the translocation and conversion of harmaline in barley leaves. The bottom layer and above layer represent the xylem and mesophyll of the leaf blade. The arrow distribution and oxidation pathways of harmaline with the leaf blade. Reproduced with permission from [23]. Copyright 2020, Elsevier.

2.3. Collection and Analysis

Processes of collecting root exudates are troublesome due to many interfering factors and uniform research methods. Different conditions have different classification standards. Thus, the correct and effective collection of plant root exudates is a key step in correctly studying the chemical composition of root exudates. Figure 3 shows commonly used methods for collecting root exudates. Among them, the root exudates collected under closed, sterile conditions can more accurately reflect the total amount of organic matter, and in situ collection under soil culture conditions can more accurately reflect the actual situation of root exudates. Thus, researchers can select appropriate collection methods in accordance with different experimental purposes [25–27]. The most commonly used method for collecting the root exudates of submerged plants is the disturbance collection method. After submerged plants are uprooted and cleaned, they are immersed in a certain amount of ultrapure water and placed in the dark for 24 h. Then, the soaking solution is filtered as the crude root exudates, followed by subsequent treatment.

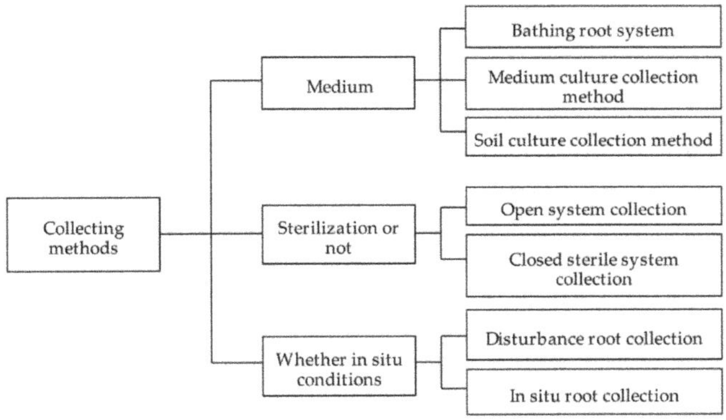

Figure 3. Common collection methods of submerged plant exudates.

Achieving accurate qualitative and quantitative analyses of all of the components of root exudates is difficult due to their complex compositions. Currently, most studies focus on analyzing high-content and important functional compounds. Among them, organic acids, amino acids, and sugars are representative root exudates that play an important role in the whole rhizosphere system. Methods of their analysis and detection are relatively mature. Chromatography is commonly used to detect these substances, including GC, GC-MS, HPLC, UPLC, and LC-MS. Table 2 summarizes the advantages and disadvantages of the methods for detecting submerged plant exudates. HPLC is mostly used in the detection and analysis of polyphenols in the exudates. Figure 4 shows the analysis process of exudates released by *Myriophyllum verticillatum*. In the early stage, Nakai used HPLC and APCI-MS to identify allelopathic polyphenols such as ellagic acid, gallic acid, pyrogallic acid, and (+)-catechin released by *Myriophyllum verticillatum*. The plants were cultured in the medium for 3 d to prepare the culture solution. Then, the solution was separated according to the polarity and molecular weight of the allelochemicals. The components were analyzed using HPLC and APCI-MS [28].

Table 2. The advantages and disadvantages of commonly used methods for analyzing and detecting submerged plant exudates.

Analysis and Test Method	Advantages	Disadvantages	Reference
GC	Analysis of the substances with a low boiling point, good thermal stability, high volatility, and stable retention time, which can directly identify the structure	Unsuitable for analyzing some substances that need pretreatment with a high boiling point and poor thermal stability via direct injection	[29,30]
GC-MS	Accurate characterization of the substances with a large database	Insufficient software for analyzing data	[31]
LC-MS	A wide range of analysis, strong separation ability, low detection limit, and high degree of automation	Lack of a standard database to identify the structure	[32]

Table 2. *Cont.*

Analysis and Test Method	Advantages	Disadvantages	Reference
UPLC	Fast analysis speed, short time, and high separation efficiency	Short service life of the chromatographic column and demanding laboratory conditions	[33–35]
HPLC	High separation efficiency, good selectivity, high detection sensitivity, automatic operation, and wide application range	High operating cost and long analysis time	[36]

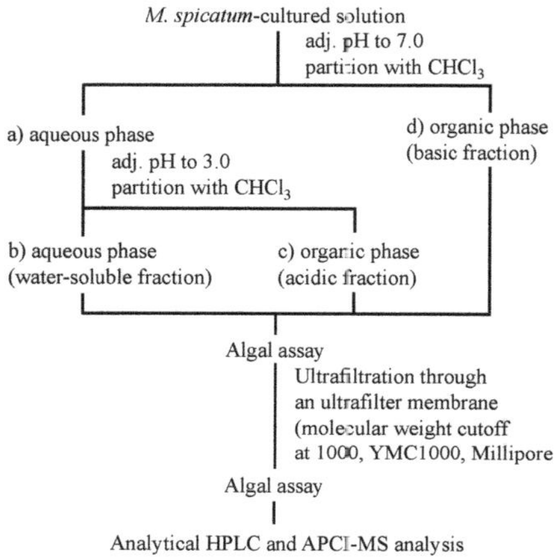

Figure 4. Study on the analysis method of exudates released by *M. spicatum*. Reproduced with permission from [28]. Copyright 2000, Elsevier.

3. Influencing Factors of Root Exudates

3.1. Plant Species and Growth Stages

Root exudates, as an inherent characteristic of plants, directly reflect the species and genetic characteristics of plants. Thus, different kinds of submerged plants release root exudates with different compositions and contents, leading to different allelopathic effects on the surrounding environment [37]. Cheng et al. found significantly different allelopathic effects on *Synechocystis* from different submerged plants. The culture water of sea cauliflower exhibited a slight promoting effect on the growth of *Synechocystis*. In contrast, the culture water of *Myriophyllum aquaticum, Ceratophyllum demersum, Myriophyllum spicatum, Hydrilla verticillata,* and *Vallisneria natans* showed different degrees of the algae-inhibition effect. This indicates that the allelochemicals released by submerged plants can affect the normal growth of *Synechocystis* [38]. Zhang et al. demonstrated that the exudates of *Phellinus linteus* and *Potamogeton malaianus* could inhibit the growth of *Microcystis aeruginosa* and *Selenastrum capricornutum*. However, the sensitivity of the two algae to the exudates of the two plants was different. A GC-MS analysis showed that the exudates of the two submerged plants contained a variety of compounds, and only three alcohols were detected in the exudates of *Potamogeton malaianus*. These specific alcohols may reflect the different degrees of sensitivity of *Microcystis aeruginosa* and *Selenastrum capricornutum* to the exudates of *Potamogeton malaianus* [39]. Xing et al. observed five organic acids detected in the root exudates of *Vallisneria natans*. Oxalic acid was the main component, accounting for 87.5%. The contents of malic acid and citric acid were 4.74% and 6.82%, respectively. Formic and ascorbic acids can be ignored when they comprise less than 1% of the total. Different types of submerged plants produce different types and contents of exudates, leading to different inhibitory effects on algae [40]. Pakdel et al. examined the allelopathic effects of

Chara australis and *Potamogeton crispus* on microalgae. All treatments exhibited significant negative effects on *A. variabilis*, with the strongest effect on *C. australis*. On the contrary, there was no significant effect on the growth of *S. quadrauda*. This result confirms that large plant allelochemicals target specific organisms [41]. In addition, the same submerged plants may have different inhibitory effects on algae at different growth stages. Xu et al. observed that the mature plant culture water of *Ottelia acuminata* had a significant effect on the growth of *P. aeruginosa* and significantly promoted the growth of *M. aeruginosa*. In contrast, the seedling culture water had no significant effect on the growth of *P. aeruginosa*. This may be because mature plants secrete nutrients or small amounts of elements that are beneficial to the growth of cyanobacteria [10]. Mulderij et al. also found different allelopathic effects of two *Chara* species on three green algae during different plant growth stages. Compared with Xu et al.'s study, their mature plants reduced the growth rate of *C. parvum*, while the effect of the young plants was the opposite [42].

3.2. Environmental Factors

The secretion of root exudates is a response characteristic of plants, and its type and quantity are affected by environmental conditions such as lighting, temperature, and nutrient levels. Abnormal conditions result in the release of abnormal root exudates, possibly leading to growth arrest or the death of plants [43]. Gross et al. found that light levels had a significant effect on the root exudates of *M. spicatum*. Specifically, bright lighting conditions increased the content of phenolic compounds secreted by *Myriophyllum spicatum*. However, the concentration of the main allelochemical, tellimagrandin II, was increased under low lighting conditions. This indicates that lighting conditions have a specific regulatory effect on the root exudates of *M. spicatum*, and the response of different types of exudates to light is different [44]. Erhard et al. observed that all related flavonoids could be detected in the exudates of *Elodea nuttallii* under different lighting conditions. However, the lighting conditions affected the quantity of specific flavonoids. In particular, high irradiance may promote the biosynthesis of luteolin diurea compounds, which was supported by field observations. They speculated that the increase in the content of luteolin diglucuronic acid is an adaptive response of plants to higher UV-B irradiation. This indicates that lighting conditions, especially the intensity of UV-B irradiation, have a significant effect on the synthesis and secretion of specific flavonoids from *Elodea nuttallii* [45]. Both strong and weak light influence the secretion of submerged plants. Martin et al. studied the effects of all-optical, continuous, and fluctuating light reduction on root exudates of three seagrasses (*Cymodocea serrulata*, *Halophila ovalis*, and *Halodule unintervis*). They found that fluctuating light exhibited the most significant effect, increasing the secretion of DOC (from the root), protein-like DOM, and humus-like DOM from the three seagrasses. This study highlights that the root exudates of seagrasses are highly correlated with light availability, and the underground environment is particularly sensitive to the reduced light reaching submerged plants [46].

Temperature is also an important factor in determining the physiological status of plant roots. Normal temperatures are conducive to the growth and physiological metabolism of plant roots. Abnormal temperatures cause adverse stress and damage the physiological metabolism of roots. Temperature can also affect the photosynthesis and respiration of plants. Therefore, temperature significantly affects the composition and content of plant root exudates. Previous studies have shown that plant roots secrete organic acids or other amino acids, enzymes, and other substances to resist high-temperature stress and adapt to environmental changes. Most root secretions increase with an increase in temperature [47]. Gu et al. found that when the water temperature was enhanced by 5 °C, the abundance of heterotrophic bacteria in seaweed exudates increased rapidly. This is because the enhancement in temperature significantly increases the assimilation rate of bacteria to the exudates, resulting in a decrease in the content of seaweed exudates [48]. Similarly, Erhard et al. observed all flavonoids in the exudates of *Elodea nuttallii* under different temperature treatments, and the temperature changed the content of individual flavonoids. For example, the content of chrysoeriol diglucuronic acid and apigenin was negatively correlated with temperature. In addition to luteolin diglucuronide, temperature has a negative impact on most phenolic compounds [45].

Plant growth is inseparable from nutrition. A lack of nutrients regulates the intensity and pathway of plant physiological and biochemical reactions and even changes the metabolic pathway of substances, thus affecting the composition and content of root exudates. Insufficient phosphorus and nitrogen can affect the production and release of chemicals by submerged plants [44]. This effect may depend on the nutrient levels of submerged plants. Most of the phosphorus required by plants reaches the root surface through diffusion. A deficiency in phosphorus occurs because phosphate always forms insoluble mineral phases with metals (such as calcium, iron, and aluminum) [49]. For example, phosphorus deficiency increases the production of polyphenols in *M. spicatum* and enhances the inhibition of cyanobacterial alkaline phosphatase [50]. In addition, many studies demonstrated that the secretion of organic acids and acid phosphatase in most plant roots increases significantly under insufficient phosphorus conditions. Phosphorus-efficient plants can promote the activation and absorption of insoluble phosphorus by increasing the secretion of organic acids. The most commonly reported organic acids are dicarboxylic and tricarboxylic acids, including oxalic, acetic, malic, fumaric, and citric acids [51]. Organic acids improve the bioavailability of P by replacing P from phosphorus-containing oxides (Fe, Al, and Ca) or complexing organic anions with metal ions in oxides [52]. Xing et al. used high-resolution dialysis and film-diffusion gradient techniques to analyze the changes in phosphorus in the rhizosphere of *Vallisneria natans*. They found that the enrichment of P and Fe in Fe patches on the rhizosphere was 5.92 and 3.12 times that of non-rhizosphere sediments, respectively. Further analysis showed five organic acids with low molecular weight in root exudates, and oxalic acid accounted for 87.5%. This finding indicates that *Vallisneria natans* significantly improves its ability to obtain rhizosphere phosphorus through the complexation of iron (III) and oxalic acid [40]. Figure 5 is a schematic diagram showing the coupling process between Fe plaque enrichment and organic acid complexation during the release of phosphorus from the rhizosphere of *Vallisneria natans*. When exploring the response of submerged plants to the nitrogen concentration in the environment, a significant phenomenon is that the released exudates will adjust with the change in the nitrogen concentration. This was confirmed by Gross et al.'s study. They found that although the nitrogen concentration within a certain range (0.6–4.8 mM NO_3^--N) had no significant effect on the total phenolic compounds in the culture water of *Myriophyllum spicatum*, the concentration of the main polyphenol, tellimagrandin II, increased significantly at low nitrogen levels. The total phenol content in the culture water of *Myriophyllum spicatum* under a low nitrogen level (0.06 mM) was much higher than that under a medium nitrogen level (0.5 mM). This result indicates that a low-nitrogen environment triggers a specific secretion pattern of plant phenols. However, ellagic acid in the plant culture water showed an opposite trend to tellimagrandin II [44]. This difference indicates that submerged plants may respond to environmental stress by secreting different chemicals, even under the same environmental conditions, which may differ significantly in function and mechanism.

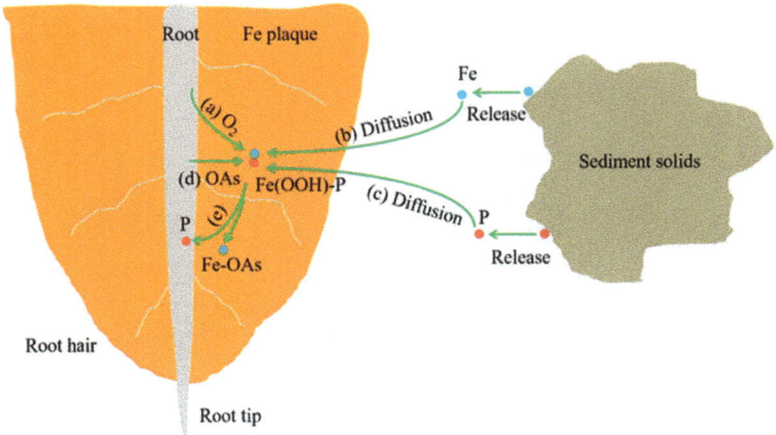

Figure 5. Schematic illustration of the coupling processes between Fe (blue color) plaque enrichment and organic acid complexation in liberating P (red color) in the *V. natans* rhizosphere. The lowercase letters show the process sequence. Reproduced with permission from [40]. Copyright 2017, Elsevier.

4. Effects of Root Exudates on Aquatic Organisms

The phenomenon of "allelopathy" is the effect exerted by one plant on its neighboring organisms by producing chemicals. This effect can be positive or negative, and it is ubiquitous in all plants. The negative effects of allelopathy include autotoxicity, soil disease, or biological invasion, while the positive effects include weed control and ecological protection [53]. In this section, the effects of allelopathic substances secreted by submerged plants on aquatic organisms will be discussed in detail.

4.1. *Effects on Phytoplankton*

The eutrophication of water bodies causes algal blooms, causing "red tide" and "bloom" phenomena. In 1969, Fitzgerald first discovered that allelochemicals secreted by submerged plants could inhibit the growth of algae, which aroused widespread interest [54]. Many scholars demonstrated that the algae content in the planting area of submerged plants was significantly lower than that in the area without submerged plants. Therefore, studies on the application of submerged plants to control algae have gradually emerged [55]. In the short term, submerged plants inhibit algae growth by secreting "algae-inhibiting substances", which are toxic to algae, rather than by nutrient competition or light shielding. As shown in Figure 6, three methods for applying allelochemicals are usually used to inhibit algae in aquatic ecosystems, including the direct cultivation of submerged plants, the release of plant residues or extracts containing allelochemicals, and the synthesis of allelochemicals [56]. This indicates a significant guide for the artificial synthesis of algae inhibitors and the application of submerged plant exudates to treat algae in water bodies.

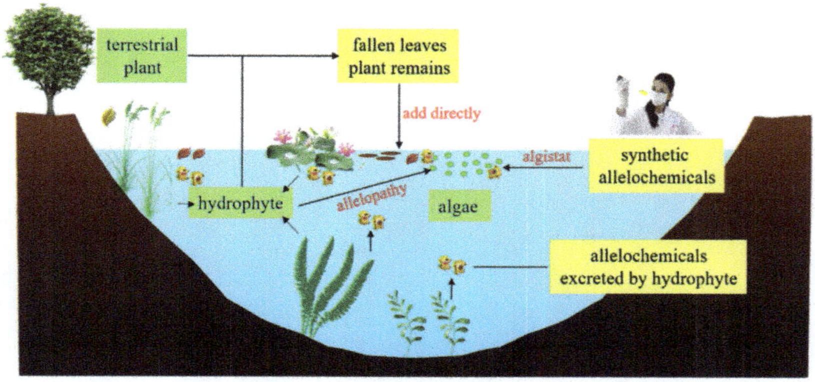

Figure 6. The application of allelochemicals in an aquatic ecosystem. Reproduced with permission from [56]. Copyright 2020, Elsevier.

Submerged plants can directly secret allelochemicals from roots to the rhizosphere, but allelochemicals have certain specificity and selectivity. A single allelochemical only affects the growth of one or several plants. When treated at higher concentrations, some root exudates, such as phenolic acids, can cause toxic effects on other plants and even themselves, inhibiting the normal physiological and metabolic activities of roots, hindering the extension of plant leaves, and affecting the normal growth of plants. Declerck et al.'s study demonstrated the strong inhibitory effect of *Elodea nuttalli* on microalgae, which lasted for more than 50 d, showing the long-term allelopathic potential [57]. Svanys et al. found that *Myriophyllum verticillatum* could effectively reduce the number of *M. aeruginosa* in eutrophic environments. The plants have a continuous negative impact on cyanobacterial biomass but a much shorter impact on other phytoplankton and green algae [58]. Wu et al.'s comparative study revealed that different submerged plants (*Pogonatherum chinense*, *Potamogeton malaianus*, and *Potamogeton crispus*) had different allelopathic effects on *P. aeruginosa* under the same conditions, emphasizing the importance of the diversity of allelochemical species and quantities in the inhibitory effect. Further studies also showed that the allelopathic activity of submerged plants may be affected by the season and growth stage [59]. Hilt et al. found the strongest allelopathic inhibitory activity of charophytes on phytoplankton in August. The growth stage of macro-submerged macrophytes may also affect allelopathic activity. Some studies have reported that young, active macrophytes exhibited greater allelopathic activity than older plants [60]. Rojo et al. tested the inhibiting efficiency of single and combined submerged plant cultures on the growth of natural phytoplankton through allelopathy. *Chara hispida*, *Chara vulgaris*, *Chara baltica*, *Nitella hyalina*, and *Myriophyllum spicatum* were used to test their single and combined allelopathic effects on environmental phytoplankton communities in the laboratory. The results showed that compared with *Myriophyllum*, *Chara* species (such as *C. hispida*) had a stronger effect. Compared with monospecific plants, combining large plants could better inhibit microalgae. Therefore, combining large plants seems to support synergistic allelopathy, directly reducing the microalgae biomass and thus improving the water quality [61]. Macro-submerged macrophytes show a significant inhibitory effect on the photosynthesis of phytoplankton by secreting specific allelochemicals, especially cyanobacteria. As shown in Figure 7, the electron transport chain may be disturbed due to abnormalities in the participating pigments, protein complexes, and electrons. For example, linoleic acid reduces the pigment content to block electron transport in *Pseudomonas aeruginosa*, while berberine inhibits photosynthesis-related gene expression and core protein synthesis [62]. In addition, some specific allelochemicals, such as tellimagrandin II, can significantly destroy the electron transport chain of cyanobacteria, which is achieved by increasing the redox midpoint potential of non-heme iron. The substance produced by *Myrophyllum spicatum* inhibits PSII of cyanobacteria by interfering with electron transfer [63]. Some allelochemicals show selective inhibition of the photosynthesis of cyanobacteria and green algae, which is attributed to the differences in their photosynthetic tissues. For example, polyphenols strongly inhibit the photosynthesis of cyanobacteria rather than green algae [64]. Similarly, the secretion of *Chara verticillata* has a significant inhibitory effect on the mutant cyanobacteria *Anabaena polymorpha* and a small effect on the growth of *Scenedesmus quadricauda* [41]. Some studies found that the exudates of large submerged plant combinations had stronger allelopathic effects on cyanobacteria and diatoms. This enhanced allelopathy is attributed to the synergistic effect of different allelochemicals produced by these plant combinations. This synergistic effect not only directly reduces the biomass of microalgae but also indirectly improves the water quality by enhancing grazing [61]. This finding provides a new theoretical basis for using multiple submerged macrophytes to control algal blooms in aquatic ecosystems. Cultivating multiple plants can effectively remove harmful cyanobacteria while retaining green algae as fish food, thereby restoring and maintaining the health of aquatic ecosystems.

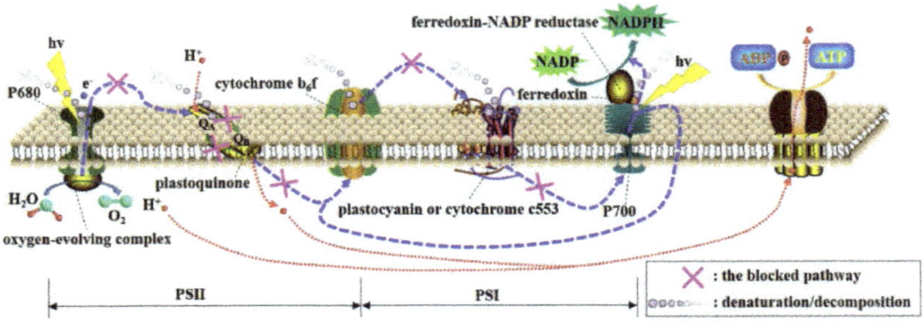

Figure 7. Schematic diagram of electron transport chain of microalgae cells. Reproduced with permission from [62]. Copyright 2020, Elsevier.

4.2. Allelopathy on Zooplankton

It is well known that submerged plants are a refuge for zooplankton, but they also inhibit the growth of zooplankton. Therefore, macro-submerged macrophytes have both positive and negative effects on zooplankton. Figure 8 shows the network of interactions between macro-submerged macrophytes and phytoplankton, zooplankton, etc. However, the effects of allelochemicals on some zooplankton species are unclear [65].

Slawomir Cerbin et al. found that the allelochemicals secreted by *Myriophyllum verticillatum* had a dual effect on *Daphnia*. In the presence of these chemicals, *Daphnia* became smaller and spawned less at maturity, but the offspring were larger. This is mainly because the allelochemicals of *Myriophyllum verticillatum* reduce the food source of water fleas and increase the energy consumption during swimming, thus limiting the growth of somatic cells. However, the increase in offspring may be an adaptation to food reduction. Despite these effects, the researchers believe that the water fleas are not directly affected by the secretion of *Myriophyllum verticillatum* [66]. Subsequently, Espinosa-Rodríguez et al. found that the allelochemicals secreted by *Egeria densa* had a positive effect on the population size of three *Simocephalus* species. The allelochemicals not only increased the age-specific reproduction yield of these zooplankton but also significantly prolonged their average life span. In the medium containing these allelochemicals, the life span, total fertility rate, and net fertility rate of zooplankton were significantly improved. This finding indicates that the biological activity and physical structure of *A. hygrophila* have a positive, stimulating effect on the population of *Daphnia* [67]. In addition, Alberto et al. explored the effects of allelochemicals secreted by *A. hygrophila* on the interaction between mendotae and three coastal clades through population growth experiments. They found that the allelochemicals increased the abundance of all measured zooplankton. In the absence of allelochemicals, the population growth rate of cladistic animals was lower than that of monoculture. However, in the presence of allelochemicals, this trend is not consistent. This further indicates that the allelochemicals of *A. philoxeroides* have a potentially positive effect on the biological populations of cladistic animals, which may increase the grazing pressure on phytoplankton [68]. In order to further study the impact of macrophytes on aquatic ecosystems, Wolters et al. compared the effects of biofilms formed on *Vallisneria spiralis* and *Egeria densa* and their artificial analogues on two large invertebrate herbivores. They found that macro-submerged macrophytes have a positive impact on large invertebrate herbivores by providing large surface areas for epiphytic algae and bacteria, improving biofilm stoichiometry and stimulating bacterial growth [69]. Finally, Bai et al. conducted a long-term observation of zooplankton biomass in the five sub-lakes of the West Lake. They observed that the zooplankton biomass showed an initial increasing trend followed by a decrease from July 2012 to April 2015. During this period, the main composition of zooplankton was dominated by rotifers, although cladocerans and copepods also accounted for a certain proportion in 2015. These studies have shown that allelochemicals secreted by submerged plants have complex and diverse effects on zooplankton, involving both positive and negative effects. These effects depend on not only the type and concentration of allelochemicals but also a variety of factors, such as environmental variables and the zooplankton species [70].

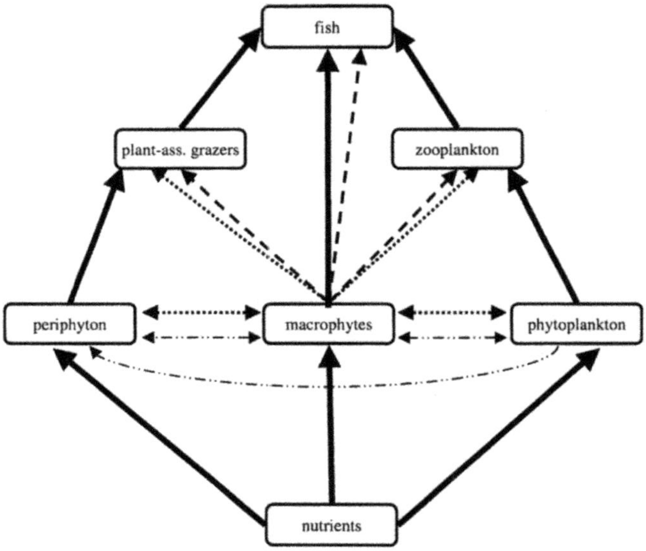

Figure 8. A schematic overview of interactions between submersed macrophytes and other constituents of shallow lake food webs. (—): food web links; (· · ·): allelopathy; (– – –): spatial refuge; (–··–··–): light conditions. Reproduced with permission from [65]. Copyright 2002, Elsevier.

4.3. *Effects on Microorganisms*

Root exudates are the main driving force for regulating rhizosphere microbial diversity and metabolic activities during plant growth [71]. Plants adjust and maintain a specific bacterial community in the rhizosphere by releasing root exudates. The bacteria produce a variety of secondary metabolites, which improve the nutrient utilization and nitrogen fixation of plants, reduce the sensitivity of plants to freezing injury, and enhance plant disease resistance by inhibiting pathogens, thus promoting their overall growth and development [72]. Roots release a variety of nutrients that are essential for microbial growth, such as vitamins, enzymes, growth regulators, and amino acids. These exudates not only affect the spatial distribution, species, and quantity of rhizosphere microorganisms but also change the physical and chemical properties of soil by promoting the formation of soil microaggregates [73]. Root exudates lead to a much higher number and species of rhizosphere microorganisms than those in non-rhizosphere areas, providing energy and good living conditions for microorganisms. Different plants release different root exudates, resulting in differences in the rhizosphere microbial community structure, affecting the water-remediation effect of plants. This natural relationship provides important inspiration for developing synthetic substances to remediate polluted water bodies.

The mechanisms of root exudates in regulating nutrient removal in water bodies are still unclear. However, root surfaces are directly affected by secretions, and REs can accumulate large amounts of organic matter and attract more microorganisms to colonize [74]. Yin et al. selected three dominant submerged plants, *Hydrilla verticillata, Potamogeton maackianus,* and *Vallisneria natans,* to evaluate their effects on the community structure and abundance of nirS-type denitrifying bacteria and anammox bacteria in the rhizosphere. They found that the concentration of organic acids in the near-root layer of submerged plants was higher than that in the root chamber and rootless layer. The concentrations of citric acid and oxalic acid were negatively correlated with the abundance of nirS-type denitrifying bacteria, and the concentration of oxalic acid was positively correlated with the abundance of anammox bacteria. These results indicate that submerged plants can reduce the abundance of nirS-type denitrifying bacteria and anammox bacteria by releasing organic acids [75]. Ma et al. also confirmed this result. As shown in Figure 9, they found that lactic acid and tartaric acid in root exudates of *Vallisneria natans* varied between 0.045–0.380 mg L^{-1} and 0.024–5.446 mg L^{-1}, respectively, which was closely related to the removal rates of TN and TP and most sediment properties. In addition, the top three relative dominant genera were *Bacillus* (0.11–17.90%), *Geobacter* (0.35–12.04%), and *Clostridium parvum* (0.14–12.05%). The results showed that lactic acid, protein, and amino acids positively correlated with *Geobacter*. This study suggests that root exudates, especially proteins, amino acids, and lactic acid, change the relative abundance and diversity of rhizosphere microorganisms, and their effects depend on bacterial species [76]. Martin et al. further emphasized the effects of lighting conditions on the root exudates and rhizosphere microorganisms of submerged plants. They found that lighting reduction affected the production of root exudates, changed the composition of seagrass root microorganisms, and reduced the abundance of potentially beneficial microorganisms. In particular, the decrease in light availability had the most significant effect on the root microorganisms of *Halophila ovalis*, which was consistent with the most significant change in the secretion pattern of the species when the light availability decreased. These results suggest that changes in root exudates are closely related to changes in the microorganisms, which play an important role in regulating seagrass–microbe relationships [77]. In addition, recent studies found that secretions of submerged plants, as an important carbon source for microorganisms, could affect Feammox activity. Although organic carbon is not necessary for Feammox, it can accelerate iron release from clay minerals involved in mediating the Feammox rate [78]. In summary, submerged macrophytes directly affect root-attached microorganisms and their surrounding environment by releasing root exudates and play a key role in nutrient removal in water bodies. The exudates can not only regulate the composition and abundance of microbial communities but also affect the metabolic activities of microorganisms, thereby affecting the nutrient-removal efficiency in water bodies. Future research should further reveal the specific mechanism and application potential of root exudates of submerged plants in water remediation.

4.4. Possibility of Using Root Exudates of Submerged Plants for Water Restoration

Submerged plants secrete "algae-inhibiting substances" to produce algae-inhibiting effects. This can provide significant guidance for the artificial synthesis of algae inhibitors in root exudates and the application of submerged plant root exudates for water restoration. Thus, the methods used to control the development of phytoplankton include introducing living plants into water bodies to prepare dry plant tissues, extracts, and natural allelochemicals or their synthetic analogues [79]. Coexistence experiments involving submerged macrophytes and target phytoplankton demonstrate that the biomass of submerged macrophytes with a 5–8 g L^{-1} wet weight can exhibit an inhibitory effect on phytoplankton [80]. The content of phenolic acids released from submerged plants to water increases gradually with an increase in the density of submerged plants. Considering the economic factors and navigation convenience of restoring submerged vegetation, a 20–50% coverage rate of the planting area may be more conducive to reconstructing submerged plant communities in shallow lakes [81]. The residence time of allelochemicals may be influenced by the evolutionary history of the donor, as microorganisms that co-evolve with allelochemicals may use them as a source of energy [82]. Although there are still some unsolved mysteries in the allelopathy of submerged plants on phytoplankton, it is possible to apply it to control harmful algal blooms or reconstruct submerged plant communities to stabilize water bodies.

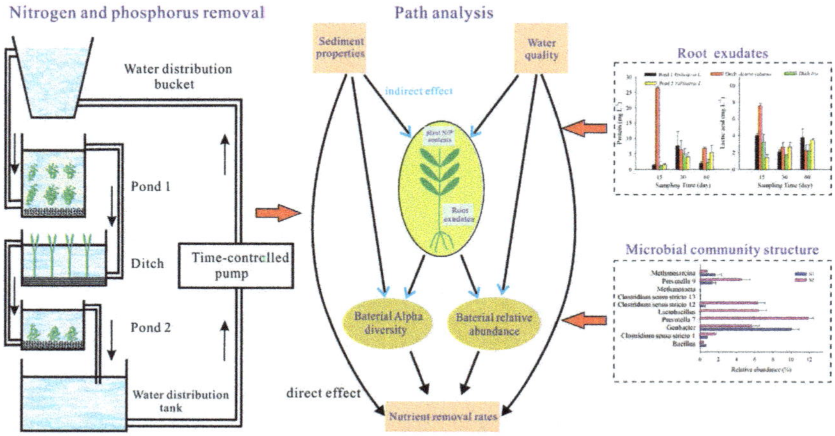

Figure 9. Interaction of root exudates, rhizosphere microorganisms, and water nutrient removal. Reproduced with permission from [76]. Copyright 2021, Elsevier.

5. Conclusions

In this study, the development history, composition, collection and detection methods, and influencing factors of root exudates released by submerged plants were reviewed. Plant species, growth stages, and environmental factors (light, temperature, and nutritional status) are crucial factors affecting root exudates. The positive or negative effects of submerged plant root exudates on phytoplankton, zooplankton, and microorganisms in water were also discussed and are crucial for clarifying the mechanisms of root exudates in water restoration by submerged plants. In particular, allelochemicals in root exudates can inhibit the growth of harmful algae, which is of great significance for maintaining the ecological balance and water restoration of water bodies. In addition, some discoveries in studies on the root exudates of submerged plants conducted in recent years revealed the screening of special bacteria (such as plant-growth-promoting rhizobia) and their application in water restoration by submerged plants, providing a new perspective and method for applying submerged plants in water restoration.

6. Prospects

Based on studies of the root exudates of submerged plants and their effects on aquatic organisms, a new type of algal inhibitor was developed and applied to water restoration. Further works should aim to (1) directly discover allelopathic substances with high algal inhibitory activity, synthesize them artificially, and realize their industrial production; and (2) conduct an in-depth and systematic study of the interactions between allelochemicals, such as synergistic and adjunctive effects. Fewer checks are required for the registration of natural compounds prepared using allelopathy. Therefore, the cost of commercialization can be reduced. However, the following issues still need to be considered in the development and application of algal suppressors: (1) increasing the ability and ease of controlling environmental conditions to produce the required allelochemicals; and (2) evaluating the environmental safety of algal inhibitors. In addition, the effects of plant exudates on other submerged plants should be further studied.

Author Contributions: Conceptualization, Y.S.; methodology, Y.S.; software, Y.S.; validation, Z.W.; formal analysis, Y.S.; investigation, Y.S. and J.G.; resources, Y.S. and J.G.; data curation, Y.S. and J.G.; writing—original draft preparation, Y.S. and J.G.; writing—review and editing, X.Z. (Xu Zhang), M.Z., and C.F.; visualization, W.G.; supervision, X.Z. (Xu Zhang) and C.F.; project administration, Y.S. and J.G.; funding acquisition, X.Z. (Xiangyong Zheng) and C.F. All authors have read and agreed to the published version of the manuscript.

Funding: This research was funded by the Wenzhou Ecological Park Research Project (grant number SY2022ZD-1002-07) and the Wenzhou Science and Technology Project for Basic Society Development (grant number S20220015).

Data Availability Statement: The data presented in this study are available on request from the corresponding author. The data are not publicly available due to funder restrictions.

Acknowledgments: The authors express their sincere gratitude for the work of the editor and the anonymous reviewers.

Conflicts of Interest: Authors Jianya Gu and Wenwen Gu were employed by the company Jiangsu B&P Testing Technology Co., Ltd. The remaining authors declare that the research was conducted in the absence of any commercial or financial relationships that could be construed as a potential conflict of interest.

Nomenclature

MC	Microcystin
PGPR	Plant-growth-promoting rhizobacteria
ABC	ATP-binding cassette
MATE	Multidrug and toxic compound extrusion
MFS	Major facilitator superfamily
ALMT	Aluminum-activated malate transporter
GC	Gas Chromatography
GC-MS	Gas Chromatography–Mass Spectrometry
HPLC	High-Performance Liquid Chromatography
UPLC	Ultra Performance Liquid Chromatography
LC-MS	Liquid Chromatography–Mass Spectrometry
APCI-MS	Atmosphere Pressure Chemical Ionization Mass Spectrometry
UV-B	Ultraviolet B
DOC	Dissolved organic carbon
DOM	Dissolved organic matter
PS	Photosynthesis System
NIRS	Nuclear Information and Resource Service

References

1. Bertin, C.; Yang, X.; Weston, L.A. The role of root exudates and allelochemicals in the rhizosphere. *Plant Soil.* **2003**, *256*, 67–83. [CrossRef]
2. Curl, E.A.; Truelove, B. *The Rhizosphere*; Springer: Berlin/Heidelberg, Germany; Springer: New York, NY, USA; Springer: Tokyo, Japan, 1986.
3. Baetz, U.; Martinoia, E. Root exudates: The hidden part of plant defense. *Trends. Plant Sci.* **2014**, *19*, 90–98. [CrossRef]
4. Hu, M.; Ma, R.; Xue, K.; Cao, Z.; Xiong, J.; Loiselle, S.A.; Shen, M.; Hou, X. Eutrophication evolution of lakes in China: Four decades of observations from space. *J. Hazard. Mater.* **2024**, *470*, 134225. [CrossRef]
5. Ceng, P. Study on the Preparation and Application of High Efficiency Phosphorus Removing Flocculant. Master's Thesis, Wuhan University of Technology, Wuhan, China, 2011.

6. Wang, X.; Song, W.; Li, N.; Lu, J.; Niu, X.; Ma, Y.; Ding, J.; Wang, M. Ultraviolet-B radiation of *Haematococcus pluvialis* for enhanced biological contact oxidation pretreatment of black odorous water in the symbiotic system of algae and bacteria. *Biochem. Eng. J.* **2020**, *157*, 107553. [CrossRef]
7. Yu, S.; Miao, C.; Song, H.; Huang, Y.; Chen, W.; He, X. Efficiency of nitrogen and phosphorus removal by six macrophytes from eutrophic water. *Int. J. Phytoremediat.* **2019**, *21*, 643–651. [CrossRef]
8. Dinakar, C.; Djilianov, D.; Bartels, D. Photosynthesis in desiccation tolerant plants: Energy metabolism and antioxidative stress defense. *Plant Sci.* **2012**, *182*, 29–41. [CrossRef]
9. Long, M.H.; McGlathery, K.J.; Zieman, J.C.; Berg, P. The role of organic acid exudates in liberating phosphorus from seagrass-vegetated carbonate sediments. *Limnol. Oceanogr.* **2008**, *53*, 2616–2626. [CrossRef]
10. Xu, R.; Wu, F.; Hilt, S.; Wu, C.; Wang, X.; Chang, X. Recovery limitation of endangered *Ottelia acuminata* by allelopathic interaction with cyanobacteria. *Aquat. Ecol.* **2015**, *49*, 333–342. [CrossRef]
11. Tamire, G.; Mengistou, S.; Degefe, G. Potential allelopathic impact of *Potamogeton schweinfurthii* on phytoplankton in Lake Ziway, Ethiopia. *Inland Waters* **2016**, *6*, 336–342. [CrossRef]
12. Wang, C.; Wang, H.; Li, Y.; Li, Q.; Yan, W.; Zhang, Y.; Wu, Z.; Zhou, Q. Plant growth-promoting rhizobacteria isolation from rhizosphere of submerged macrophytes and their growth-promoting effect on *Vallisneria natans* under high sediment organic matter load. *Microb. Biotechnol.* **2021**, *14*, 726–736. [CrossRef]
13. Myrold, D.D. The rhizosphere: Biochemistry and organic substances at the soil-plant interface, second edition. *Soil Sci. Soc. Am. J.* **2008**, *72*, 339–353. [CrossRef]
14. Rovira, A.D. Plant root exudates. *Bot. Rev.* **1969**, *35*, 35–57. [CrossRef]
15. Meng, H.; Yan, Z.; Li, X. Effects of exogenous organic acids and flooding on root exudates, rhizosphere bacterial community structure, and iron plaque formation in *Kandelia obovata* seedlings. *Sci. Total Environ.* **2022**, *830*, 154695. [CrossRef]
16. Dietz, S.; Herz, K.; Doell, S.; Haider, S.; Jandt, U.; Bruelheide, H.; Scheel, D. Semi-polar root exudates in natural grassland communities. *Ecol. Evol.* **2019**, *9*, 5526–5541. [CrossRef]
17. Haichar, F.E.; Santaella, C.; Heulin, T.; Achouak, W. Root exudates mediated interactions belowground. *Soil Biol. Biochem.* **2014**, *77*, 69–80. [CrossRef]
18. Ma, Y.; Oliveira, R.S.; Freitas, H.; Zhang, C. Biochemical and molecular mechanisms of plant-microbe-metal interactions: Relevance for phytoremediation. *Front. Plant Sci.* **2016**, *7*, 918. [CrossRef]
19. Huang, X.F.; Chaparro, J.M.; Reardon, K.F.; Zhang, R.; Shen, Q.; Vivanco, J.M. Rhizosphere interactions: Root exudates, microbes, and microbial communities. *Botany* **2014**, *92*, 267–275. [CrossRef]
20. Weston, L.A.; Ryan, P.R.; Watt, M. Mechanisms for cellular transport and release of allelochemicals from plant roots into the rhizosphere. *J. Exp. Bot.* **2012**, *63*, 3445–3454. [CrossRef]
21. Selmar, D.; Abouzeid, S.; Radwan, A.; Hijazin, T.; Kleinwchter, M. Horizontal Natural Product Transfer: A Novel Attribution in Allelopathy. In *Co-Evolution of Secondary Metabolites*, 3rd ed.; Springer: Cham, Switzerland, 2020; pp. 429–439.
22. Hijazin, T.; Lewerenz, L.; Yahyazadeh, M.; Selmar, D. Horizontal Natural Product Transfer: A Phenomenon Which Is Responsible for the Widespread Alkaloidal Contaminations of Herbal Products. In *Environmental Challenges and Medicinal Plants*, 3rd ed.; Springer: Cham, Switzerland, 2022; pp. 183–201.
23. Lewerenz, L.; Hijazin, T.; Abouzeid, S.; Haensch, R.; Selmar, D. Pilot study on the uptake and modification of harmaline in acceptor plants: An innovative approach to visualize the interspecific transfer of natural products. *Phytochemistry* **2020**, *174*, 112362. [CrossRef]
24. Hijazin, T.; Radwan, A.; Lewerenz, L.; Abouzeid, S.; Selmar, D. The uptake of alkaloids by plants from the soil is determined by rhizosphere pH. *Rhizosphere* **2020**, *15*, 100234. [CrossRef]
25. Tu, S.; Ma, L.; Luongo, T. Root exudates and arsenic accumulation in arsenic hyperaccumulating *Pteris vittata* and non-hyperaccumulating *Nephrolepis exaltata*. *Plant Soil.* **2004**, *258*, 9–19. [CrossRef]
26. Kudoyarova, G.R.; Melentiev, A.I.; Martynenko, E.V.; Timergalina, L.N.; Arkhipova, T.N.; Shendel, G.V.; Kuz'mina, L.Y.; Dodd, I.C.; Veselov, S.Y. Cytokinin producing bacteria stimulate amino acid deposition by wheat roots. *Plant Physiol. Bioch.* **2014**, *83*, 285–291. [CrossRef] [PubMed]
27. Pierre, P.; Alex, W.; Anne, C.; Alexander, E.M.; Stephen, A.R.; Jurriaan, T. Metabolite profiling of non-sterile rhizosphere soil. *Plant J.* **2017**, *92*, 147–162.
28. Nakai, S.; Inoue, Y.; Hosomi, M.; Murakami, A. *Myriophyllum spicatum*-released allelopathic polyphenols inhibiting growth of blue-green algae *Microcystis aeruginosa*. *Water Res.* **2000**, *34*, 3026–3032. [CrossRef]
29. Herz, K.; Dietz, S.; Gorzolka, K.; Haider, S.; Jandt, U.; Scheel, D.; Bruelheide, H. Linking root exudates to functional plant traits. *PLoS ONE* **2019**, *14*, e0204128. [CrossRef] [PubMed]
30. Casas, M.E.; Matamoros, V. Analytical challenges and solutions for performing metabolomic analysis of root exudates. *Trends Environ. Anal. Chem.* **2021**, *31*, e00130. [CrossRef]
31. Robinson, M.D.; De Souza, D.P.; Keen, W.W.; Saunders, E.C.; McConville, M.J.; Speed, T.P.; Likic, V.A. A dynamic programming approach for the alignment of signal peaks in multiple gas chromatography-mass spectrometry experiments. *BMC Bioinform.* **2007**, *8*, 419. [CrossRef] [PubMed]
32. Song, Y.; Song, Q.; Liu, Y.; Li, J.; Wan, J.B.; Wang, Y.; Jiang, Y.; Tu, P. Integrated work-flow for quantitative metabolome profiling of plants, *Peucedani Radix* as a case. *Anal. Chim. Acta* **2017**, *953*, 40–47. [CrossRef] [PubMed]

33. Michopoulos, F.; Lai, L.; Gika, H.; Theodoridis, G.; Wilson, I. UPLC-MS-based analysis of human plasma for metabonomics using solvent precipitation or solid phase extraction. *J. Proteome Res.* **2009**, *8*, 2114–2121. [CrossRef]
34. Plumb, R.S.; Johnson, K.A.; Rainville, P.; Smith, B.W.; Wilson, I.D.; Castro-Perez, J.M.; Nicholson, J.K. UPLC/MS^E; a new approach for generating molecular fragment information for biomarker structure elucidation. *Rapid Commun. Mass Spectrom.* **2006**, *20*, 1989–1994. [CrossRef]
35. Zheng, X.B.; Liu, S.H.; Wen, B.Y.; Peng, W.; Li, H.M.; Kathiresan, M.; Zhang, Y.J.; Jin, S.Z.; Li, J.F. Rapid detection and whole class control of quinolone antibiotics in pork based on surface-enhanced Raman spectroscopy. *J. Raman Spectrosc.* **2023**, *54*, 468–476. [CrossRef]
36. Oledzka, I.; Baczek, T. Urinary steroids measured by modern separation techniques and applied as biomarkers in stress studies. *Curr. Pharm. Anal.* **2010**, *6*, 151–163. [CrossRef]
37. Hu, L.; Robert, C.A.M.; Cadot, S.; Zhang, X.; Ye, M.; Li, B.; Manzo, D.; Chervet, N.; Steinger, T.; van der Heijden, M.G.A.; et al. Root exudate metabolites drive plant-soil feedbacks on growth and defense by shaping the rhizosphere microbiota. *Nat. Commun.* **2018**, *9*, 2738. [CrossRef] [PubMed]
38. Wu, C.; Chang, X.X.; Wu, F.; Liu, J.Y.; Zheng, G.L. Studies on allelopathy of aquatic macrophytes on *Synahocystis* sp. *J. Yunnan Univ.* **2008**, *30*, 535–540.
39. Zhang, S.H.; Sun, P.S.; Ge, F.J.; Wu, Z.B. Different sensitivities of *Selenastrum capricornutum* and toxic strain *Microcystis aeruginosa* to exudates from two *Potamogeton* species. *Pol. J. Environ. Stud.* **2011**, *20*, 1359–1366.
40. Xing, X.; Ding, S.; Liu, L.; Chen, M.; Yan, W.; Zhao, L.; Zhang, C. Direct evidence for the enhanced acquisition of phosphorus in the rhizosphere of aquatic plants: A case study on *Vallisneria natans*. *Sci. Total Environ.* **2018**, *616*, 386–396. [CrossRef] [PubMed]
41. Pakdel, F.M.; Sim, L.; Bearda, J.; Davis, J. Allelopathic inhibition of microalgae by the freshwater stonewort, *Chara australis*, and a. submerged angiosperm, *Potamogeton crispus*. *Aquat. Bot.* **2013**, *110*, 24–30. [CrossRef]
42. Mulderij, G.; Van, D.E.; Roelofs, J.G.M. Differential sensitivity of green algae to allelopathic substances from Chara. *Hydrobiologia* **2003**, *491*, 261–271. [CrossRef]
43. Wang, W.; Jia, T.; Qi, T.; Li, S.; Degen, A.A.; Han, J.; Bai, Y.; Zhang, T.; Qi, S.; Huang, M.; et al. Root exudates enhanced rhizobacteria complexity and microbial carbon metabolism of toxic plants. *iScience* **2022**, *25*, 105243. [CrossRef]
44. Gross, E.M. Differential response of tellimagrandin II and total bioactive hydrolysable tannins in an aquatic angiosperm to changes in light and nitrogen. *Oikos* **2003**, *103*, 497–504. [CrossRef]
45. Erhard, D.; Gross, E.M. Do environmental factors influence composition of potential allelochemicals in the submersed freshwater macrophyte *Elodea nuttallii* (*Hydrocharitaceae*)? *Int. Assoc. Theor. Appl. Limnoloy* **2005**, *29*, 287–291. [CrossRef]
46. Martin, B.C.; Statton, J.; Siebers, A.R.; Grierson, P.F.; Ryan, M.H.; Kendrick, G.A. Colonizing tropical seagrasses increase root exudation under fluctuating and continuous low light. *Limnol. Oceanogr.* **2018**, *63*, S381–S391. [CrossRef]
47. Bressan, M.; Roncato, M.A.; Bellvert, F.; Comte, G.; Haichar, F.E.Z.; Achouak, W.; Berge, O. Exogenous glucosinolate produced by *Arabidopsis thaliana* has an impact on microbes in the rhizosphere and plant roots. *ISME J.* **2009**, *3*, 1243–1257. [CrossRef] [PubMed]
48. Gu, L.Y.; Wyatt, K.H. Light availability regulates the response of algae and heterotrophic bacteria to elevated nutrient levels and warming in a northern boreal peatland. *Freshw. Biol.* **2016**, *61*, 1442–1453. [CrossRef]
49. Gu, S.; Qian, Y.; Jiao, Y.; Li, Q.; Pinay, G.; Gruau, G. An innovative approach for sequential extraction of phosphorus in sediments: Ferrous iron P as an independent P fraction. *Water Res.* **2016**, *103*, 352–361. [CrossRef] [PubMed]
50. Gross, E.M.; Meyer, H.; Schilling, G. Release and ecological impact of algicidal hydrolysable polyphenols in *Myriophyllum spicatum*. *Phytochemistry* **1996**, *41*, 133–138. [CrossRef]
51. Jones, D.L. Organic acids in the rhizosphere. A critical review. *Plant Soil.* **1998**, *205*, 25–44. [CrossRef]
52. Taghipour, M.; Jalali, M. Effect of low-molecular-weight organic acids on kinetics release and fractionation of phosphorus in some calcareous soils of western Iran. *Environ. Monit. Assess.* **2013**, *185*, 5471–5482. [CrossRef] [PubMed]
53. Bachheti, A.; Sharma, A.; Bachheti, R.K.; Husen, A.; Pandey, D.P. Plant Allelochemicals and Their Various Applications. In *Co-Evolution of Secondary Metabolites*; Springer: Cham, Switzerland, 2020.
54. Filzgerald, G.P. Some factors in the competition or antagonism among bacteria, algae, and aquatic weeds. *J. Phycol.* **1969**, *5*, 351–359. [CrossRef]
55. Hilt, S.; Gross, E.M.; Hupfer, M.; Morscheid, H.; Maehlmann, J.; Melzer, A.; Poltz, J.; Sandrock, S.; Scharf, E.M.; Schneider, S.; et al. Restoration of submerged vegetation in shallow eutrophic lakes—A guideline and state of the art in Germany. *Limnologica* **2006**, *36*, 155–171. [CrossRef]
56. Li, B.; Yin, Y.; Kang, L.; Feng, L.; Liu, Y.; Du, Z.; Tian, Y.; Zhang, L. A review: Application of allelochemicals in water ecological restoration--algal inhibition. *Chemosphere* **2021**, *267*, 128869. [CrossRef] [PubMed]
57. Declerck, S.; Vanderstukken, M.; Pals, A.; Muylaert, K.; de Meester, L. Plankton biodiversity along a gradient of productivity and its mediation by macrophytes. *Ecology* **2007**, *88*, 2199–2210. [CrossRef] [PubMed]
58. Svanys, A.; Paskauskas, R.; Hilt, S. Effects of the allelopathically active macrophyte *Myriophyllum spicatum* on a natural phytoplankton community: A mesocosm study. *Hydrobiologia* **2014**, *737*, 57–66. [CrossRef]
59. Wu, Z.; Liu, A.; Zhang, S.; Cheng, S.; Wu, X. Short-term effects of drawing water for connectivity of rivers and lakes on zooplankton community structure. *J. Environ. Sci.* **2008**, *20*, 419–423. [CrossRef]

60. Hilt, S.; Ghobrial, M.G.N.; Gross, E.M. In situ allelopathic potential of *Myriophyllum verticillatum* (*Haloragaceae*) against selected phytoplankton species. *J. Phycol.* **2006**, *42*, 1189–1198. [CrossRef]
61. Rojo, C.; Segura, M.; Rodrigo, M.A. The allelopathic capacity of submerged macrophytes shapes the microalgal assemblages from a recently restored coastal wetland. *Ecol. Eng.* **2013**, *58*, 149–155. [CrossRef]
62. Zhu, X.; Dao, G.; Tao, Y.; Zhan, X.; Hu, H. A review on control of harmful algal blooms by plant-derived allelochemicals. *J. Hazard. Mater.* **2021**, *401*, 123403. [CrossRef] [PubMed]
63. Leu, E.; Krieger-Liszkay, A.; Goussias, C.; Gross, E.M. Polyphenolic allelochemicals from the aquatic angiosperm *Myriophyllum spicatum* inhibit photosystem II. *Plant Physiol.* **2002**, *130*, 2011–2018. [CrossRef] [PubMed]
64. Zhu, J.; Liu, B.; Wang, J.; Gao, Y.; Wu, Z. Study on the mechanism of allelopathic influence on cyanobacteria and chlorophytes by submerged macrophyte (*Myriophyllum spicatum*) and its secretion. *Aquat. Toxicol.* **2010**, *98*, 196–203. [CrossRef]
65. Van Donk, E.; van de Bund, W.J. Impact of submerged macrophytes including charophytes on phyto- and zooplankton communities: Allelopathy versus other mechanisms. *Aquat. Bot.* **2002**, *72*, 261–274. [CrossRef]
66. Cerbin, S.; van Donk, E.; Gulati, R.D. The influence of *Myriophyllum verticillatum* and artificial plants on some life history parameters of *Daphnia magna*. *Aquat. Ecol.* **2007**, *41*, 263–271. [CrossRef]
67. Espinosa-Rodriguez, C.A.; Rivera-De la Parra, L.; Martinez-Tellez, A.; Gomez-Cabral, G.C.; Sarma, S.S.S.; Nandini, S. Allelopathic interactions between the macrophyte *Egeria densa* and plankton (alga, *Scenedesmus acutus* and cladocerans, *Simocephalus* spp.): A laboratory study. *J. Limnol.* **2016**, *75*, 151–160. [CrossRef]
68. Espinosa-Rodriguez, C.A.; Sarma, S.S.S.; Nandini, S. Effect of the allelochemicals from the macrophyte *Egeria densa* on the competitive interactions of pelagic and littoral cladocerans. *Chem. Ecol.* **2017**, *33*, 247–256. [CrossRef]
69. Wolters, J.W.; Reitsema, R.E.; Verdonschot, R.C.M.; Schoelynck, J.; Verdonschot, P.F.M.; Meire, P. Macrophyte-specific effects on epiphyton quality and quantity and resulting effects on grazing macroinvertebrates. *Freshwater. Biol.* **2019**, *64*, 1131–1142. [CrossRef]
70. Bai, G.; Liu, Y.; Liu, Z.; Kong, L.; Tang, Y.; Ding, Z.; Zou, Y.; Wang, C.; Zhang, C.; Chen, D.; et al. Effects of lake geo-engineering on plankton in a typical shallow urban lake: Evidence based on 10-year data. *ACS EST Eng.* **2022**, *3*, 105–120. [CrossRef]
71. Vanderstukken, M.; Declerck, S.A.J.; Decaestecker, E.; Muylaert, K. Long-term allelopathic control of phytoplankton by the submerged macrophyte *Elodea nuttallii*. *Freshwater Biol.* **2014**, *59*, 930–941. [CrossRef]
72. Scheffer, M.; Hosper, S.H.; Meijer, M.L.; Moss, B.; Jeppesen, E. Alternative equilibria in shallow lakes. *Trends Ecol. Evol.* **1993**, *8*, 275–279. [CrossRef] [PubMed]
73. de Vries, F.T.; Williams, A.; Stringer, F.; Willcocks, R.; McEwing, R.; Langridge, H.; Straathof, A.L. Changes in root-exudate-induced respiration reveal a novel mechanism through which drought affects ecosystem carbon cycling. *New Phytol.* **2019**, *224*, 132–145. [CrossRef]
74. Landi, L.; Valori, F.; Ascher, J.; Renella, G.; Falchini, L.; Nannipieri, P. Root exudate effects on the bacterial communities, CO_2 evolution, nitrogen transformations and ATP content of rhizosphere and bulk soils. *Soil Biol. Biochem.* **2006**, *38*, 509–516. [CrossRef]
75. Yin, X.; Lu, J.; Wang, Y.; Liu, G.; Hua, Y.; Wan, X.; Zhao, J.; Zhu, D. The abundance of *nirS*-type denitrifiers and anammox bacteria in rhizospheres was affected by the organic acids secreted from roots of submerged macrophytes. *Chemosphere* **2020**, *240*, 124903. [CrossRef]
76. Ma, L.; Yang, L.; Liu, W.; Zhang, Y.; Zhou, Q.; Wu, Z.; He, F. Effects of root exudates on rhizosphere bacteria and nutrient removal in pond-ditch circulation systems (PDCSs) for rural wastewater treatment. *Sci. Total Environ.* **2021**, *785*, 147282. [CrossRef] [PubMed]
77. Martin, B.C.; Gleeson, D.; Statton, J.; Siebers, A.R.; Grierson, P.; Ryan, M.H.; Kendrick, G.A. Low light availability alters root exudation and reduces putative beneficial microorganisms in seagrass roots. *Front. Microbiol.* **2018**, *8*, 2667. [CrossRef] [PubMed]
78. Sun, S.; Zhang, M.; Gu, X.; Yan, P.; He, S.; Chachar, A. New insight and enhancement mechanisms for Feammox process by electron shuttles in wastewater treatment-A systematic review. *Bioresour. Technol.* **2023**, *369*, 128495. [CrossRef] [PubMed]
79. Gao, Y.; Dong, J.; Fu, Q.; Wang, Y.; Chen, C.; Li, J.; Li, R.; Zhou, C. Allelopathic effects of submerged macrophytes on phytoplankton. *Allelopath. J.* **2017**, *40*, 1–22. [CrossRef]
80. Eigemann, F. Allelopathic Effects of Submerged Macrophytes on Phytoplankton: Determining the Factors of Phytoplankton Sensitivity and Detection of New Modes of Action. Ph.D. Thesis, Freie Universität Berlin, Berlin, Germany, 2013; pp. 77–100.
81. Peng, X.; Lin, Q.; Liu, B.; Huang, S.; Yan, W.; Zhang, L.; Ge, F.; Zhang, Y.; Wu, Z. Effect of submerged plant coverage on phytoplankton community dynamics and photosynthetic activity in situ. *J. Environ. Manag.* **2022**, *301*, 113822. [CrossRef]
82. Müller, N.; Hempel, M.; Philipp, B.; Gross, E.M. Degradation of gallic acid and hydrolysable polyphenols is constitutively activated in the freshwater plant-associated bacterium *Matsuebacter* sp. FB25. *Aquat. Microb. Ecol.* **2007**, *47*, 83–90. [CrossRef]

water

MDPI

Article

Impact of Gas-to-Water Ratio on Treatment Efficiency of Submerged-Macrophyte Constructed Wetland Systems

Hongming Mao [1], Shiwen Lu [1,2], Guiying Huang [3], Xianle Jia [4], Chenqian Bao [1], Xinyi Yan [1], Xiangyong Zheng [1], Min Zhao [1] and Zhan Jin [1,*]

1 National and Local Joint Engineering Research Center of Ecological Treatment Technology for Urban Water Pollution, Institute for Eco-Environmental Research of Sanyang Wetland, Wenzhou University, Wenzhou 325035, China; 18857150334@163.com (H.M.); 18770825248@163.com (S.L.); averyccq@163.com (C.B.); 17816259001@163.com (X.Y.); 00121051@wzu.edu.cn (X.Z.); zmzjcn@sina.com.cn (M.Z.)

2 Jiujiang Branch of Jiangsu Hongrun Biomass Energy Technology Co., Ltd., Jiujiang 332000, China

3 Taishun Branch of Wenzhou Municipal Bureau of Ecology and Environment, Wenzhou 325500, China; huanggy5715@163.com

4 Wenzhou Municipal Bureau of Ecology and Environment, Wenzhou 325000, China; 18758749968@163.com

* Correspondence: 20150080@wzu.edu.cn

Abstract: Constructed wetland systems employing submerged macrophytes are increasingly utilized for treating municipal and industrial wastewater, as well as odoriferous and eutrophic water bodies. However, the pollutant removal efficiency of these systems needs further enhancement. In this study, we examined the impact of the gas-to-water ratio on the treatment efficiency of the constructed wetland of Vallisneria. We also examined the extracellular polymeric substances (EPSs) of the floating biofilm and the structure of the microbial community in this system. Our findings showed that the gas-to-water ratio significantly affects the total nitrogen (TN) removal rate within the Vallisneria wetlands, with an optimum removal at a gas-to-water ratio of 15:1, while the removal efficiencies for chemical oxygen demand (COD), NH_4^+-N, and total phosphorus (TP) remain relatively unaffected. Increased gas-to-water ratios corresponded to a notable decrease in biofilm EPSs. High-throughput sequencing analysis demonstrated a shift in biofilm-denitrifying bacteria from anoxic heterotrophic to aerobic denitrifiers, alongside a significant rise in the abundance of denitrifying bacteria, whereas excessively high gas-to-water ratios inhibited the growth of these bacteria. A gas-to-water ratio of 15:1 constituted the optimal condition for ecological restoration of the water body within the Vallisneria wetland systems. These results could contribute to the optimization of submerged-macrophyte constructed wetland system design and the enhancement of treatment efficiency.

Keywords: constructed wetlands; Vallisneria; gas-to-water ratio; microbial community; extracellular polymeric substances

1. Introduction

High levels of nitrogen and phosphorus in water bodies can threaten the biodiversity of aquatic environments, disturb the stability of ecosystems, and lead to eutrophication [1]. The current denitrification methods used to solve the prevalent problem of water eutrophication can be classified as physicochemical or biological methods. Among them, biological methods (such as constructed wetlands) have attracted extensive attention due to their effective denitrification performance and non-toxic by-products and their potential application in wastewater treatment [2,3].

Constructed wetlands are extensively utilized in the treatment of domestic sewage [1], industrial wastewater [4], and black and malodorous water [5] due to their cost-effectiveness, ease of operation and maintenance, and environmentally friendly nature [6]. As a crucial component of constructed wetlands, plants, particularly submerged plants, are fully immersed beneath the water surface and come into direct contact with both sewage and microorganisms. The roots, stems, leaves, and attached biofilms of these plants play a significant role in the removal of nutrients from the water column [7,8]. However, previous studies have indicated that the impact of submerged plants on nutrient removal during water purification is limited, with the removal of nitrogen and phosphorus attributed to plant-attached microbial communities [9]. Submerged plants primarily serve as habitats and provide fixed substrates for the attached biota [10], while also releasing DO for other microorganisms [11,12]. This influences the abundance, activity, and reaction processes of the functional bacteria in the plants' epiphytic biofilms, thereby affecting the efficacy of the ecosystem treatment.

However, plants have a limited secretory capacity, and their efficiency in reducing water pollution requires further improvement [13]. Studies have revealed that in constructed wetlands dominated by emergent aquatic plants, enhanced measures such as aeration are often employed to augment the removal capacity of constructed wetlands. Specifically, a higher gas-to-water ratio can lead to improved effectiveness in the removal of pollutants [14]. Currently, there is a greater emphasis on comprehensive investigations of constructed wetlands with submerged plants, both domestically and internationally [10]. Conversely, research on submerged-plant-based constructed wetlands primarily focuses on the removal mechanisms of pollutants by attached biofilms [15,16]; intramembrane microbial community structures [10]; external environmental conditions such as microcystins [17], microplastics [18], antibiotics [19], and harmful algal bloom harvests [20]; and other factors in the structure of biofilm microbial communities. However, limited attention has been paid to measures aimed at enhancing the removal efficiency of submerged-plant-based constructed wetlands, which hinders their widespread adoption and application.

The Vallisneria exhibits well-developed root tissue, strong pollution resistance, high reproductive capacity, low light tolerance, and effective removal of ammonia nitrogen [21]. Vallisneria natans is a common submerged macrophyte in most eutrophic lakes in China that can tolerate and purify polluted water [17]. Therefore, this study used the epiphytic biofilm on grass leaves as its research object, focusing on investigating the wastewater treatment efficiency of a constructed wetland with submerged plants and different air–water ratios. Additionally, the microbial community structure within the epiphytic biofilm was characterized to explore the influence mechanism of the air–water ratio on the submerged plant system.

2. Materials and Methods

2.1. System Construction

Three PVC setups measuring 50 × 30 × 40 cm, with a 5 cm layer of river sand at their base and 15 L water, were constructed. Forty-eight Vallisneria plants of similar growth were selected from a laboratory-acclimated biofilm culture and planted in each sept using sixteen plants per setup. Three gas-to-water ratios of 10:1 (A, 2 h of aeration), 15:1 (B, 3 h of aeration), and 20:1 (C, 4 h of aeration) were investigated for their effects on the water quality, biofilm EPS content, and microbial community structure. A 3-day water renewal cycle and a 25-day experimental run were adopted. COD, TN, TP, NH_4^+-N, NO_3^--N, NO_2^--N, DO, and pH of the influent and effluent were measured. High-throughput sequencing analysis of the biofilm was performed at the end of the experiments using plant samples from each setup (Shanghai Meji Biomedical Technology Co., Ltd.; Shanghai, China).

2.2. Determination of EPS Protein and Polysaccharide in Biofilm

Polysaccharides and proteins in EPSs were extracted by thermal digestion: 3.0 g of leaves was first removed from the reactor and then placed in a 30 mL centrifuge tube and shaken with 20 mL of deionized water for 1 min to separate the leaves from the biofilm. We removed the leaves, placed the remaining solution in a 70 °C water bath for 30 min, and then centrifuged at 2500× g for 15 min [20] and the supernatant was used for EPS analysis. Phenol-sulfuric acid method and BCA assay (Beyotime, P0012;

Shanghai Biyuntian Biotechnology Co., Ltd., Shanghai, China) were utilized for quantifying polysaccharides and proteins, respectively.

2.3. High-Throughput Sequencing

High-throughput sequencing was used to analyze microbial communities in leaf epiphytic biofilms. The collected plant leaf samples were transported on ice to the lab, where the biofilm was detached via repeated ultrasonication and vortexing in phosphate-buffered saline (pH = 8.0). The samples were then filtered through a 0.22 μm membrane. Membranes were flash-frozen in liquid nitrogen and stored at −80 °C. The biofilm was cut into 50 μm slices using a cryostat, and the DNA was extracted from the biofilm samples for subsequent gel electrophoresis and NanoDrop-2000 spectrophotometry to check the quality and quantity of the DNA. Amplification of the 16S V3-V4 region was performed on a Veriti FAST thermal cycler using 338F and 806R universal primers. After the sequencing data were spliced, quality-controlled, and de-spliced, the optimized sequences were obtained. Based on the optimized sequences, OUT clustering was performed to obtain the OUT abundance for subsequent biological information analysis. The whole sequencing experiment was completed by Shanghai Meiji Biomedical Technology Co., Ltd., Shanghai, China.

2.4. Data Processing and Analysis

Data processing and analyses were performed using SPSS 24.0 software (IBM, Armonk, NY, USA) for ANOVA or Student's *t*-test to discern significant differences across treatment groups ($p < 0.05$), with plotting and further analyses conducted in Origin 2018b (OriginLab Corporation, Northampton, MA, USA).

3. Results and Discussion

3.1. Effect of Different Gas-to-Water Ratios on Water Quality Treatment

Different gas-to-water ratios were used to assess the effectiveness of the water treatment using Vallisneria. These gas-to-water ratios were 10:1 (A), 15:1 (B), and 20:1 (C). The removal rates for the various indicators were as follows: A—COD, 51.92 ± 15.37%; TN, 13.90 ± 7.08%; TP, 30.32 ± 6.42%, NH_4^+-N, 20.13 ± 10.71%; B—COD, 57.51 ± 19.08%%; TN, 23.49 ± 8.86%; TP, 36.49 ± 7.81%, NH_4^+-N, 26.97 ± 7.31; and C—COD, 57.51 ± 19.08%; TN, 20.24 ± 10.49%; TP, 20.51 ± 8.56%, NH_4^+-N, 23.64 ± 9.89%. Significant differences in TN removal between groups A and B were noted ($p < 0.05$), with no significant differences found for other parameters. The DO in the outflow water for groups A, B, and C averaged 1.64 ± 1.31 mg/L, 1.89 ± 1.20 mg/L, and 2.46 ± 0.90 mg/L, respectively, while the average pH values in these groups were 7.58 ± 0.08, 7.56 ± 0.90, and 7.46 ± 0.19, respectively. All three gas-to-water ratios resulted in an increase in the concentration of DO in the effluent (Figure 1a). The COD removal rates increased, whereas the pH decreased with rising gas-to-water ratios, with a significant difference in pH value between conditions A and C. Increases in COD removal rates and decreases in pH values upon increasing gas-to-water ratios could possibly be a result of a positive correlation between the DO and COD and a negative correlation between the DO and pH [22].

3.2. Effects of Different Gas-to-Water Ratios on EPS

As shown in Figure 2, the contents of proteins and polysaccharides in group A increased with time, while the contents of polysaccharides and proteins in supernatants from group B and group C decreased with treatment time. On the one hand, the proteins and polysaccharides in the EPSs were biodegradable, and the increase in gas-to-water ratio increased the concentration of DO in the system. As the microorganisms in the system adapted to the environment, their activities became more intense. The accumulated EPS in the system is often used as a substrate for microbial consumption and decomposition [23]. On the other hand, the decrease in the amount of proteins and polysaccharides in groups B and C may be due to the increase in the gas-to-water ratio, oxygen deoxygenation, and oxygen transfer rate in the system, so that the microbial membrane has a higher level of metabolic activity, and the polysaccharide production rate is less than the consumption rate [24]. As polysaccharide is the main source of EPS stickiness [25], a decrease in polysaccharide content and EPS stickiness may lead to a decrease in the shedding of the attached biofilm on the leaf surface, consequently leading to a decrease in the removal rate of ammonium nitrogen. At the same time, the pH value was higher under the condition of a low gas-to-water ratio (Figure 1c), and the alkaline environment would cause the aggregation of polysaccharide sticky substances [26], which may be the reason for the higher EPS content in group A. Although the content of polysaccharides and protein in group A increased, in the case of a gas-to-water ratio of 15:1, the system has the best nitrogen removal effect (Figure 1a). A low aeration rate may affect the nitrification efficiency and fail to provide sufficient electron acceptors for denitrification, resulting in poor TN removal, while a high gas-to-water ratio may lead to excessive oxidation of carbon sources and affect the denitrification effect, resulting in a poor TN removal effect.

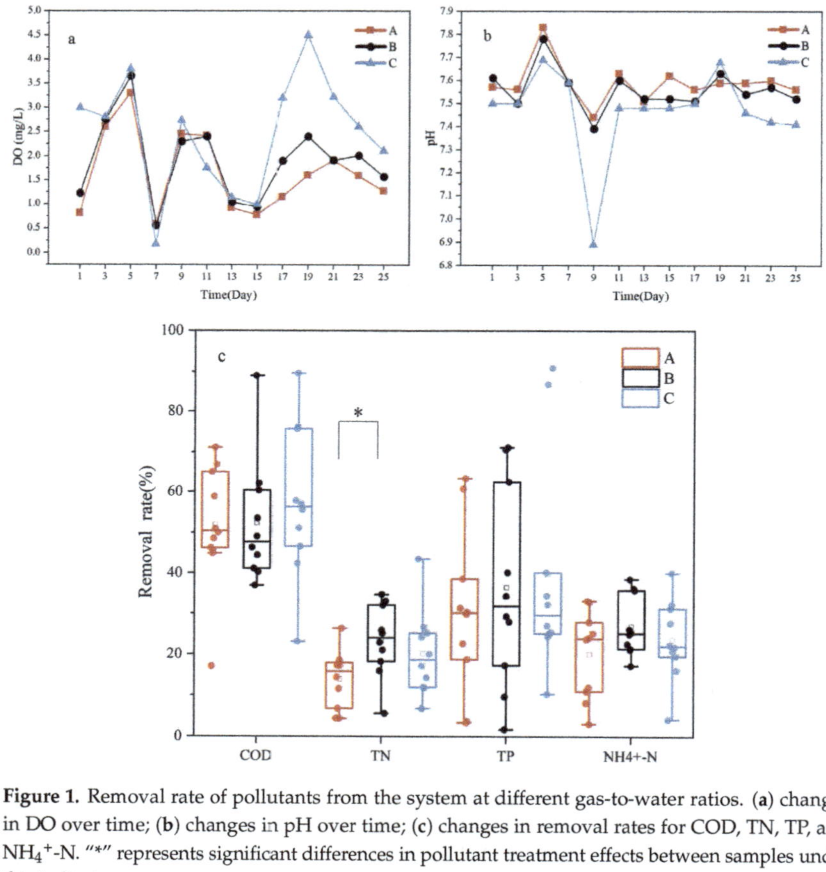

Figure 1. Removal rate of pollutants from the system at different gas-to-water ratios. (**a**) changes in DO over time; (**b**) changes in pH over time; (**c**) changes in removal rates for COD, TN, TP, and NH_4^+-N. "*" represents significant differences in pollutant treatment effects between samples under this indicator.

3.3. Alpha Diversity Analysis and Microbial Community Composition

As depicted in Table 1, the alpha diversity analysis identified an average of 1042, 789 and 1608 observed OTU sequences in groups A, B, and C. This indicates that group C bacteria have a higher abundance of microbial species. The coverage indices for all samples exceeded 0.96, indicating a sufficient sequencing depth. The Shannon and Chao indexes are closely related to bacterial community diversity, while the Simpson index reflects the most prevalent species within communities, and the Ace indicator can be used to measure the sample coverage. No statistical differences in the Simpson index were found among the three groups ($p = 0.0509$), whereas significant differences were observed among the three groups for the following three indices: Shannon ($p = 0.0273$), Chao indices ($p = 0.0273$), and Ace indices ($p = 0.0273$) (Table 1). Based on these findings, it was anticipated that the gas-to-water ratio could significantly influence microbial community diversity. However, it does not affect the uniformity of the species, supporting the results of significant differences between groups.

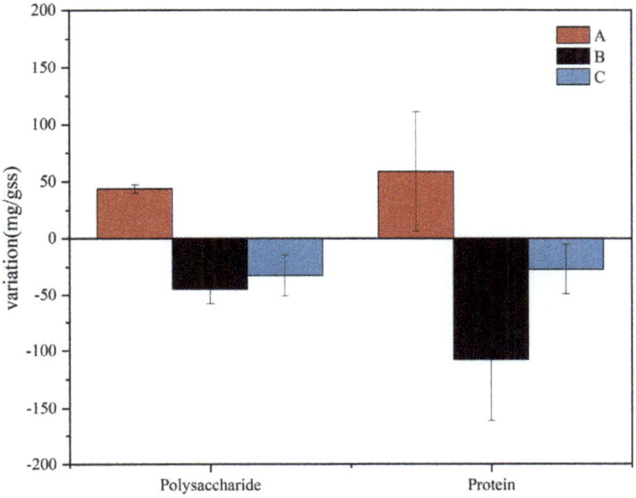

Figure 2. Polysaccharide and protein mass variation chart.

Table 1. Statistical table of microbial diversity analysis.

Sample	Sobs *	Shannon *	Simpson	Ace *	Chao *	Coverage
A1	1037	5.28	0.013	1472.07	1508.50	0.9772
A2	1017	5.30	0.012	1454.19	1472.63	0.9776
A3	1073	5.29	0.013	1539.70	1525.28	0.9761
B1	784	4.22	0.040	1781.48	1364.31	0.9772
B2	779	4.28	0.040	1683.49	1287.37	0.9780
B3	805	4.43	0.031	1661.36	1257.76	0.9782
C1	1541	5.44	0.021	2130.90	2106.45	0.9661
C2	1655	5.73	0.013	2393.24	2352.90	0.9612
C3	1627	5.67	0.013	2415.20	2374.84	0.9604

Note: '*' indicates that the indicators of samples under the different gas-to-water ratios have significant differences (A1 vs. B1).

3.4. Microbial Community Diversity

At the genus level, the top 10 abundant bacteria accounted for 35.92%, 47.09%, and 37.54% of the total bacteria in groups A, B, and C, respectively, with the dominant genera being *Dechloromonas*, *unclassified_f__Rhodocyclaceae*, *unclassified_f__Comamonadaceae*, *Zoogloea*, *Gemmobacter*, *Azohydromonas*, *env.OPS_17*, and *Ramlibacter* (Figure 3).

Group A was significantly enriched with genera such as *Azohydromonas* (4.557%), *unclassified_f__Comamonadaceae* (6.841%), *Aquabacterium* (3.667%), *norank_f__Polyangiaceae* (3.178%), *Thauera* (3.204%), *Ideonella* (2.925%), and *norank_f__Roseiflexaceae* (1.784%). *Aquabacterium* and *unclassified_f__Comamonadaceae* are known as heterotrophic anoxic denitrifiers [27,28], *Ideonella* as heterotrophic nitrifying-aerobic denitrifiers [29], and *Thauera* as facultative heterotrophic denitrifiers [30,31]. In group B, significantly enriched genera included *Dechloromonas* (12.13%), *unclassified_f__Rhodocyclaceae* (8.092%), *Ramlibacter* (6.503%),

Arenimonas (4.499%), *Limnothrix* (4.559%), *Thermomonas* (3.992%), and *Azospira* (1.412%). *Dechloromonas*, *Rhodocyclaceae*, *Azospira*, and *Arenimonas* are heterotrophic nitrifying aerobic denitrifiers [32–35], and *Thermomonas* are autotrophic nitrifying aerobic denitrifiers [36]. In group C, *Flavobacterium* (4.454%) was significantly enriched as a heterotrophic nitrifying aerobic denitrifier (Figure 4).

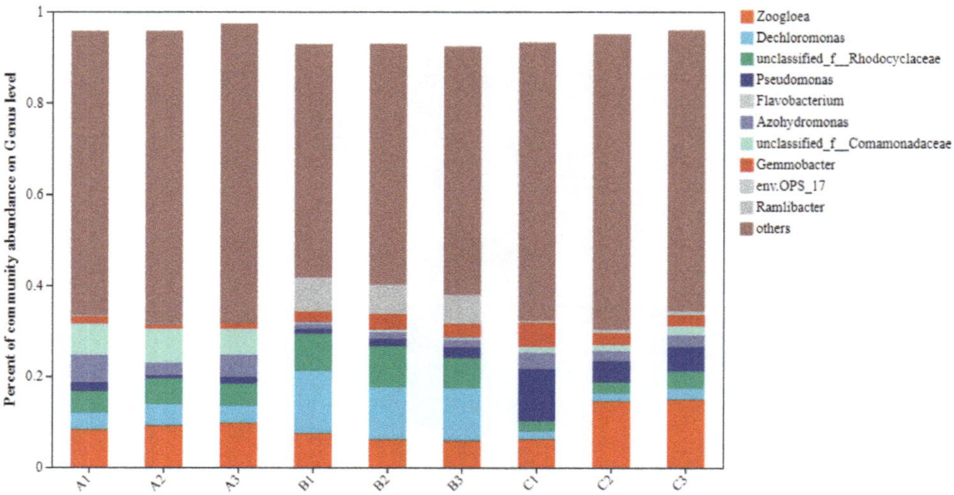

Figure 3. The proportion of TOP10 bacteria in each group. Note: the horizontal/vertical coordinates indicate the sample name. The vertical/horizontal coordinates show the proportion of species in the sample. The different colors of the columns represent different species, and the lengths of the columns represent the size of the proportion of species.

Further analysis identified that group A was predominantly enriched with heterotrophic denitrifying bacteria, with a total abundance of 13.71%, which was significantly higher than the 0.58% of group B and 3.21% of group C. Group B was mainly enriched with aerobic denitrifying bacteria, with a total abundance of 42.76%, which was significantly higher than the 25.03% and 23.92% of groups A and C, respectively (Figure 5). It is evident that with increasing gas-to-water ratios, the dominant denitrifying bacteria in the epiphytic biofilm changed from anoxic heterotrophic to aerobic denitrifiers. A further increase in the gas-to-water ratio will also inhibit the aerobic denitrifying bacteria, possibly as a result of over-oxidation of the carbon source from the inflow COD, which suppressed the growth of heterotrophic/aerobic denitrifiers, leading to a significantly lower abundance of total denitrifying bacteria in group C than in groups A and B. Coupled with wastewater TN removal analysis, group B significantly outperformed group A (Figure 1a), which is consistent with its higher denitrifier abundance. Aerobic denitrifiers such as *Thermomonas* in group B were more competitive in the denitrification of biodegradable organic matter than anoxic heterotrophic denitrifiers like *Thauera* [37]. No significant difference in TN removal rate was observed between groups A and C, possibly because of the low gas-to-water ratio in group A, thereby restricting the nitrification process and failing to supply sufficient nitrate electron acceptors for subsequent denitrification. The main phosphate-removing bacteria across the systems were *Flavobacterium* and *unclassified_f__Comamonadaceae*, with *Flavobacterium* identified as a denitrifying phosphate remover. The total proportion of polyphosphate-accumulating bacteria in the biofilm decreased ($p < 0.05$) as the gas-to-water ratio increased. However, no significant difference in TP removal rate was observed among the three groups, likely because systemic phosphorus removal requires residual sludge exclusion, and the experimental constructed wetland with Vallisneria had no sludge discharge.

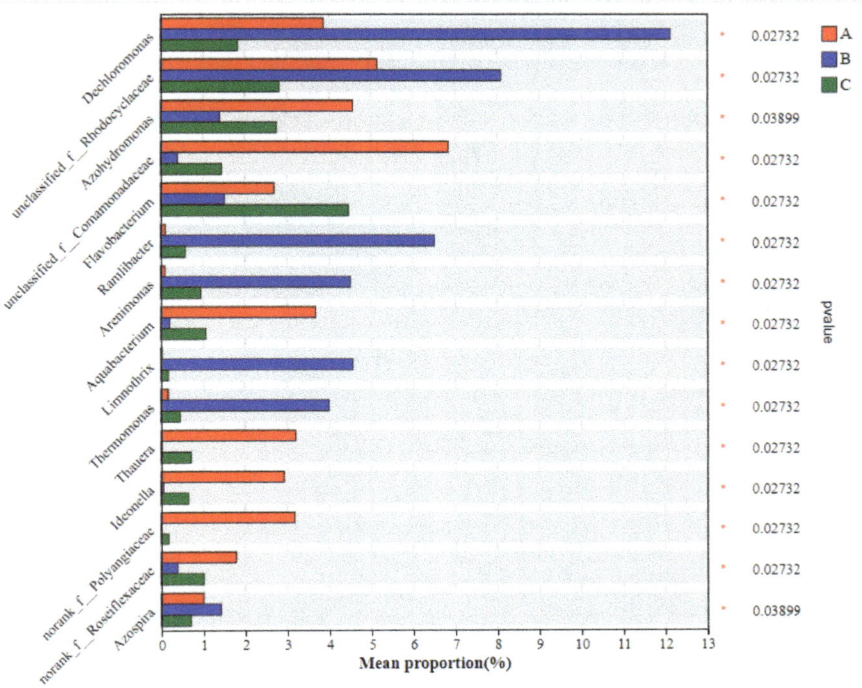

Figure 4. Differences in the mean relative abundance of the same species among different groups. Note: the y-axis represents the species name at a certain taxonomic level; the x-axis represents the average relative abundance of the species in different groups; the colored bars represent different groups; and the column of numbers on the right-hand shows the p-values. The "*" on the right axis of the figure represents significant differences in the genus of bacteria among different groups.

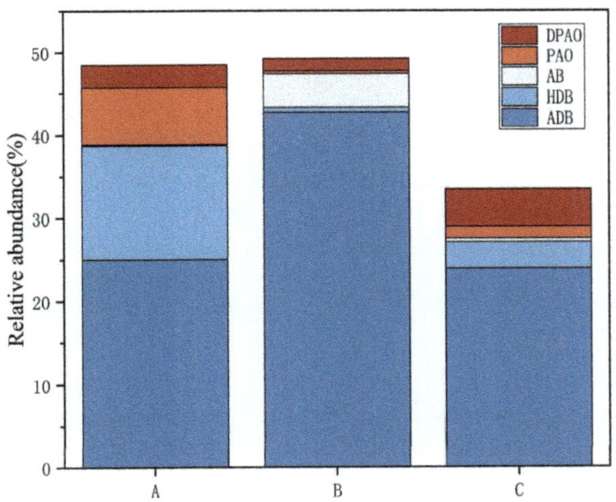

Figure 5. Proportion of functional bacteria for nitrogen and phosphorus removal in each group under different gas-to-water ratios. Note: DPAO stands for denitrifying polyphosphate bacteria, PAO stands for polyphosphate bacteria, AB stands for autotrophic denitrifying bacteria, HDB stands for heterotrophic denitrifying bacteria, and ADB stands for aerobic denitrifying bacteria.

Additionally, within group A, the relative abundance of *Zoogloea* and *Azohydromonas* was relatively high. *Zoogloea* can promote the production of EPSs [38], and *Azohydromonas* mainly functions in alkalinization, with alkaline conditions being favorable for EPS growth. The pH value in group A was indeed significantly higher than in the other two groups, and this may be the main reason for the significant initial rise in EPS content under conditions of a low gas-to-water ratio (Figures 1c and 2). In systems B and C, both *Ramlibacter* and *Limnothrix* were significantly higher than in group A ($p < 0.05$), and these bacteria have been reported to contribute to EPS accumulation [39–41]. However, the EPS content in systems B and C decreased from the initial state, possibly because these systems promoted the growth and metabolism of bacteria such as *Pseudomonas* and *Flavobacterium* that can decompose several polysaccharides. *Flavobacterium* also produces weak acids, which might be detrimental to EPS accumulation [42,43].

4. Conclusions

This study explored the impact of the gas-to-water ratio on the removal rate of pollutants and the epiphytic biofilm community structure in constructed wetland systems with Vallisneria. The findings clearly indicated that the gas-to-water ratio could significantly affect the TN removal rate, with the optimal conditions being a gas to water ratio of 15:1. Such a gas-to-water ratio had minimal effect on the influences on the removal efficiencies for COD, NH_4^+-N, and TP. An increase in the gas-to-water ratio led to a marked decline in biofilm EPSs. High-throughput sequencing analysis demonstrated that as the gas-to-water ratio increased, the content of denitrifying bacteria changed from anoxic heterotrophic denitrifiers to aerobic denitrifiers, with a significant rise in the abundance of denitrifier, although excessive ratios could suppress the growth of denitrifiers. Overall, a gas-to-water ratio of 15:1 presented the optimal condition for ecological restoration in Vallisneria wetland systems.

Author Contributions: Conceptualization, Z.J.; methodology, Z.J., H.M., C.B. and X.Y.; software, H.M., C.B. and X.J.; formal analysis, G.H.; investigation, S.L. and Z.J.; resources, X.Z. and M.Z.; writing—original draft preparation, H.M. and Z.J.; writing—review and editing, X.Z. and M.Z.; supervision, G.H. and X.J.; funding acquisition, Z.J. All authors have read and agreed to the published version of the manuscript.

Funding: This work was supported by the National Natural Science Foundation of China (No. 52100197) and the Major Program of Institute for Eco-environmental Research of Sanyang Wetland, Wenzhou University (SY2022ZD-1002-06).

Data Availability Statement: The data presented in this study are available on request from the corresponding author.

Acknowledgments: We thank Alan K. Chang (Wenzhou University) for his kind help with revising the language of the manuscript.

Conflicts of Interest: Author Shiwen Lu is employed in Jiujiang Branch of Jiangsu Hongrun Biomass Energy Technology Co., Ltd. The remaining authors declare that the research was conducted in the absence of any commercial or financial relationships that could be construed as a potential conflict of interest.

References

1. Angeles-Núñez, J.G.; Cruz-Acosta, T. Aislamiento, caracterización molecular y evaluación de cepas fijadoras de nitrógeno en la promoción del crecimiento de frijol X1-Isolation, molecular characterization and evaluation of nitrogen-fixing strains in promoting the growth of beans. *Rev. Mex. Cienc. Agrícolas* **2015**, *6*, 929–942.
2. Chen, H.; Zhang, S.; Lv, X.; Guo, S.; Ma, Y.; Han, B.; Hu, X. Interactions between suspended sediments and submerged macrophytes-epiphytic biofilms under water flow in shallow lakes. *Water Res.* **2022**, *222*, 118911. [CrossRef]
3. Cheng, F.; Zhang, H.; Sun, S.; Li, Z. Cooperative denitrification in biocathodes under low carbon to nitrogen ratio conditions coupled with simultaneous degradation of ibuprofen in photoanodes. *Bioresour. Technol.* **2022**, *351*, 126988. [CrossRef] [PubMed]
4. Fan, Y.; Sun, S.; He, S. Iron plaque formation and its effect on key elements cycling in constructed wetlands: Functions and outlooks. *Water Res.* **2023**, *235*, 119837. [CrossRef]

5. Fu, F.; Huang, S.; Hu, H.; Lu, Y.; Wang, Y.; Yuan, J.; Gong, Z.; Wu, J.; Zhang, Y. Transformation of N and S pollutants and characterization of microbial communities in constructed wetlands with *Vallisneria natans*. *J. Water Process Eng.* **2021**, *42*, 102186. [CrossRef]
6. Gonzalez-Gil, G.; Sougrat, R.; Behzad, A.R.; Lens, P.N.; Saikaly, P.E. Microbial Community Composition and Ultrastructure of Granules from a Full-Scale Anammox Reactor. *Microb. Ecol.* **2015**, *70*, 118–131. [CrossRef] [PubMed]
7. Gu, P.; Wu, H.; Zhang, Z.; Li, Q.; Zhang, W.; Zheng, Z.; Yang, K.; Miao, H.; Xu, J. Biological effects of harvesting harmful algal blooms on submerged macrophytes and leaf biofilms: A mesocosm experiment. *J. Clean. Prod.* **2022**, *361*, 132256. [CrossRef]
8. Han, S.-F.; Jin, W.; Qu, F.; Hanelt, D.; Abomohra, A. Integrated municipal wastewater treatment and lipid accumulation by a self-flocculating/floating microalga *Limnothrix* sp. *Bioresour. Technol.* **2024**, *394*, 130165. [CrossRef] [PubMed]
9. Hussein, A. Textile Wastewater Treated by Constructed Wetlands–A Critical Review. *J. Ecol. Eng.* **2023**, *24*, 256–275. [CrossRef]
10. Jiang, X.; Wang, M.; He, D.; Zhu, J.; Yang, S.; Fang, F.; Yang, L. Submerged macrophyte promoted nitrogen removal function of biofilms in constructed wetland. *Sci. Total Environ.* **2024**, *914*, 169666. [CrossRef]
11. Kumar, S.; Sangwan, V.; Kumar, M.; Deswal, S. A survey on constructed wetland publications in the past three decades. *Environ. Monit. Assess.* **2023**, *195*, 992. [CrossRef] [PubMed]
12. Li, Q.; Tian, X.; Gu, P.; Yang, G.; Deng, H.; Zhang, J.; Zheng, Z. Transcriptomic analysis reveals phytohormone and photosynthetic molecular mechanisms of a submerged macrophyte in response to microcystin-LR stress. *Aquat. Toxicol.* **2022**, *245*, 106119. [CrossRef] [PubMed]
13. Li, T.; Cao, X.; Wu, Z.; Liu, J.; Hu, B.; Chen, H.; Li, B. Biotransformation of nitrogen and tetracycline by counter-diffusion biofilm system: Multiple metabolic pathways, mechanism, and slower resistance genes enrichment. *Chem. Eng. J.* **2023**, *474*, 145637. [CrossRef]
14. Lin, Q.; Fan, M.; Peng, X.; Ma, J.; Zhang, Y.; Yu, F.; Wu, Z.; Liu, B. Response of *Vallisneria natans* to aluminum phytotoxicity and their synergistic effect on nitrogen, phosphorus change in sediments. *J. Hazard. Mater.* **2020**, *400*, 123167. [CrossRef] [PubMed]
15. Liu, M.Z.; Hu, X.B.; Yan, J.; Mei, L.Y.; Li, X.T. Effects of intermittent aeration under different operation conditions on the pollutant removal of a pilot-scale constructed wetland. *Fresenius Environ. Bull.* **2019**, *28*, 1312–1318.
16. Liu, Y.; Feng, B.; Yao, Y. Research Trends and Future Prospects of Constructed Wetland Treatment Technology in China. *Water* **2024**, *16*, 738. [CrossRef]
17. Meng, F.; Lyu, Y.; Zhao, H.; Lyu, F.; Bie, X.; Lu, Y.; Zhao, M.; Chen, Y.; Lu, Z. LsrR-like protein responds to stress tolerance by regulating polysaccharide biosynthesis in *Lactiplantibacillus plantarum*. *Int. J. Biol. Macromol.* **2023**, *225*, 1193–1203. [CrossRef]
18. Nájera, A.F.; Serwecińska, L.; Szklarek, S.; Mankiewicz-Boczek, J. Seasonal and spatial changes of N-transforming microbial communities in sequential sedimentation-biofiltration systems—Influence of system design and environmental conditions. *Int. Biodeterior. Biodegrad.* **2021**, *159*, 105203. [CrossRef]
19. Ren, X.; Wang, Y.; Wan, J.; Yan, Z.; Ma, Y.; Zhang, G.; Zhu, B. The Nitrogen Removal Performance and Functional Bacteria in Heterotrophic Denitrification and Mixotrophic Denitrification Process. *Water* **2022**, *14*, 3603. [CrossRef]
20. Seviour, R.J.; Mino, T.; Onuki, M. The microbiology of biological phosphorus removal in activated sludge systems. *FEMS Microbiol. Rev.* **2003**, *27*, 99–127. [CrossRef]
21. Tian, J.; Han, Y.; Yin, P.; Zhang, J.; Guo, T.; Li, H.; Hou, Y.; Song, Y.; Guo, J. Response of dissimilatory perchlorate reducing granular sludge (DPR-GS) system to high-strength perchlorate and starvation stress in UASB reactor: Performance, kinetics and recovery mechanism. *J. Environ. Chem. Eng.* **2023**, *11*, 109414. [CrossRef]
22. Tian, X.; Zhao, J.; Huang, J.; Chen, G.; Zhao, Y. The metabolic process of aerobic granular sludge treating piggery wastewater: Microbial community, denitrification genes and mathematical model calculation. *J. Environ. Chem. Eng.* **2021**, *9*, 105392. [CrossRef]
23. Tomás-Martínez, S.; Zwolsman, E.J.; Merlier, F.; Pabst, M.; Lin, Y.; van Loosdrecht, M.C.; Weissbrodt, D.G. Turnover of the extracellular polymeric matrix of granules performing biological phosphate removal. *Appl. Microbiol. Biotechnol.* **2023**, *107*, 1997–2009. [CrossRef] [PubMed]
24. Tsolcha, O.N.; Tekerlekopoulou, A.G.; Akratos, C.S.; Aggelis, G.; Genitsaris, S.; Moustaka-Gouni, M.; Vayenas, D.V. Agroindustrial Wastewater Treatment with Simultaneous Biodiesel Production in Attached Growth Systems Using a Mixed Microbial Culture. *Water* **2018**, *10*, 1693. [CrossRef]
25. Wan, L.; Zhang, S.; Zhou, Z.; Chen, S. Plant Compartments Shape the Assembly and Network of Vallisneria natans-Associated Microorganisms. *Diversity* **2023**, *15*, 676. [CrossRef]
26. Wang, C.; Sun, Q.; Wang, L.Y. Cadmium Toxicity and Phytochelatin Production in a Rooted-Submerged Macrophyte *Vallisneria spiralis* Exposed to Low Concentrations of Cadmium. *Environ. Toxicol.* **2009**, *24*, 271–278. [CrossRef] [PubMed]
27. Wang, X.; Wu, Y.; Chen, N.; Piao, H.; Sun, D.; Ratnaweera, H.; Maletskyi, Z.; Bi, X. Characterization of Oxidation-Reduction Potential Variations in Biological Wastewater Treatment Processes: A Study from Mechanism to Application. *Processes* **2022**, *10*, 2607. [CrossRef]
28. Wang, Z.; Zhang, Y.; Li, X.; Li, J.; Zhao, Z.; Hou, X. Mixed culture of plants improved nutrient removal in constructed wetlands: Response of microbes and root exudates. *Environ. Sci. Pollut. Res.* **2022**, *30*, 5861–5872. [CrossRef] [PubMed]
29. Wu, C.; Zhou, L.; Zhou, C.; Zhou, Y.; Xia, S.; Rittmann, B.E. Co-removal of 2,4-dichlorophenol and nitrate using a palladized biofilm: Denitrification-promoted microbial mineralization following catalytic dechlorination. *J. Hazard. Mater.* **2022**, *422*, 126916. [CrossRef]

30. Wu, M.; Hao, H.; Ge, Y.; Pu, T.; He, Z.; Ge, D.; Rene, E.R.; Huang, Z. Treatment of Black-Odorous Water Using Submerged Plants: The Physiological Response of *Vallisneria natans*. *Water* **2023**, *15*, 653. [CrossRef]
31. Wu, W.; Yang, L.; Wang, J. Denitrification performance and microbial diversity in a packed-bed bioreactor using PCL as carbon source and biofilm carrier. *Appl. Microbiol. Biotechnol.* **2013**, *97*, 2725–2733. [CrossRef]
32. Xiao, R.; Zheng, Y. Overview of microalgal extracellular polymeric substances (EPS) and their applications. *Biotechnol. Adv.* **2016**, *34*, 1225–1244. [CrossRef]
33. Xing, W.; Li, J.; Li, P.; Wang, C.; Cao, Y.; Li, D.; Yang, Y.; Zhou, J.; Zuo, J. Effects of residual organics in municipal wastewater on hydrogenotrophic denitrifying microbial communities. *J. Environ. Sci.* **2018**, *65*, 262–270. [CrossRef]
34. Xu, P.; Xiao, E.; Wu, J.; He, F.; Zhang, Y.; Wu, Z. Enhanced nitrate reduction in water by a combined bio-electrochemical system of microbial fuel cells and submerged aquatic plant *Ceratophyllum demersum*. *J. Environ. Sci.* **2019**, *78*, 338–351. [CrossRef]
35. Yan, L.; Zhang, S.; Lin, D.; Guo, C.; Yan, L.; Wang, S.; He, Z. Nitrogen loading affects microbes, nitrifiers and denitrifiers attached to submerged macrophyte in constructed wetlands. *Sci. Total Environ.* **2018**, *622–623*, 121–126. [CrossRef]
36. Yu, G.; Chen, H.; Chen, J.; Chen, S.; Long, Y.; Huang, J.; Wang, Y.; He, S. Enhanced nitrogen removal through aerobic denitrifying bacteria in horizontal subsurface flow constructed wetlands: Influencing factors and microbial community structure. *Chem. Eng. J.* **2024**, *481*, 148654. [CrossRef]
37. Yue, F.; Zhang, F.; Qu, Q.; Wang, C.; Qin, Y.; Ma, L.; Jia, Y.; Ismael, M.; Jiang, Y.; Sun, T.; et al. Effects of ageing time on the properties of polysaccharide in tangerine peel and its bacterial community. *Food Chem.* **2023**, *417*, 135812. [CrossRef]
38. Zhang, H.; Ge, Z.; Li, Y.; Huang, S.; Zhang, J.; Zheng, Z. Response of submerged macrophytes and leaf biofilms to different concentrations of oxytetracycline and sulfadiazine. *Chemosphere* **2022**, *308*, 136098. [CrossRef]
39. Zhang, L.-J.; Xie, Y.; Ding, L.-Y.; Qiao, X.-J.; Tao, H.-C. Highly efficient ammonium removal through nitrogen assimilation by a hydrogen-oxidizing bacterium, Ideonella sp. TH17. *Environ. Res.* **2020**, *191*, 110059. [CrossRef]
40. Zhang, L.; Huang, X.; Zhou, J.; Ju, F. Active predation, phylogenetic diversity, and global prevalence of myxobacteria in wastewater treatment plants. *ISME J.* **2023**, *17*, 671–681. [CrossRef]
41. Zhao, D.; Chen, C.; Yang, J.; Zhou, S.; Du, J.; Zhang, M.; An, S. Mutual promotion of submerged macrophytes and biofilms on artificial macrophytes for nitrogen and COD removal improvement in eutrophic water. *Environ. Pollut.* **2021**, *277*, 116718. [CrossRef] [PubMed]
42. Zhao, T.; Hu, H.; Chow, A.T.; Chen, P.; Wang, Y.; Xu, X.; Gong, Z.; Huang, S. Evaluation of organic matter and nitrogen removals, electricity generation and bacterial community responses in sediment microbial fuel cell coupled with *Vallisneria natans*. *J. Environ. Chem. Eng.* **2023**, *11*, 110058. [CrossRef]
43. Zheng, S.; Liu, X.; Yang, X.; Zhou, H.; Fang, J.; Gong, S.; Yang, J.; Chen, J.; Lu, T.; Zeng, M.; et al. The nitrogen removal performance and microbial community on mixotrophic denitrification process. *Bioresour. Technol.* **2022**, *363*, 127901. [CrossRef] [PubMed]

water

MDPI

Review

A Comprehensive Review on Ecological Buffer Zone for Pollutants Removal

Dongsheng Wang [1,2,3,†], Xing Gao [1,2,3,†], Suqing Wu [1,2,3], Min Zhao [1,2,3], Xiangyong Zheng [1,2,3], Zhiquan Wang [1,2,3], Yejian Zhang [1,2,*] and Chunzhen Fan [1,2,3,*]

[1] College of Life and Environmental Science, Wenzhou University, Wenzhou 325000, China; wang18072058562@163.com (D.W.); gx18344630052022@163.com (X.G.); wusuqing@wzu.edu.cn (S.W.); zmzjcn@sina.com (M.Z.); x.zheng@wzu.edu.cn (X.Z.); wangzhiquan@wzu.edu.cn (Z.W.)

[2] National and Local Joint Engineering Research Center of Ecological Treatment Technology for Urban Water Pollution, Wenzhou University, Wenzhou 325000, China

[3] Institute for Eco-Environmental Research of Sanyang Wetland, Wenzhou University, Wenzhou 325014, China

* Correspondence: yjzhang@wzu.edu.cn (Y.Z.); fanchunzhen@wzu.edu.cn (C.F.)

† These authors contributed equally to this work.

Abstract: The issue of agricultural non-point source pollution has attracted global attention. A buffer zone is an effective, eco-friendly, and economically feasible remediation ecosystem to reduce the impact of agricultural non-point source pollution on water bodies. They can effectively remove pollutants in agricultural drainage through physical processes (infiltration, filtration, deposition, etc.), plant absorption and assimilation, and microbial processes, improving the water quality of water bodies. This article provides a comprehensive review of the current studies on using buffer zones to remediate agricultural non-point source pollution, with a focus on the key affecting factors for pollutant removal efficiencies. The main factors included buffer zone width, vegetation type, slope, seasonal variation, soil variation, and vegetation density. The influencing mechanisms of these factors on the pollutant removal efficiencies of buffer zones were also discussed. This review can serve as a reference for a deep understanding of buffer zones and help optimize their design and management in real ecological remediation projects.

Keywords: buffer zone; agriculture; non-point source pollution; ecological functions; influencing factors

1. Introduction

With the rapid development of the social economy, an increasing level of attention is being paid to the issues of water environments. At present, eutrophication has become one of the global water pollution issues [1,2]. According to relevant research [3,4], the usage of chemical pesticides increased from 0.73 million tons in 1990 to 1.66 million tons in 2017 in China. The usage of mineral fertilizers increased from 8.84 million tons in 1978 to 58.59 million tons in 2017. Obviously, agricultural growth relies on intensive inputs of production factors, leading to serious non-point source pollution in water environments [5]. In the United States, agricultural activities are also a major source of surface water pollution, including excessive nutrients from fertilizers and pesticides, as well as an increase in water turbidity caused by soil erosion [6], accounting for approximately 55% of surface water pollution from non-point sources. Additionally, global agriculture production releases around 31 million tons of nitrogen and 2.9 million tons of phosphorus into freshwater bodies per year [7].

Agricultural non-point source pollution is characterized by its extensive dispersion, complex migration routes, hidden nature, and cumulative effects, leading to challenges in effectively controlling such pollution [8]. Although significant efforts have been conducted to reduce fertilizer application and adopt optimal land management practices [9], nutrient pollution in water bodies persists. This is partly due to the continuous loss of function of natural riparian wetlands [10]. Extensive studies have been focused on the prevention and controlling of agricultural non-point source pollution to reduce the impact on aquatic ecosystems. At present, the commonly used technologies include developing precision agriculture, ecological ditches, buffer zones, compost technology, and soil microbial fertilizers [11–13]. Among them, buffer zone technology has become a widely accepted and effective technology due to its pollutant removal capabilities. It is widely used to control non-point source pollution and support the development of more sustainable agriculture [14,15]. The detailed arrangement and functionality of buffer zones are shown in Figure 1 [16].

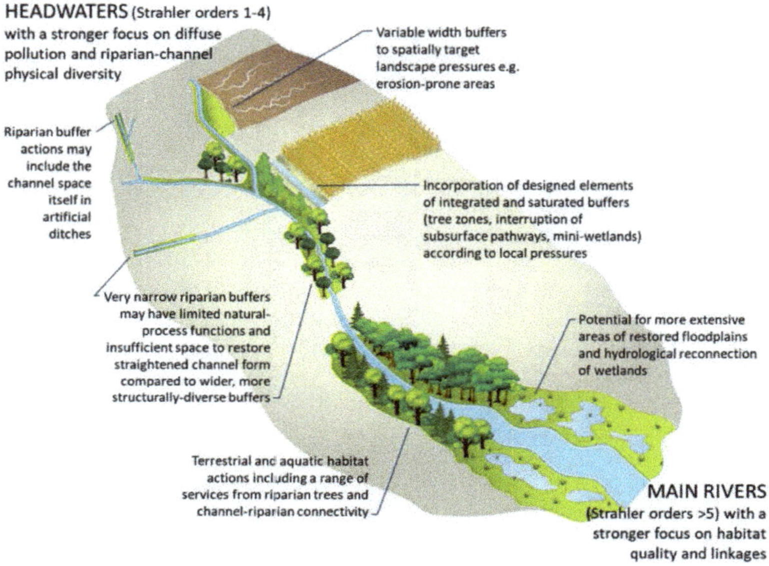

Figure 1. A schematic diagram of the resistance and control of non-point source pollution in buffer zones. Reproduced with permission from ref. [16], Copyright 2019, ASA/CSSA/SSSA.

Buffer zones serve as critical interfaces between surface water and groundwater systems [17], aiming to improve water quality by capturing pollutants from surface water and shallow groundwater and absorbing excess pollutants [18]. In addition, buffer zones exhibit complex bio-geochemistry processes that play an important role in maintaining the river balance of nature, promoting biodiversity conservation and providing a variety of ecological services [19]. As shown in Figure 2, after pesticides and nutrients are discharged from agricultural fields, they enter buffer zones and can be effectively removed through processes such as soil filtration, plant absorption, and microbial degradation, significantly reducing their impact on water bodies and ecosystems [20]. Liu et al. [21] studied the impact of buffer zones on controlling non-point source pollution in Chaohu lake. They found that the pollutant removal efficiency of the buffer zone was significantly better than that of the constructed wetland and the permeable pedestrian pathways. The reduction rates of non-point source pollution load for total nitrogen and total phosphorus were 15.29% and 15.03%, respectively. Further studies demonstrated that buffer zones could reduce nitrogen fluxes by up to 90% through a series of complex processes, including plant absorption, denitrification, and storage [22]. In summary, buffer zones have been demonstrated as an effective, eco-friendly, and economically viable method for trapping runoff and sediment. Therefore, it is becoming a popular non-point measure of soil and water conservation [23,24].

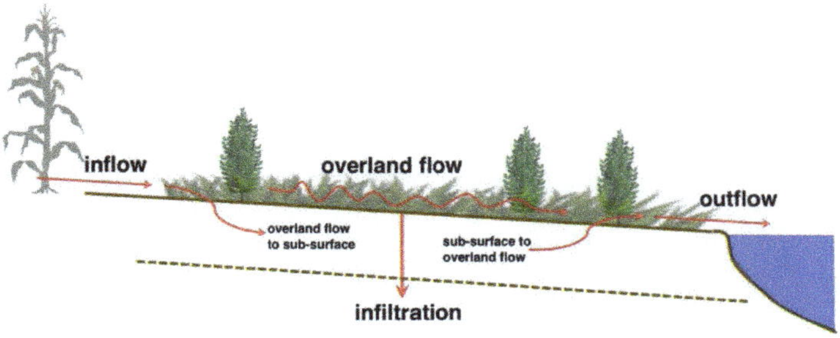

Figure 2. Pathways of pesticide and nutrient movement from an agricultural field through a vegetated buffer strip to an aquatic ecosystem and major pathways of retention. Reproduced with permission from ref. [20], Copyright 2020, Academic Press Inc.

Previous studies mainly focused on the removal effect of buffer zones on agricultural pollutants, the restrictive factors of buffer zones, and the development of models or technical methods for evaluating buffer zones in the field [4,25–30]. During the processes of pollutant removal via buffer zones, it is necessary to clarify the mechanism of nitrogen and phosphorus removal. The pollutant removal processes are influenced by many factors, such as the buffer zone width, vegetation type, slope, seasonal variation, soil composition, and vegetation density. Therefore, the mechanisms of pollutant removal via buffer zones are quite complex and require further research. However, to the best of our knowledge, few comprehensive reviews were focused on pollutant removal via buffer zones, which prompted us to write this critical and comprehensive review. The specific objectives are the following: (1) provide insights on recent study trends and the developing progress of buffer zones, aiming to demonstrate the importance of buffer zones in reducing the introduction of pollutants from agricultural activities into water bodies; (2) clarify the main retention process and mechanisms of pollutant removal in buffer zones; and (3) explore the influencing factors for the performance of buffer zones in pollutant removal. This review can provide important references for the design and construction of buffer zones, improving the development and application of buffer zones in water pollution control and ecosystem protection.

2. Methods

2.1. Literature Acquisition Sources

To search for the relevant literature, the Web of Science (WoS) and ScienceDirect (SD) databases were used, which cover the main academic journals and published papers with a high degree of authority and credibility in almost all major subject areas. The search process is summarized in Table 1. The initial search was conducted based on specific keywords, and 565 papers published from 2010 up to 30 June 2024 were obtained, including 502 papers in WoS and 63 papers in SD.

Table 1. Literature search strings.

Database	Retrieval String	Number	Search Data
Web of Science	First search string: buffer zone, second search string: water pollution, third search string: agriculture	364	30 June 2024
	First search string: buffer zone, second search string: water pollution, third search string: mechanism	138	30 June 2024
ScienceDirect	Keywords in the title or abstract: buffer zone, pollution	28	30 June 2024
	Keywords in the title or abstract: buffer zone, agriculture	35	30 June 2024
Total	—	565	30 June 2024

2.2. Literature Selection Criteria and Classification

According to the topic of this review, the relevant literature was further screened out by titles, abstracts, keywords, and full-text articles in turn, and the duplicate and irrelevant articles were eliminated manually, followed by intensive reading to determine the eligible articles. The detailed screening process is shown in Figure 3. Eventually, 318 relevant articles were identified.

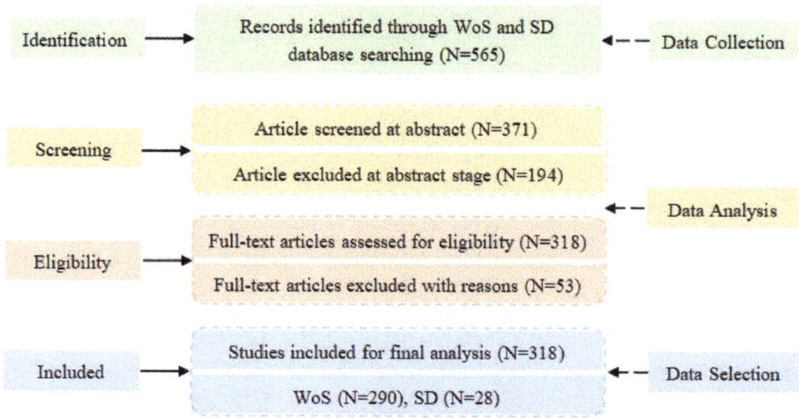

Figure 3. Data selection process. (WoS represents Web of Science, SD represents ScienceDirect).

On the basis of the above searching results, further classification was carried out according to the corresponding retrieval string, as shown in Table 2. Among them, when the articles involved multiple influencing factors, they were separately counted in each influencing factor.

Table 2. Literature classification.

Retrieval String	Number of Articles in the Initial Searching	Number of Relevant Articles Based on the Exacting Screening
Mechanism	52	28
Buffer zone width	83	56
Vegetation type	67	45
Slope	58	37
Seasonal variation	48	38
Soil composition	201	28
Vegetation density	22	15
Runoff intensity	20	12
Others	—	—

3. Mechanism of Buffer Zones in Removing Pollutants

Buffer zones can reduce the concentration of pollutants in the water through a series of complex physical, chemical, and biological processes [31]. The nitrogen and phosphorus removal processes are closely related to some factors, such as buffer zone width, vegetation type, slope, seasonal variation, soil composition, and vegetation density. A combination of these internal and external factors is essential for complete understanding and effective management of the pollutant removal capacity of buffer zones. Nitrogen and phosphorus removal via buffer zones mainly involve physical adsorption, vegetation absorption and assimilation, soil adsorption, and microbial absorption and transformation, as well as denitrification processes [32]. Although many researchers have studied the migration and transformation processes of nitrogen and phosphorus in different ways using many experimental methods, there is still much controversy regarding the relative importance of each process [33].

3.1. Physical Processes

Buffer zones protect aquatic ecosystems by effectively reducing nitrogen and phosphorus in water through physical processes, such as precipitation, filtration, infiltration, absorption, and degradation [34]. Firstly, vegetation within buffer zones plays a crucial role. Vegetation covering the soil surface can effectively increase runoff resistance and slow down surface runoff velocity, allowing more surface runoff to infiltrate through soil pores and become subsurface flow [35]. As a result, most solid particles carrying pollutants gradually settle, and particulate pollutants or suspended solids in runoff are effectively filtered and intercepted [36]. At the same time, soluble pollutants in the subsurface infiltrate into deeper soil layers through the relatively loose soil in the buffer zone. The transport capacity of surface runoff for soluble pollutants is decreased [20], thus reducing the loss of total nitrogen and total phosphorus [37]. The root system of vegetation can penetrate the deep layer of soil and increase the structural stability, water permeability, and aeration of the soil. It also plays a role in filtering and adsorbing nutrients such as nitrogen and phosphorus, thus purifying groundwater and surface water. Goloran et al. [38] found that plant roots could absorb and intercept nutrients from underground runoff, reducing the loss of total nitrogen and total phosphorus from underground runoff. The more developed root system and the higher biomass of the buffer zone can promote root absorption and microbial degradation of plant roots, thus increasing the efficiency of pollutant interception in runoff [39].

Wu et al. [40] found that the nitrogen and phosphorus removal efficiency of a buffer zone exceeded 60% at 2–5% slopes, which was significantly higher than that of runoff. Alemu et al. [41] observed that 99% of total phosphorus and 85% of nitrate nitrogen could be reduced through approximately 10 m of herbaceous buffer zone. This result suggests that the establishment of herbaceous buffer zones on both sides of riverbanks can reduce the entry of nitrogen and phosphorus into the water body. It was also found that the populus buffer zone in Taihu significantly reduced sediment and nitrogen loss from surface runoff and the loss flux decreased with the increasing plant density in the buffer zone [42].

Furthermore, the soil layer of the buffer zone also possesses the ability to adsorb, filter, and immobilize nutrients. When the soil contains large clay, small silt, and sand particles, the runoff velocity of water slows down [43]. Meanwhile, soil pores serve as channels for water transport, retaining pollutants in the soil through water absorption. Moreover, the infiltration capacity of soil is increased and the erosion of surface runoff on the soil surface is reduced under the action of gravity. In addition, soil texture, soil–water interaction, organic matter content, and soil nutrient coverage were found to affect phosphorus release [44]. The retention and transformation of phosphorus in buffer zones are primarily driven by physical processes [45], including the adsorption effect by particles in surface runoff and the infiltration effect of the soil, thus the retained phosphorus can be absorbed and utilized by plant roots. These physical processes lead to the reduction in phosphorus in the surface runoff.

In addition, the types and forms of pollutants in surface runoff also have a significant influence on the interception effect of buffer zones. When surface runoff containing particulate nitrogen and phosphorus and dissolved nitrogen flows through a buffer zone, the vegetation in the buffer zone effectively suppresses soil erosion due to the runoff, increases the roughness of the surface, and effectively decreases the runoff velocity. It also improves the hydraulic permeability of the soil and assists in the effective removal of nitrogen and phosphorus from runoff [16,46]. Besides, the width, slope, and other factors of buffer zones also play an important role in their removal of pollutants, improving the exchange and transformation of substances among the vegetation, soil, and water.

In summary, buffer zones effectively intercept and transform nitrogen and phosphorus in surface runoff through physical processes involving vegetation, soil, and topography.

The amount of nitrogen and phosphorus entering water bodies can be reduced, thus protecting the health of aquatic ecosystems.

3.2. Absorption and Assimilation Process by Plants

Absorption and assimilation via plants play an important role in the removal of nitrogen, phosphorus, and other pollutants in buffer zones. Firstly, root systems absorb pollutants like nitrogen and phosphorus from the soil and convert them into nutrients for plant growth [37]. Pollutants from water and soil are effectively transferred into plants through this absorption process, reducing their pollution levels in water bodies and the soil. When dissolved nitrogen (usually in the form of nitrate or ammonia) infiltrates into the root zone, it is absorbed by plant roots [25]. Then, it is converted into organic nitrogen through a series of biochemical reactions, mainly existing in the forms of amino acids and proteins stored in plants [47]. This process involves the participation of various enzymes, such as nitrate reductase and glutamine synthetase, which catalyze the transformation and synthesis of nitrogen. Inorganic phosphorus in the soil is absorbed by plant roots and converted into organic phosphorus, such as phospholipids and nucleic acids. This process involves biochemical reactions such as phosphorylation and esterification. Organic phosphorus plays important biological functions in plants, such as energy conversion and cell metabolism.

Secondly, plants absorb CO_2 and release O_2 during their growth through photosynthesis. Plant roots also release O_2, which can boost O_2 content in water bodies. The release of O_2 is essential for the survival of aquatic organisms in water bodies. It maintains the balance of aquatic ecosystems and promotes the decomposition and degradation of organic matter, further reducing the concentration of pollutants in water bodies. At the same time, it can promote the growth and activity of soil microorganisms and root-associated microorganisms, accelerating the degradation and removal of pollutants. As displayed in Figure 4 [48], oxygen and root exudates are transferred from the upper biomass to the microbial communities on the root surface. With the action of rhizosphere microorganisms and enzymes, the ability of plants to remove pollutants is significantly enhanced.

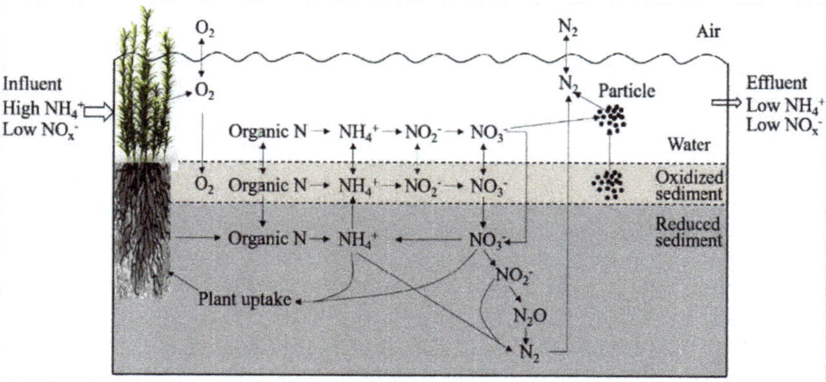

Figure 4. The major nitrogen transformations in the Daniaopi Constructed Wetland. Major nitrogen pathways illustrated are nitrification, denitrification, and dissimilatory nitrate reduction to ammonia and anammox. Reproduced with permission from ref. [48], Copyright 2017, Elsevier.

Moreover, it was found that the majority of nitrogen and phosphorus in plants was returned to the soil through plant aging or leaf litter [49]. Peterjohn et al. also found that nitrogen in leaf litter returned to the soil accounted for up to 80% of nitrogen absorption in a deciduous forest-dominated buffer zone [50]. This phenomenon is similar to Yang et al.'s study [51]. Nevertheless, plant assimilation remains an important mechanism for nitrogen and phosphorus removal via buffer zones [47]. It can alter the forms of nitrogen and phosphorus in soil. The mineral decomposition of plant residue can produce a lot of inorganic salts and available carbon sources, thus creating favorable conditions for microbial activity.

In summary, absorption and assimilation via plants play an important role in removing nitrogen and phosphorus in buffer zones, involving the absorption of pollutants, release of oxygen, and promotion of microbial activity. This contributes to the improvement of environmental quality in both the water and soil.

3.3. Microbiological Effects

Microorganisms are an important driving factor for nitrogen and phosphorus removal in buffer zones. They promote the transformation and absorption of nitrogen and phosphorus through biodegradation, phosphorus dissolution, and precipitation, as well as symbiotic interactions with plants, improving the removal efficiency of buffer zones and maintaining the health of aquatic ecosystems [52].

In buffer zones, microorganisms biodegrade organic matter to release organic nitrogen and phosphorus, converting complex organic molecules into different forms of inorganic compounds such as ammonia nitrogen, nitrates, and phosphates, which can be more easily absorbed by soil or water. These microbial metabolites and activities influence the cycling processes of nitrogen and phosphorus, promoting their removal efficiency in buffer zones and maintaining healthy aquatic ecosystems. The intensity of microbial processes varies in different types of vegetative buffer zones [53,54].

Ammonification, nitrification, and denitrification processes by microorganisms play important roles in nitrogen transport and transformation [25,26]. Organic nitrogen is converted to ammonia nitrogen through ammonification. Ammonia nitrogen can be absorbed and assimilated by microorganisms, as well as converted to nitrate nitrogen by nitrifying bacteria and nitroso bacteria. Studies have shown that the main mechanisms of nitrogen removal in buffer zones are plant absorption and microbially mediated denitrification in soil [25,26]. As shown in Figures 4 and 5, nitrate nitrogen can be more easily removed via the denitrification effect of denitrifying bacteria, reducing it to N_2 and releasing it into the atmosphere, which can be completely removed from the buffer zone [48,55]. Additionally, a portion of nitrate nitrogen is reduced to ammonia nitrogen via nitrate reductase and is further synthesized into amino acids and proteins. Denitrification and microbial assimilation are important processes for nitrogen removal in buffer zones [53,54]. Gu et al. [55] argued that microbial nitrogen removal processes play much larger roles than plant uptake, and microbial fixation plays a minor role in the nitrogen removal processes [56]. Different types of buffer zones also exhibit varying degrees of denitrification intensity. Studies have demonstrated that denitrification can remove 20–1600 kg of nitrogen per hectare of buffer zone annually [57].

Microorganisms also play an important role in the migration and transformation of phosphorus in buffer zones. Soluble phosphorus can be assimilated and absorbed by plant roots and microorganisms and subsequently converted into organic phosphorus in plants, effectively reducing phosphorus in water [34]. In addition, microorganisms are involved in phosphorus cycling processes such as desorption, adsorption, and mineralization, which affect the availability of phosphorus in soil [58].

Moreover, the microbial communities around plant roots symbiotically enhance the absorption capacity of plants for nitrogen and phosphorus, thereby promoting the effectiveness of buffer zones in removing nitrogen and phosphorus. In summary, microorganisms provide strong support for nitrogen and phosphorus removal in buffer zones through their diverse metabolic pathways and interactions with plants, contributing to the maintenance of the health and stability of aquatic ecosystems.

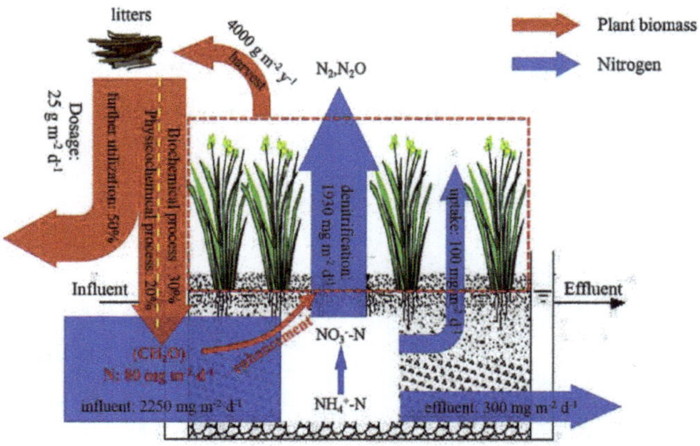

Figure 5. The model and mechanism of secondary wastewater treatment plants (WWTPs) effluent treatment via Iris pseudacorus self-consumed subsurface flow constructed wetlands (SSF CWS). Reproduced with permission from ref. [55], Copyright 2021, Elsevier.

4. Affecting Factors for Pollutants Removal

As shown in Figure 6a, buffer zones have been increasingly applied in agricultural pollution control and have attracted more and more attention from researchers from 2010 to 2023. The migration and transformation of nitrogen and phosphorus in buffer zones is highly complex, involving many physical, chemical, and biological processes [25–27]. These processes are influenced by many factors, such as the width of the buffer zone, vegetation type, seasonal variation, soil composition, vegetation density, slope, and runoff intensity, which in turn affect the effectiveness of buffer zones in removing pollutants such as nitrogen and phosphorus [28,59–78]. The number of published articles on different affecting factors are summarized from 2010 to June 2024 as shown in Figure 6b. Table 3 displays the key factors for buffer zones. These factors should be carefully considered when a buffer zone is applied for protecting agricultural ecosystems. The following sections assess the important affecting factors for pollutant removal in buffer zones.

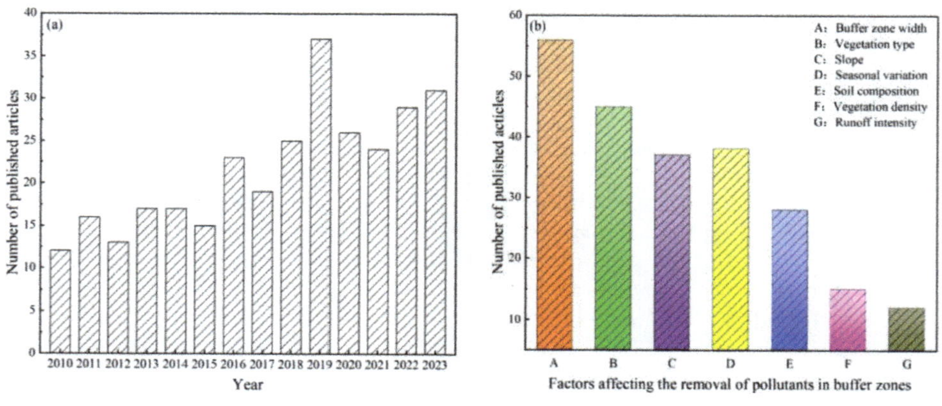

Figure 6. (**a**) The number of published articles from 2010 to 2023 on the impact of buffer zones on agricultural pollution control. (**b**) The number of published articles from 2010 to June 2024 on influencing factors for pollutant removal in buffer zones.

Table 3. The key factors for buffer zones.

Buffer Zone Width	Vegetation Type	Slope	Soil Composition	Average Annual Precipitation	Average Temperature	Reference
1, 3, 7 m	Grass	<3%	—	665 mm	8.2 °C	[59]
5, 9, 13 m	Weeds, sweet clover, and sweet clover/Chinese wingnut	10–20%	Sand, clay, and silt	665 mm	Minus 13.7 °C to 23.7 °C	[60]
10, 30 m	White clover, meadow fescue, and timothy	1–14%	Fine sand	—	—	[61]
10, 15, 30 m	Grass, deciduous trees, and trees	—	Hagerstown and Opequon	1050 mm	—	[62]
12, 24, 36, 48, 60 m	Woody vegetation, shrubs, or grass	8–9%	Clay	1650 mm	18 °C	[63]
0, 10, 20 m	Willows and poplars	—	Loess soil	350–600 mm	8 °C	[64]
12, 36, 60 m	Grass vegetation, shrubs, and woody vegetation	8–9%	—	1650 mm	18 °C	[65]
2, 4, 8 m	Native tallgrass prairie grasses and forbs	5–10%	Sand or clay	1035.8 mm	8 °C	[66]
25, 45 m	Tall forbs or swamp non-forest communities	0.5–3.0%	Soil, fine sands, grits, and coarse sand	600 mm	8.6 °C	[67]
3.05, 6.1, 9.14 m	Ordeum vulgare, medicago sativa, bromus marginatus, and pascopyrum smithii	0.5–2.0%	Loamy sand, sandy loam, loam, and silty loam	150 mm	0–23.1 °C	[68]
0–200, 200–500, 500–1000 m	Agriculture, forest, grassland, and urban	0–5%	—	1900 mm	17 °C	[69]
1, 3, 4, 6 m	Forest and tillage crops	8%	Loamy soils	450–700 mm	6.25 °C	[70]
100–700 m	Forest, paddy field, and tea field	0–80.30°	Ultisols, anthrosols, and inceptisols	1340 mm	17.5 °C	[71]
500, 1000 m	Site buffer, riparian buffer, and catchment buffer	—	—	—	—	[72]
500, 800, 1000, 1200, 1500, 1800 m	Forest land, water area, agricultural land, bare land, construction land	—	—	1680 mm	17.5 °C	[73]
—	Silvopastoral systems, silvoarable agroforestry, and linear tree plantings	—, —	—	—	—	[74]
—	Rice plant and grass samples	—	Ponds, rice fields, and natural wetlands	1200 mm	20 °C	[75]
—	Arboraceous, herbaceous, and aerenchymous	—	Organic and mineral	—	—	[14]
4.5 m	Trees and grasses	1–5%, 5–9%	Putnam silt loam soil and armstrong loam soil	978 mm	—	[76]
—	Phragmites australis	—	Gravel, gravel + biochar, ceramsite + biochar, and modified ceramsite + biochar	1191.5 mm	16.1 °C	[77]
30 m	—	≤10%, >10%	—	—	—	[78]

4.1. Buffer Zone Width

The width of buffer zones plays an important role in the removal of pollutants such as nitrogen and phosphorus [28]. Scholars have explored the impact of buffer zone width on pollutant removal efficiency through field experiments, numerical simulations, and modeling studies. The optimal width of buffer zones was also determined to optimize the design of buffer zones and enhance their pollutant removal capacity.

The width of a buffer zone determines whether it can completely exhibit its ecological service function [4]. Chen et al. [79] found that the removal capacity of a buffer zone for pollutants such as nitrogen, phosphorus, and others depended highly on its width.

Jiang et al. [80] discovered that the total nitrogen in water was reduced by 23.21, 50.39, and 56.20% for buffer zones 20, 40, and 60 m wide, with the removal efficiency of total phosphorus of 18.16, 45.93, and 52.14%, respectively. Clearly, wider buffer zones result in better efficiency in trapping and transforming pollutants. Wang et al. [81] also found that the optimal widths of buffer zones in Dianchi Lake, Erhai Lake, and Fuxian Lake were 450, 100, and 150 m, respectively. The findings of these studies indicate that the width of buffer zones is positively correlated with their efficiency in improving water quality and protecting aquatic environments, and a wider buffer zone plays an important role. It could be attributed to the increased vegetation area provided by the wider buffer zone, which in turn increases the contact area between pollutants, soil, and vegetation and promotes the removal and degradation of pollutants [82]. Aguiar et al. [63] studied the effects of buffer zone width (12, 24, 36, 48, and 60 m) on nutrient removal. They found that the vegetation strip with a 60 m width exhibited the optimum removal efficiency, especially for nitrogen. The finding differs from Johnson et al.'s study, which suggests that additional buffer zone width does not necessarily produce proportional groundwater water quality benefits [83].

Mayer et al. [84] conducted a meta-analysis on 89 buffer zones with different widths. They found that the reduction rate of nitrate nitrogen in water was significantly enhanced as the width increased from 0 to 25 m. However, increasing the width from 25 to 50 m did not significantly enhance the removal rate of nitrate nitrogen. Lv and Wu observed a similar phenomenon [85]. The highest nitrogen removal was achieved at a width of 15 m. When the width exceeded 15 m, the increasing trend in nitrogen removal rate noticeably slowed down and even decreased. However, Valkama et al. [86] used a meta-analysis and found no effect of width on nitrogen removal efficiency, which is contrary to the model predictions of Zhang et al. [78]. Wanyama et al. [87] found that there was no linear relationship between buffer zone width and phosphorus removal efficiency in water. As the width increased (7.5 and 15 m), the pollutant removal efficiency by unit width of buffer zone continuously decreased [88]. Based on previous studies, it can be concluded that buffer zones cannot be expanded indefinitely, which will not only increase the cost and complexity, but also may lead to resource wasting and performance reduction. Therefore, it is necessary to discuss the optimal width of buffer zones.

The optimal width of buffer zones has been extensively studied by many researchers through field investigations or mathematical models, which suggested suitable widths [64]. Fischer et al. [89] suggested that different widths of buffer zones can meet the requirements of various types of ecological and environmental protection. In general, a 3–10 m buffer zone can be used for removing organic matter, a 10–20 m buffer zone is suitable for stabilizing streams, a 5–30 m buffer zone is suggested for water quality protection, a 20–150 m buffer zone can be applied for flood controlling, and a 30–500 m buffer zone can provide riparian habitats. Additionally, some studies suggested that buffer zone width should exceed 500 m, aiming to observe the ecological status of the forest riparian zone. The buffer width was recommended to be between 0.9 and 30.5 m and the planting gap should not exceed 7.6 cm, which is more conducive to the removal of pollutants [36]. It was also found that the optimal width of buffer zones ranged from 5 to 12 m in the small-scale research, whereas it was wider than 15 m in field-scale research [90]. The buffer zone width used in the United States is usually 30 m, but the removal of pollutants is still significant when the buffer zone width exceeds 30 m [91]. As shown in Table 3, it is clear that although various studies were focused on the optimal width of buffer zones, especially on small slopes, there is no consensus on the optimal width of buffer zones. This may be due to differences in natural conditions such as geographical location, types of pollutants, composition and structure of soils, plant communities, and climate change. These factors increase the difficulty in establishing a unified optimal width for buffer zones. Moreover, the fixed-width buffer zone may fail to achieve the desired objectives in certain areas. As shown in Figure 7, it is recommended that the buffer zone width should be extended by 5–15 m compared to the practical width for enhancing their function [92]. Thus, more in-depth studies are needed. As shown in Figures 8 and 9, the optimum buffer zone width should be determined according to the specific environmental conditions and pollutant types, aiming to effectively improve the performance of buffer zones and the water quality and protect the aquatic environment [81,93].

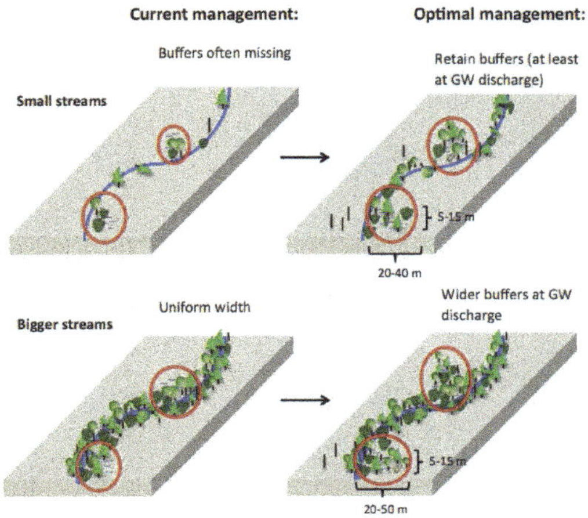

Figure 7. Optimal (site-specific) riparian buffer management in comparison to today's practice. GW stands for groundwater. Reproduced with permission from ref. [92], Copyright 2014, Elsevier.

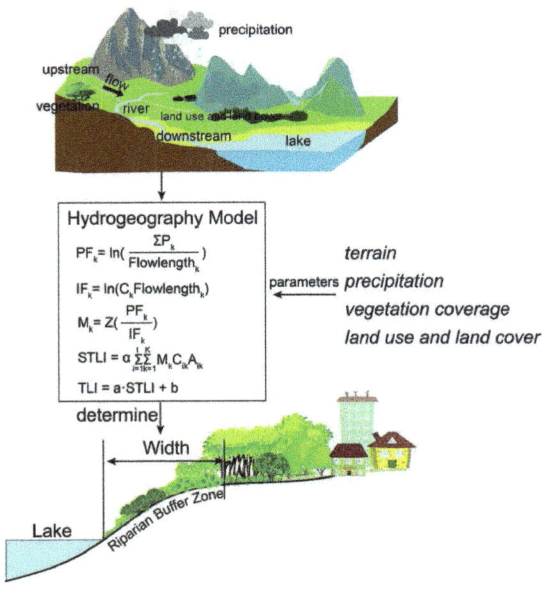

Figure 8. Influence factors and calculation of optimum width of buffer zone. (Where k is a certain point in the basin, PF_k is the pollutant producing factor of point k, ΣP_k is the cumulative water-collecting amount from the upstream area of point k, $Flowlength_k$ is the length of the flow path from point k to the lake or reservoir, IF_k is the interception factor of point k, C_k is the cumulative vegetation coverage on the flow path k, M_k is the weight factor of point k in the range of 0–1, STLI is the simulated comprehensive trophic level index (dimensionless), and α is a coefficient to revise the approximate equation. The TLI of a lake can be considered as the sum of the contributions of all LULC in the basin). Reproduced with permission from ref. [81], Copyright 2020, Elsevier.

Figure 9. A visual representation of the quantified ecological and economic outcomes when altering riparian buffer width on agricultural properties. Reproduced with permission from ref. [93], Copyright 2023, Elsevier.

4.2. Vegetation Type

Vegetation types play an important role in intercepting and removing pollutants in buffer zones, including wetland, aquatic, riparian, and grassland vegetation, as well as wetland trees. The influencing factors for the purification effectiveness of vegetation buffer zones mainly include vegetation type and the structure and characteristics of pollutants.

Different types of vegetation have different root structures, growth characteristics, and absorption capacities, leading to differences in their effectiveness in removing pollutants. Compared with non-vegetated buffer zones, vegetated buffer zones have better runoff stagnation capacity, effectively enhancing soil hydraulic permeability and improving removal efficiencies of nitrogen, phosphorus, and other pollutants [94]. At the same time, plant roots can effectively increase soil porosity, and aboveground portions of plants can enhance water storage and conductivity of soil through transpiration and root absorption, facilitating the transformation of dissolved nitrogen and plant absorption [95]. Trees have certain advantages in protecting groundwater, stabilizing riverbanks, and resisting floods [96]. A grass-based buffer zone is suitable for absorbing pollutants, improving plant and animal habitats, and increasing agricultural biodiversity [97]. Forest-based buffer zones can more effectively intercept rainfall [98]. *Arundinaria gigantea* is excellent at increasing water penetration rates, controlling surface runoff, and reducing total suspended sediment and total phosphorus concentrations [99]. Selection of suitable vegetation types is essential to minimize nutrient loss and maximize nutrient removal efficiency.

Aguiar et al. [63] found that, under the same buffer width, the interception capacity of woody vegetation for nitrogen, phosphorus, and nitrate was 100%, while removal rates by shrub vegetation were 83%, 66%, and 80%, and the rates for grassland vegetation were 61%, 53%, and 52%, respectively. Lv and Wu discovered that the order of removal efficiency of total nitrogen was as follows: *Taxodium hybrid 'Zhongshanshan'* + poplar (Nanlin-95) (65.57%) > poplar (Nanlin-95) (62.67%) > *Taxodium hybrid 'Zhongshanshan'* (60.63%). However, the order of removal efficiency for nitrate and ammonium nitrogen was as follows: poplar (Nanlin-95) > *Taxodium hybrid 'Zhongshanshan'* > *Taxodium hybrid 'Zhongshanshan'* + poplar (Nanlin-95) [85]. Dunn et al. [100] studied the effects of different vegetation on runoff and sediment loss. They found that the order of runoff reduction was as follows: willow (49%) > deciduous woodland (46%) > grass (33%). The decreasing order of suspended substance loss was as follows: willow (44%) > deciduous woodland (30%) > grass (29%). Apparently, the willow-based buffer zone showed a strong capacity for removing pollutants. Stutter et al. [13] reported that there was an increasing interest in willow (*Salix* spp.) due to its potential as a biomass energy source and its effectiveness as a barrier to prevent the flowing of soil and nutrients from agricultural land to rivers. Additionally, it possesses the ability for rapid regeneration after coverage and a high adaptability for various growing conditions. The differences among different types of riparian buffer zones are mainly attributed to the quality and quantity of organic carbon, aerobic and anaerobic conditions, and the composition of microbial communities [101]. However, some studies suggest that there is no difference in the maximum pollutant removal capacity among different vegetation types, which is possibly due to vegetation coverage exceeding 80%, resulting in an insignificant effect of vegetation composition on the buffer zone [102]. Therefore, in the designing process of buffer zones, suitable vegetation types for the local environment and water features should be selected and combined with the function of vegetation and ecological benefits to effectively improve water quality and protect the ecological environment.

Besides considering buffer zone width, it is difficult to summarize the effects of different vegetation coverage. Under the same herbaceous vegetation and external environment, a higher biomass of vegetation usually exhibits a stronger ability for reducing pollutants [103]. Hou et al. also considered that there was a positive correlation between vegetation coverage and the purification capacity index, and the optimal vegetation coverage should be higher than 84% [8]. However, some studies indicate that the impact of different degrees of vegetation coverage on pollution control is up to 20% with the same buffer width [41]. It is also suggested that the impact of vegetation coverage may be very limited or nonexistent [104].

Due to the importance of plant communities in the buffer zone, particular attention has been paid to their effects on pollutants entering buffer zones [93]. Xu et al. [34] found that *Platycladus orientalis* was more effective in removing nitrogen, phosphorus, and other pollutants from rivers, while *Pinus tabuliformis* could effectively intercept pollutants. Different vegetation types usually lead to differences in plant composition, root system type, and microbial community composition, thus affecting the absorption and transformation of nitrogen, phosphorus, and other pollutants via different types of buffer zones. Therefore, further studies are needed to deeply explore the effects of different vegetation types on pollutants. Quantitative analysis of the impact of vegetation types, biomass, and morphology on pollutant removal and their potential mechanisms should be emphasized.

4.3. Slope and Runoff Intensity

The slope and runoff intensity of buffer zones are important influencing factors for the removal of agricultural non-point source pollutants [105]. They significantly affect the transport rate and extent of soluble pollutants in runoff and sediments [106]. The slow-moving surface runoff in a gently sloping riparian zone can provide a longer contact and buffer time, and the sediment deposition can be utilized to increase the interception and degradation efficiency of pollutants. In contrast, the steep slope accelerates the speed of runoff, following pre-existing channels and bypassing vegetation and greatly diminishing the effectiveness of the buffer zone. Rainfall intensity has a significant effect on runoff, which increases with rainfall intensity [107]. Therefore, previous studies have focused on slope and runoff intensity as key factors for studying the effectiveness of rainwater runoff pollutant retention. It is worth noting that the slope of a buffer zone may not have a significant impact on the removal of pollutants without exceeding a certain threshold of rainfall duration and intensity.

Wu et al. [40] conducted a quantitative study on the removal loads of nitrogen and phosphorus using a constructed buffer zone and runoff hydrological measuring devices in fields with different slopes (2, 3, 4, and 5%). The initial runoff outflow time for a buffer zone with a 2% slope was 16.4 min, whereas it was only 9.1 min with a 5% slope. This indicates that the capacity for hindering runoff decreases with the increase in slope, likely since buffer zones with gentler slopes can enhance soil hydraulic permeability, significantly slowing down runoff. It was also found that the infiltration removal rates with the slopes of 2, 3, 4, and 5% reached 71.66, 68.14, 64.39, and 61.93%, respectively. The result indicates that lower slopes result in higher infiltration rates and infiltration ability for removing pollutants. This finding is consistent with Hille et al.'s study [46]. They attributed the higher pollutant removal rates to soil retention, filtration, microbial degradation, and root absorption [46]. However, Zhang et al. [78] found that a 10% slope was a crucial turning point for the water quality protection function of ecological buffer zones. The different results may be due to the sensitivity of slope parameters. The slight increase in slope will greatly increase the runoff velocity, which will then affect the retention and absorption of pollutants. These results provide a scientific basis for the designing and construction of buffer zones to effectively control non-point source pollution.

Additionally, some studies indicate that the total amount and intensity of rainfall should be considered along with buffer zone slope when determining the recommended width of a buffer zone [20]. In summary, slope plays a crucial role in determining buffer zone width. Furthermore, it is unsuitable to construct a wide enough buffer zone due to the limitations of construction costs and land resource. Therefore, it is necessary to conduct comparative analyses of the differences in pollutant removal effectiveness between runoff and infiltration and comprehensively consider buffer zone slopes and widths to improve the removal capacity for agricultural non-point source pollution via buffer zones.

4.4. Seasonal Variation

Seasonal variation has a significant influence on the pollutant removal ability of buffer zones. During the growing season, plant absorption is one of the main pathways for nitrogen removal. However, plant absorption may significantly decline or even cease in winter. Salazar et al. [108] pointed out that nitrogen fixation mainly occurs in the autumn and winter, whereas nitrogen mineralization primarily leads to approximately 70% absorption of nitrogen during the spring and summer.

Seasonal variation primarily affects nutrient absorption by plants through changes in temperature, light, precipitation, and plant communities. As shown in Figure 10, buffer zones exhibit good performance in reducing nutrient concentrations in warm or temperate climates [28]. However, several issues occur in colder climates, such as vegetation flattening or dying due to ice and snow and reduced infiltration due to frozen soil, resulting in poorer nutrient removal via buffer zones [37]. Additionally, Kumwimba et al. also found that many biological activities in buffer zones decreased during the spring snowmelt period, reducing plant uptake and assimilation of nutrients [18,28]. At the same time, some plants may be flattened or submerged by snow/ice during melting, these plants may decay and decompose to release nitrogen and phosphorus into surface runoff, consequently leading to an increase in pollutant concentrations in water. Duan et al. [97] also found that the removal rate of ammonia nitrogen via buffer zones was higher than that of total nitrogen in all experimental groups. This is mainly due to the absorption function of plant roots and the denitrification process in soil caused by water saturation and anoxic conditions during the warm growing season [109].

In addition, vegetation can be periodically harvested during the winter to reduce the loss of nutrients through runoff or leaching [110]. Vegetation in buffer zones absorbs the phosphorus in soil throughout its growing season. But vegetation begins to transport phosphorus from the branches to the roots in late autumn. Harvesting vegetation in autumn is beneficial for removing phosphorus and reducing the phosphorus releasing capacity of vegetation. However, more phosphorus is released from vegetation in completely unmanaged buffer zones, which may lead to a higher phosphorus content in soil and surface runoff. Zhang et al.'s study also suggests that harvesting vegetation before the end of October can avoid backflow and reduce agricultural non-point source pollution [49]. Therefore, it is necessary to understand the growth rate, life history, and community structure of vegetation in different seasons to fully utilize the function of buffer zone vegetation [111], providing a scientific basis for buffer zone management.

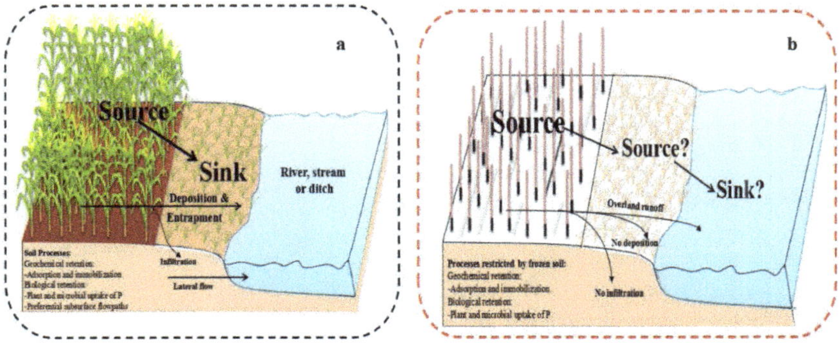

Figure 10. A conceptual diagram exhibiting a number of the processes through which buffer zones decrease pollutants in (**a**) warm and (**b**) cold climate regions. Reproduced with permission from ref. [28], Copyright 2023, Academic Press Inc.

4.5. Soil Composition

Physical and chemical properties of soil, such as particle size, organic carbon content, texture, structure, and moisture status, have a certain impact on the pollutant removal efficiency of buffer zones. Soil serves as a sink for pollutants like nitrogen and phosphorus, providing important functions such as interception, adsorption, and degradation. However, as phosphorus accumulates in soil, its fixing capacity for phosphorus gradually decreases, eventually leading to soil phosphorus entering water bodies through surface runoff, becoming a significant source of phosphorus in rivers. As shown in Figure 11, there are differences in phosphorus pools and their distribution between farmland and buffer zones, indicating that components of soil phosphorus and phosphorus stocks can be used to assess changes in the behavioral characteristics of soil phosphorus [112]. Compared with the paddy field without a soil plant buffer zone, the effluent concentration of each indicator in the paddy field with the operation mode of a soil plant buffer system is significantly reduced, and the interception rates of total dissolved nitrogen and phosphorus are 64.28% and 83.73%, respectively [113]. Thus, soil plant buffer zones can effectively reduce non-point source pollution in paddy fields and enhance yield and fertilizer utilization.

Walton et al. [14] studied the impact of organic and mineral soil on buffer zones, as depicted in Figure 12. They found that the average removal efficiency of nitrate nitrogen in organic soil and mineral soil was 52% and 51%, while that of total nitrogen was 45% and 36%, respectively. Clearly, wetland buffer zones with organic soil showed better pollutant removal effectiveness under higher loading rates, which may be related to their higher content of organic matter and stronger capacity for water infiltration. The combination of industrial by-products and buffer zone was also studied, which provides a potential strategy for improving the removal of soluble phosphorus from agricultural runoff. Some studies used industrial by-products containing large amounts of activated aluminum, iron, and calcium to reduce the release of soluble reactive phosphorus through adsorption or precipitation reactions [114,115]. Additionally, the adsorption capacity could be maintained under a wide pH range. Moreover, it has been found that phosphorus that adsorbed onto these materials was not easily desorbed. This study suggests that the combination of appropriate industrial by-products with buffer zones is an economical, efficient, and environmentally friendly method that can minimize the loss of soluble phosphorus.

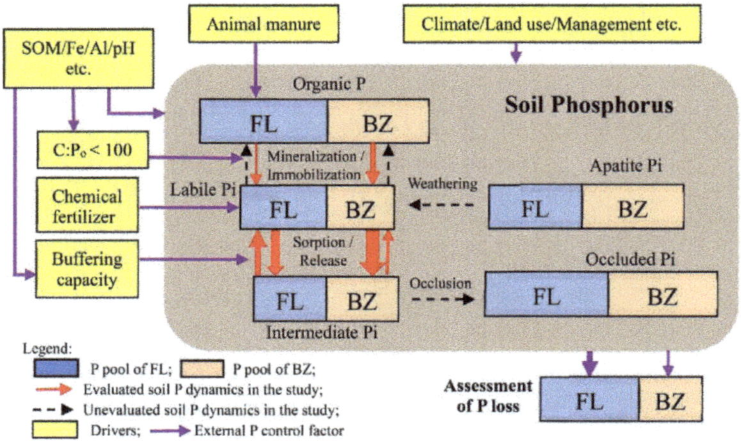

Figure 11. A conceptual framework of soil P fraction responses to different land use and how soil physicochemical characteristics affect soil P behavior in the lakeside area. FL, farmland; BZ, buffer zone. The size of the boxes reflects the size of the P pools. The thickness of the red arrows represents the amount of flux between the P fraction pools. Reproduced with permission from ref. [112], Copyright 2021, Elsevier.

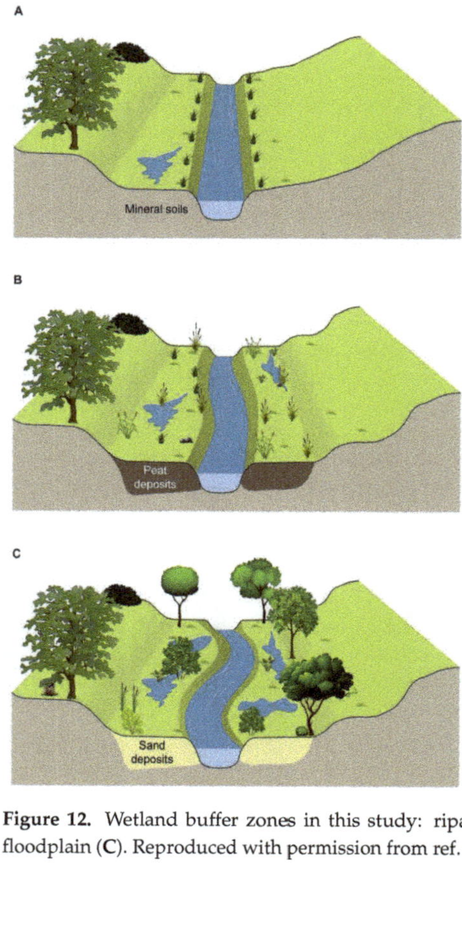

Figure 12. Wetland buffer zones in this study: riparian mineral soil wetland (**A**), fen (**B**), and floodplain (**C**). Reproduced with permission from ref. [14], Copyright 2020, Elsevier.

4.6. Vegetation Density

Vegetation density is one of the important influencing factors for the pollutant retention efficiency of buffer zones. Currently, there are not many studies on the pollutant retention efficiency of buffer zones with appropriate vegetation density. Tang et al. [116] suggested that increasing vegetation coverage was an important approach to reduce soil erosion in aquatic ecosystems. Yang et al. [117] also found that the recharge depth of soil water increased by 0.1 m after planting grass, which significantly improved rainwater infiltration and reduced the spatiotemporal changes in soil water porosity compared with non-vegetated sloping land. Liang et al. [107] found that a negative correlation between runoff and vegetation coverage (10, 30, and 50%) existed under the same rainfall intensity. This is due to the increase in vegetation coverage enhancing the roughness, with leaves and roots blocking more runoff. When the vegetation coverage increased from 0 to 50%, the surface runoff decreased by 4.10, 12.32, and 19.10%, respectively, indicating that increasing vegetation density can effectively reduce surface runoff and improve water resource management and soil conservation. Liang et al. [107] suggested that vegetation coverage should be higher than 50% to produce significant benefits in soil and water conservation.

Jin and Römkens found that when the vegetation density of buffer zones increased from 2500 to 10,000 clumps/m^2, the runoff decreased significantly, and the removal rate of suspended solids increased by 45% [118]. Lv and Wu studied the effect of different vegetation densities (400, 1000, and 1600 plants/hm^2) in buffer zones on nitrogen removal. They found that when the buffer zone width was 30–40 m, the average removal rates for total nitrogen, nitrate nitrogen, and ammonia nitrogen gradually increased with the increase in vegetation density [85]. However, the optimum nitrogen removal efficiency of the buffer zone with a plant density of 1000 plants/hm^2 was achieved with a width of 5 m. This may be due to a higher amount of litter in the buffer zone with a greater vegetation density, resulting in more nitrogen released from litter to the soil through litter decomposition [115], as shown in Figure 5. Approximately 46% of the nutrient returns to the soil through decomposition of plant litter in the whole Dinghushan forest area [119]. Similarly, suitable vegetation density facilitates the removal of phosphorus from water bodies via buffer zones [42]. Hénault-Ethier et al. [120] also found an insignificant difference in buffering capacity between a 3 m wide buffer zone planted with willow and a buffer zone with natural regeneration herb coverage, regardless of density. Therefore, the different buffer zone configuration has a significant impact on its pollution interception effect in the process of buffer zone construction, which is also the main reason for the difference in the pollutant removal effect of buffer zones.

4.7. Other Factors

Besides the above six factors, the types and contents of pollutants, microbial activity, and other factors can also have a certain influence on the ability of buffer zones to remove pollutants.

4.7.1. Types and Contents of Pollutants

One of the main factors that determine the change in nitrogen removal efficiency is the pollution source [86]. Agricultural non-point source pollution originates from various sources, primarily including wastewater discharged from livestock farming, runoff of pesticides and fertilizers, and soil erosion and nutrient loss during heavy rainfall [103]. As shown in Figure 13, different types of pollution correspond to different types of buffer zones. Within the 0–50 m range, the riparian buffer zone is predominant, where natural vegetation can be effectively utilized to intercept and absorb nutrients during runoff. The 300–1000 m range typically comprises agricultural zones, where implementing a mulberry and rapeseed intercropping system can effectively control nitrogen and phosphorus loss. In forested or grassland areas, or orchards beyond 1000 m, reducing nitrogen and phosphorus into rivers can be achieved through soil and water conservation projects (such as increasing understory vegetation cover) and afforestation efforts [121]. The efficiency of buffer zones is different based on the source, type, and form of pollutants. Among them, nitrogen usually exists in the form of soluble nitrogen such as nitrate nitrogen and ammonia nitrogen. Phosphorus mainly consists of soluble, particulate, and organic phosphorus [28]. When the pollutants in the runoff pass through the buffer zone, the removal rate of the particulate adsorbed pollutants is the highest, while that of the dissolved pollutants is the lowest. However, buffer zones can improve groundwater quality to the same extent regardless of the source of contamination. In addition, it was found that nitrogen retention in surface runoff and groundwater was linearly correlated with the initial nitrogen concentration entering buffer zones by performing a robust weighted meta-analysis. The higher initial nitrogen concentration led to the greater amount of nitrogen retained in the buffer zone [86].

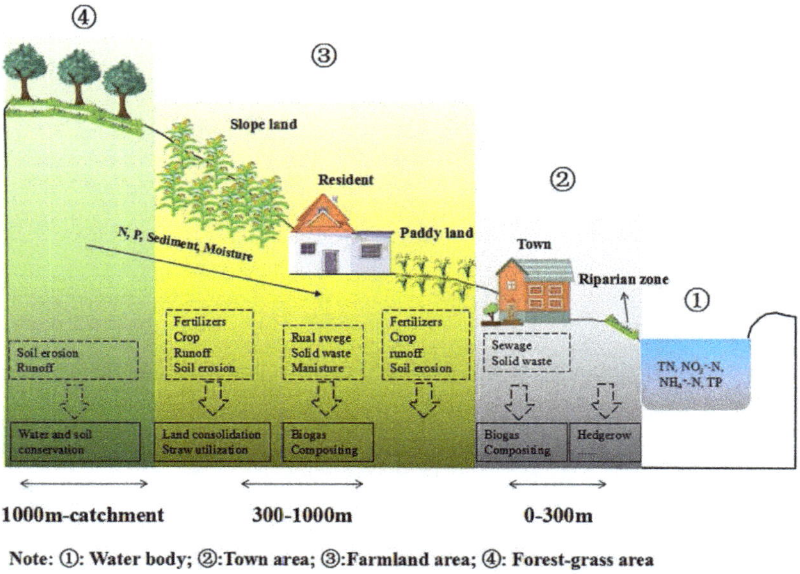

Figure 13. Multi-scale control system of nutrients in Qi river basin. TN stands for total nitrogen, NO_3^--N stands for nitrate nitrogen, NH_4^+-N stands for ammonia nitrogen, and TP stands for total phosphorus. Reproduced with permission from ref. [121], Copyright 2023, Elsevier.

4.7.2. Microbial Activity

The presence of vegetation root systems in buffer zones promotes soil biodiversity and provides a habitat for the growth and reproduction of microorganisms [122]. Microorganisms form a symbiotic relationship with vegetation, mutually enhancing their growth and development. Microorganisms are abundant and diverse components of buffer zone systems, which influence the ecological function and water quality of buffer zones via biodegradation, nitrogen cycling, and degradation of toxic substances [28]. These processes not only degrade pollutants such as nitrogen and phosphorus, but also improve soil texture, increase microbial diversity, and reduce ecological damage [123]. At the same time, the presence and management of buffer zones can also affect microbial diversity, with specific impacts on particular environments. Microorganisms are influenced by the physical characteristics of the buffer zone (such as topography and width), vegetation structure (such as type and density), and physical attributes of the river (such as width and hydrology) [124,125]. Therefore, more attention should be paid to maintaining the diversity and activity of microorganisms in buffer zones, which is of great significance for the protection of the ecological environment and the improvement of water quality.

5. Conclusions and Prospects

A buffer zone is one of the most commonly implemented management methods for controlling agricultural non-point source pollution. This review summarized the main affecting factors for buffer zones, including width, vegetation type, slope, seasonal variation, soil composition, vegetation density, types and forms of pollutants, and microbial activity. All these factors have an appropriate range to achieve maximum pollutant removal efficiency. Their range is closely related to the types of pollutants and their external environment. The inappropriate ranges for these factors may lead to the weakening or invalidation of buffer zones, consequently affecting the removal effect for agricultural non-point source pollution.

So far, great achievements have been focused on the pollutant removal efficiency of buffer zones, but few reviews were conducted on the remediation of agricultural non-point source pollution via buffer zones. Many issues require further research. Due to insufficient dynamic studies, the mechanisms through which single factors affect the pollutant removal efficiency of buffer zones is unclear. More attention should be paid to the combination of multiple factors. At the same time, studies on the mechanisms of pollutant removal are not deep enough, which may lead to insufficient understanding of the specific processes of buffer zones for removing pollutants. More research on this topic should be carried out to provide effective and scientific guidance for the practical application of buffer zones.

Author Contributions: Conceptualization, D.W. and X.G.; methodology, D.W. and X.G.; software, D.W. and X.G.; validation, Z.W.; formal analysis, D.W. and X.G.; investigation, D.W. and X.G.; resources, D.W. and X.G.; data curation, D.W. and X.G.; writing—original draft preparation, D.W., X.G. and S.W.; writing—review and editing, S.W., M.Z., Y.Z. and C.F.; visualization, Z.W.; supervision, Y.Z. and C.F.; project administration, D.W. and X.G.; funding acquisition, X.Z. and C.F. All authors have read and agreed to the published version of the manuscript.

Funding: This research was funded by the Wenzhou Ecological Park Research Project (grant number SY2022ZD-1002-07) and the Wenzhou Science and Technology Project for Basic Society Development (grant number S20220015).

Data Availability Statement: The data presented in this study are available on request from the corresponding author. The data are not publicly available due to funder restrictions.

Acknowledgments: The authors express their sincere gratitude for the warm work of the editor and the anonymous reviewers.

Conflicts of Interest: The authors declare no conflicts of interest.

Nomenclature

TN Total nitrogen	TP Total phosphorus
NH_4^+-N Ammonia nitrogen	PO_4^{3-}-P Phosphate
NO_3^--N Nitrate nitrogen	WWTPs Wastewater treatment plants
NO_2^--N Nitrite nitrogen	SSF CWS Subsurface flow constructed wetlands
N_2O Nitrous oxide	GW Groundwater
NO_X Nitrogen oxides	CH_2O Formaldehyde
N_2 Nitrogen	

References

1. Yang, C.; Shen, X.; Wu, J.; Shi, X.; Cui, Z.; Tao, Y.; Lu, H.; Li, J.; Huang, Q. Driving forces and recovery potential of the macrophyte decline in East Taihu Lake. *J. Environ. Manag.* **2023**, *342*, 118154. [CrossRef]
2. Xia, W.; Zhu, B.; Zhang, S.; Liu, H.; Qu, X.; Liu, Y.; Rudstam, L.G.; Anderson, J.T.; Ni, L.; Chen, Y. Climate, hydrology, and human disturbance drive long-term (1988–2018) macrophyte patterns in water diversion lakes. *J. Environ. Manag.* **2022**, *319*, 115726. [CrossRef]
3. Zou, L.; Liu, Y.; Wang, Y.; Hu, X. Assessment and analysis of agricultural non-point source pollution loads in China: 1978–2017. *J. Environ. Manag.* **2020**, *263*, 110400. [CrossRef]
4. Wu, S.; Bashir, M.A.; Raza, Q.U.A.; Rehim, A.; Geng, Y.; Cao, L. Application of riparian buffer zone in agricultural non-point source pollution control—A review. *Front. Sustain. Food Syst.* **2023**, *7*, 985870. [CrossRef]

5. Zhang, Y.; Long, H.; Li, Y.; Tu, S.; Jiang, T. Non-point source pollution in response to rural transformation development: A comprehensive analysis of China's traditional farming area. *J. Rural Stud.* **2021**, *83*, 165–176. [CrossRef]
6. Jabbar, F.K.; Grote, K. Statistical assessment of nonpoint source pollution in agricultural watersheds in the Lower Grand River watershed, MO, USA. *Environ. Sci. Pollut. Res.* **2018**, *26*, 1487–1506. [CrossRef]
7. Wang, Y.; Yang, J.; Liang, J.; Qiang, Y.; Fang, S.; Gao, M.; Fan, X.; Yang, G.; Zhang, B.; Feng, Y. Analysis of the environmental behavior of farmers for non-point source pollution control and management in a water source protection area in China. *Sci. Total Environ.* **2018**, *633*, 1126–1135. [CrossRef]
8. Hou, G.; Zheng, J.; Cui, X.; He, F.; Zhang, Y.; Wang, Y.; Li, X.; Fan, C.; Tan, B. Suitable coverage and slope guided by soil and water conservation can prevent non-point source pollution diffusion: A case study of grassland. *Ecotoxicol. Environ. Saf.* **2022**, *241*, 113804. [CrossRef]
9. Li, X.; Qiao, L.; Huang, Y.; Li, D.; Xu, M.; Ge, T.; Meersmans, J.; Zhang, W. Manuring improves soil health by sustaining multifunction at relatively high levels in subtropical area. *Agric. Ecosyst. Environ.* **2023**, *353*, 108539. [CrossRef]
10. Zak, D.; Goldhammer, T.; Cabezas, A.; Gelbrecht, J.; Gurke, R.; Wagner, C.; Reuter, H.; Augustin, J.; Klimkowska, A.; McInnes, R.; et al. Top soil removal reduces water pollution from phosphorus and dissolved organic matter and lowers methane emissions from rewetted peatlands. *J. Appl. Ecol.* **2017**, *55*, 311–320. [CrossRef]
11. Lind, L.; Hasselquist, E.M.; Laudon, H. Towards ecologically functional riparian zones: A meta-analysis to develop guidelines for protecting ecosystem functions and biodiversity in agricultural landscapes. *J. Environ. Manag.* **2019**, *249*, 109391. [CrossRef]
12. Wang, J.; Chen, G.; Fu, Z.; Song, X.; Yang, L.; Liu, F. Application performance and nutrient stoichiometric variation of ecological ditch systems in treating non-point source pollutants from paddy fields. *Agric. Ecosyst. Environ.* **2020**, *299*, 106989. [CrossRef]
13. Srivastav, A.L.; Patel, N.; Rani, L.; Kumar, P.; Dutt, I.; Maddodi, B.S.; Chaudhary, V.K. Sustainable options for fertilizer management in agriculture to prevent water contamination: A review. *Environ. Dev. Sustain.* **2023**, *26*, 8303–8327. [CrossRef]
14. Walton, C.R.; Zak, D.; Audet, J.; Petersen, R.J.; Lange, J.; Oehmke, C.; Wichtmann, W.; Kreyling, J.; Grygoruk, M.; Jabłońska, E.; et al. Wetland buffer zones for nitrogen and phosphorus retention: Impacts of soil type, hydrology and vegetation. *Sci. Total Environ.* **2020**, *727*, 138709. [CrossRef]
15. Han, H.; Yan, X.; Xie, H.; Qiu, J.; Li, X.; Zhao, D.; Li, X.; Yan, X.; Xia, Y. Incorporating a new landscape intensity indicator into landscape metrics to better understand controls of water quality and optimal width of riparian buffer zone. *J. Hydrol.* **2023**, *625*, 130088. [CrossRef]
16. Stutter, M.; Kronvang, B.; Ó hUallacháin, D.; Rozemeijer, J. Current Insights into the Effectiveness of Riparian Management, Attainment of Multiple Benefits, and Potential Technical Enhancements. *J. Environ. Qual.* **2019**, *48*, 236–247. [CrossRef]
17. Ghimire, S.R.; Nayak, A.C.; Corona, J.; Parmar, R.; Srinivasan, R.; Mendoza, K.; Johnston, J.M. Holistic Sustainability Assessment of Riparian Buffer Designs: Evaluation of Alternative Buffer Policy Scenarios Integrating Stream Water Quality and Costs. *Sustainability* **2022**, *14*, 12278. [CrossRef]
18. Kieta, K.A.; Owens, P.N.; Lobb, D.A.; Vanrobaeys, J.A.; Flaten, D.N. Phosphorus dynamics in vegetated buffer strips in cold climates: A review. *Environ. Rev.* **2018**, *26*, 255–272. [CrossRef]
19. Graziano, M.P.; Deguire, A.K.; Surasinghe, T.D. Riparian Buffers as a Critical Landscape Feature: Insights for Riverscape Conservation and Policy Renovations. *Diversity* **2022**, *14*, 172. [CrossRef]
20. Prosser, R.S.; Hoekstra, P.F.; Gene, S.; Truman, C.; White, M.; Hanson, M.L. A review of the effectiveness of vegetated buffers to mitigate pesticide and nutrient transport into surface waters from agricultural areas. *J. Environ. Manag.* **2020**, *261*, 110210. [CrossRef]
21. Liu, H.; Chen, J.; Zhang, L.; Sun, K.; Cao, W. Simulation effects of clean water corridor technology on the control of non-point source pollution in the Paihe River basin, Chaohu lake. *Environ. Sci. Pollut. Res.* **2021**, *28*, 23534–23546. [CrossRef] [PubMed]
22. Dlamini, J.C.; Cardenas, L.M.; Tesfamariam, E.H.; Dunn, R.M.; Loick, N.; Charteris, A.F.; Cocciaglia, L.; Vangeli, S.; Blackwell, M.S.A.; Upadhayay, H.R.; et al. Riparian buffer strips influence nitrogen losses as nitrous oxide and leached N from upslope permanent pasture. *Agric. Ecosyst. Environ.* **2022**, *336*, 108031. [CrossRef]
23. Buchanan, S.W.; Baskerville, M.; Oelbermann, M.; Gordon, A.M.; Thevathasan, N.V.; Isaac, M.E. Plant Diversity and Agroecosystem Function in Riparian Agroforests: Providing Ecosystem Services and Land-Use Transition. *Sustainability* **2020**, *12*, 568. [CrossRef]
24. Jager, H.I.; Nair, S.S.; Efroymson, R.A.; DeRolph, C.R.; Parish, E.S.; Wang, G. Ecosystem services from partially harvested riparian buffers can offset biomass production costs. *Sci. Total Environ.* **2023**, *889*, 164199. [CrossRef] [PubMed]
25. Lyu, C.; Li, X.; Yuan, P.; Song, Y.; Gao, H.; Liu, X.; Liu, R.; Yu, H. Nitrogen retention effect of riparian zones in agricultural areas: A meta-analysis. *J. Clean. Prod.* **2021**, *315*, 128143. [CrossRef]
26. Cui, R.; Zhang, D.; Liu, G.; Wang, P.; Chen, A.; Wang, H. Shift of lakeshore cropland to buffer zones greatly reduced nitrogen loss from the soil profile caused by the interaction of lake water and shallow groundwater. *Sci. Total Environ.* **2022**, *803*, 150093. [CrossRef] [PubMed]
27. Stutter, M.; Costa, F. The interactions of site-specific factors on riparian buffer effectiveness across multiple pollutants: A review. *Sci. Total Environ.* **2021**, *798*, 149238. [CrossRef] [PubMed]
28. Nsenga Kumwimba, M.; Huang, J.; Dzakpasu, M.; De Silva, K.; Ohore, O.E.; Ajibade, F.O.; Li, X.; Jing, J.S.; Muyembe, D.K.; Kaixuan, H. An updated review of the efficacy of buffer zones in warm/temperate and cold climates: Insights into processes and drivers of nutrient retention. *J. Environ. Manag.* **2023**, *336*, 117646. [CrossRef] [PubMed]

29. Xie, W.F.; Li, J.K.; Liu, Y.W.; Peng, K.; Zhang, K. Evaluation of ecological buffer zone based on landscape pattern for non-point source pollution control: A case study in Hanjiang River basin, China. *J. Hydrol.* **2023**, *626*, 130341. [CrossRef]
30. Lisandro, A.; Angel, V.G. Spatial modeling tool to assess and rank peri-urban land use in an agricultural region of the Midwestern United States. *Ecol. Inform.* **2024**, *81*, 102587. [CrossRef]
31. Oshunsanya, S.O.; Li, Y.; Yu, H. Vetiver grass hedgerows significantly reduce nitrogen and phosphorus losses from fertilized sloping lands. *Sci. Total Environ.* **2019**, *661*, 86–94. [CrossRef] [PubMed]
32. Gundersen, P.; Laurén, A.; Finér, L.; Ring, E.; Koivusalo, H.; Sætersdal, M.; Weslien, J.O.; Sigurdsson, B.D.; Högbom, L.; Laine, J.; et al. Environmental Services Provided from Riparian Forests in the Nordic Countries. *Ambio* **2010**, *39*, 555–566. [CrossRef]
33. Baulch, H.M.; Elliott, J.A.; Cordeiro, M.R.C.; Flaten, D.N.; Lobb, D.A.; Wilson, H.F. Soil and water management: Opportunities to mitigate nutrient losses to surface waters in the Northern Great Plains. *Environ. Rev.* **2019**, *27*, 447–477. [CrossRef]
34. Xu, K.; Mo, L.; Zhang, Z.; Zhang, M. Water quantity and quality changes from forested riparian buffer in Beijing. *Environ. Sci. Pollut. Res.* **2019**, *26*, 29041–29051. [CrossRef] [PubMed]
35. Weatherill, J.J.; Krause, S.; Ullah, S.; Cassidy, N.J.; Levy, A.; Drijfhout, F.P.; Rivett, M.O. Revealing chlorinated ethene transformation hotspots in a nitrate-impacted hyporheic zone. *Water Res.* **2019**, *161*, 222–231. [CrossRef]
36. Gene, S.M.; Hoekstra, P.F.; Hannam, C.; White, M.; Truman, C.; Hanson, M.L.; Prosser, R.S. The role of vegetated buffers in agriculture and their regulation across Canada and the United States. *J. Environ. Manag.* **2019**, *243*, 12–21. [CrossRef]
37. Wang, R.; Wang, Y.; Sun, S.; Cai, C.; Zhang, J. Discussing on "source-sink" landscape theory and phytoremediation for non-point source pollution control in China. *Environ. Sci. Pollut. Res.* **2020**, *27*, 44797–44806. [CrossRef] [PubMed]
38. Goloran, J.B.; Phillips, I.R.; Chen, C. Forms of Nitrogen Alter Plant Phosphorus Uptake and Pathways in Rehabilitated Highly Alkaline Bauxite Processing Residue Sand. *Land Degrad. Dev.* **2016**, *28*, 628–637. [CrossRef]
39. Moset, V.; Hille, S.; Rubæk, G.H.; Møller, H.B.; Wahid, R.; Baattrup-Pedersen, A. Indicators of biomass and methane yields in vegetated buffer strips. *J. Clean. Prod.* **2019**, *210*, 907–915. [CrossRef]
40. Wu, J.Q.; Xiong, L.J.; Sha, C.Y. Removal of N, P from seepage and runoff by different vegetated and slope buffer strips. *Water Sci. Technol.* **2020**, *82*, 351–363. [CrossRef]
41. Alemu, T.; Bahrndorff, S.; Alemayehu, E.; Ambelu, A. Agricultural sediment reduction using natural herbaceous buffer strips: A case study of the east African highland. *Water Environ. J.* **2017**, *31*, 522–527. [CrossRef]
42. Bu, X.; Xue, J.; Zhao, C.; Wu, Y.; Han, F.; Zhu, L. Sediment and nutrient removal by integrated tree-grass riparian buffers in Taihu Lake watershed, eastern China. *J. Soil Water Conserv.* **2016**, *71*, 129–136. [CrossRef]
43. Gong, Y.; Tian, R.; Li, H. Coupling of surface charges, adsorbed counterions and particle-size distribution on soil water infiltration and transport. *Eur. J. Soil Sci.* **2018**, *69*, 1008–1017. [CrossRef]
44. Sharpley, A.N.; Kleinman, P.J.A.; Flaten, D.N.; Buda, A.R. Critical source area management of agricultural phosphorus: Experiences, challenges and opportunities. *Water Sci. Technol.* **2011**, *64*, 945–952. [CrossRef]
45. Dorioz, J.M.; Wang, D.; Poulenard, J.; Trévisan, D. The effect of grass buffer strips on phosphorus dynamics—A critical review and synthesis as a basis for application in agricultural landscapes in France. *Agric. Ecosyst. Environ.* **2006**, *117*, 4–21. [CrossRef]
46. Hille, S.; Andersen, D.K.; Kronvang, B.; Baattrup-Pedersen, A. Structural and functional characteristics of buffer strip vegetation in an agricultural landscape—High potential for nutrient removal but low potential for plant biodiversity. *Sci. Total Environ.* **2018**, *628–629*, 805–814. [CrossRef]
47. Hefting, M.M.; Clement, J.C.; Bienkowski, P.; Dowrick, D.; Guenat, C.; Butturini, A.; Topa, S.; Pinay, G.; Verhoeven, J.T.A. The role of vegetation and litter in the nitrogen dynamics of riparian buffer zones in Europe. *Ecol. Eng.* **2005**, *24*, 465–482. [CrossRef]
48. Tan, E.; Hsu, T.C.; Huang, X.F.; Lin, H.J.; Kao, S.J. Nitrogen transformations and removal efficiency enhancement of a constructed wetland in subtropical Taiwan. *Sci. Total Environ.* **2017**, *601–602*, 1378–1388. [CrossRef]
49. Zhang, J. *Forestry Measures for Ecologically Controlling Non-Point Source Pollution in Taihu-Lake, China*; Springer: Singapore, 2016.
50. Peterjohn, W.T.; Correll, D.L. Nutrient dynamics in an agricultural watershed: Observations on the role of a riparian forest. *Ecology* **1984**, *65*, 1466–1475. [CrossRef]
51. Yang, J.; Wu, F.; Wei, X.; Zhang, X.; Wu, Q.; Yue, K.; Ni, X. Global Positive Effects of Litter Inputs on Soil Nitrogen Pools and Fluxes. *Ecosystems* **2022**, *26*, 860–872. [CrossRef]
52. Saadani, O.; Jebara, S.H.; Fatnassi, I.C.; Chiboub, M.; Mannai, K.; Zarrad, I.; Jebara, M. Effect of *Vicia faba* L. var. minor and *Sulla coronaria* (L.) *Medik* associated with plant growth-promoting bacteria on lettuce cropping system and heavy metal phytoremediation under field conditions. *Environ. Sci. Pollut. Res.* **2019**, *26*, 8125–8135.
53. Qin, Y.; Chen, Z.; Ding, B.; Li, Z. Impact of sand mining on the carbon sequestration and nitrogen removal ability of soil in the riparian area of Lijiang River, China. *Environ. Pollut.* **2020**, *261*, 114220. [CrossRef]
54. Wang, S.; Wang, W.; Zhao, S.; Wang, X.; Hefting, M.M.; Schwark, L.; Zhu, G. Anammox and denitrification separately dominate microbial N-loss in water saturated and unsaturated soils horizons of riparian zones. *Water Res.* **2019**, *162*, 139–150. [CrossRef]
55. Gu, X.; Chen, D.; Wu, F.; He, S.; Huang, J. Recycled utilization of *Iris pseudacorus* in constructed wetlands: Litters self-consumption and nitrogen removal improvement. *Chemosphere* **2021**, *262*, 127863. [CrossRef]
56. Martin, T.L.; Kaushik, N.K.; Trevors, J.T.; Whiteley, H.R. Review: Denitrification in Temperate Climate Riparian Zones. *Water Air Soil Pollut.* **1999**, *111*, 171–186. [CrossRef]
57. Brüsch, W.; Nilsson, B. Nitrate transformation and water movement in a wetland area. *Hydrobiologia* **1993**, *251*, 103–111. [CrossRef]

58. Satchithanantham, S.; English, B.; Wilson, H. Seasonality of Phosphorus and Nitrate Retention in Riparian Buffers. *J. Environ. Qual.* **2019**, *48*, 915–921. [CrossRef]
59. Janssen, M.; Frings, J.; Lennartz, B. Effect of grass buffer strips on nitrate export from a tile-drained field site. *Agric. Water Manag.* **2018**, *208*, 318–325. [CrossRef]
60. Tang, J.; Zhu, Y.; Wei, Z.; Feng, L.; Yang, N.; Sun, Z.; Luo, Q. Effectiveness of Riparian Vegetated Filter Strips in Removing Agricultural Nonpoint Source Pollutants in Agricultural Runoff from the Liao River Area, China. *Pol. J. Environ. Stud.* **2021**, *30*, 4709–4718. [CrossRef]
61. Dunn, A.M.; Julien, G.; Ernst, W.R.; Cook, A.; Doe, K.G.; Jackman, P.M. Evaluation of buffer zone effectiveness in mitigating the risks associated with agricultural runoff in Prince Edward Island. *Sci. Total Environ.* **2011**, *409*, 868–882. [CrossRef]
62. Jiang, F.; Preisendanz, H.E.; Veith, T.L.; Cibin, R.; Drohan, P.J. Riparian buffer effectiveness as a function of buffer design and input loads. *J. Environ. Qual.* **2020**, *49*, 1599–1611. [CrossRef] [PubMed]
63. Aguiar, T.R., Jr.; Rasera, K.; Parron, L.M.; Brito, A.G.; Ferreira, M.T. Nutrient removal effectiveness by riparian buffer zones in rural temperate watersheds: The impact of no-till crops practices. *Agric. Water Manag.* **2015**, *149*, 74–80. [CrossRef]
64. Liu, C.; Qu, L.; Clausen, J.; Lei, T.; Yang, X. Impact of Riparian Buffer Zone Design on Surface Water Quality at the Watershed Scale, a Case Study in the Jinghe Watershed in China. *Water* **2023**, *15*, 2696. [CrossRef]
65. Aguiar, T.R.; Bortolozo, F.R.; Hansel, F.A.; Rasera, K.; Ferreira, M.T. Riparian buffer zones as pesticide filters of no-till crops. *Environ. Sci. Pollut. Res.* **2015**, *22*, 10618–10626. [CrossRef] [PubMed]
66. Noble, D.T.; MacDougall, A.S.; Levison, J. Impacts of soil, climate, and phenology on retention of dissolved agricultural nutrients by permanent-cover buffers. *Sci. Total Environ.* **2023**, *860*, 160532. [CrossRef] [PubMed]
67. Izydorczyk, K.; Michalska-Hejduk, D.; Jarosiewicz, P.; Bydałek, F.; Frątczak, W. Extensive grasslands as an effective measure for nitrate and phosphate reduction from highly polluted subsurface flow—Case studies from Central Poland. *Agric. Water Manag.* **2018**, *203*, 240–250. [CrossRef]
68. Campo-Bescós, M.A.; Muñoz-Carpena, R.; Kiker, G.A.; Bodah, B.W.; Ullman, J.L. Watering or buffering? Runoff and sediment pollution control from furrow irrigated fields in arid environments. *Agric. Ecosyst. Environ.* **2015**, *205*, 90–101. [CrossRef]
69. Jiang, Y.; Xie, Z.; Zhang, H.; Xie, H.; Cao, Y. Effects of land use types on dissolved trace metal concentrations in the Le'an River Basin, China. *Environ. Monit. Assess.* **2017**, *189*, 633. [CrossRef] [PubMed]
70. Lin, X.; Tang, J.; Li, Z.; Li, H. Finite Element Simulation of Total Nitrogen Transport in Riparian Buffer in an Agricultural Watershed. *Sustainability* **2016**, *8*, 288. [CrossRef]
71. Liu, H.; Meng, C.; Wang, Y.; Liu, X.; Li, Y.; Li, Y.; Wu, J. Multi-spatial scale effects of multidimensional landscape pattern on stream water nitrogen pollution in a subtropical agricultural watershed. *J. Environ. Manag.* **2022**, *321*, 115962. [CrossRef]
72. Wang, W.; Zhang, F.; Zhao, Q.; Liu, C.; Jim, C.Y.; Johnson, V.C.; Tan, M.L. Determining the main contributing factors to nutrient concentration in rivers in arid northwest China using partial least squares structural equation modeling. *J. Environ. Manag.* **2023**, *343*, 118249. [CrossRef] [PubMed]
73. Huang, W.; Mao, J.; Zhu, D.; Lin, C. Impacts of Land Use and Land Cover on Water Quality at Multiple Buffer-Zone Scales in a Lakeside City. *Water* **2019**, *12*, 47. [CrossRef]
74. Zhu, X.; Liu, W.; Chen, J.; Bruijnzeel, L.A.; Mao, Z.; Yang, X.; Cardinael, R.; Meng, F.R.; Sidle, R.C.; Seitz, S. Reductions in water, soil and nutrient losses and pesticide pollution in agroforestry practices: A review of evidence and processes. *Plant Soil* **2019**, *453*, 45–86. [CrossRef]
75. Uwimana, A.; van Dam, A.A.; Irvine, K. Effects of conversion of wetlands to rice and fish farming on water quality in valley bottoms of the Migina catchment, southern Rwanda. *Ecol. Eng.* **2018**, *125*, 76–86. [CrossRef]
76. Salceda-Gonzalez, M.; Udawatta, R.P.; Anderson, S.H. Agroforestry on runoff nitrogen and phosphorus losses from three paired watersheds after 25 years of implementation. *Agrofor. Syst.* **2023**, *98*, 603–617. [CrossRef]
77. Wang, J.; Fu, Z.; Liu, F.; Qiao, H.; Bi, Y. Effects of substrate improvement on winter nitrogen removal in riparian reed (*Phragmites australis*) wetlands: Rhizospheric crosstalk between plants and microbes. *Environ. Sci. Pollut. Res.* **2023**, *30*, 95931–95944. [CrossRef]
78. Zhang, X.; Liu, X.; Zhang, M.; Dahlgren, R.A.; Eitzel, M. A Review of Vegetated Buffers and a Meta-analysis of Their Mitigation Efficacy in Reducing Nonpoint Source Pollution. *J. Environ. Qual.* **2010**, *39*, 76–84. [CrossRef]
79. Chen, H.; Zhou, X.; Wang, Y.; Wu, W.; Cao, L.; Zhang, X. Study on the planning and influential factors of the safe width of riparian buffer zones in the upper and middle reaches of the Ziwu River, China. *Environ. Sci. Pollut. Res.* **2023**, *30*, 103703–103717. [CrossRef] [PubMed]
80. Jiang, K.; Li, Z.; Luo, C.; Wu, M.; Chao, L.; Zhou, Q.; Zhao, H. The reduction effects of riparian reforestation on runoff and nutrient export based on AnnAGNPS model in a small typical watershed, China. *Environ. Sci. Pollut. Res.* **2019**, *26*, 5934–5943. [CrossRef]
81. Wang, M.; Duan, L.; Wang, J.; Peng, J.; Zheng, B. Determining the width of lake riparian buffer zones for improving water quality base on adjustment of land use structure. *Ecol. Eng.* **2020**, *158*, 106001. [CrossRef]
82. Sirabahenda, Z.; St-Hilaire, A.; Courtenay, S.C.; Van Den Heuvel, M.R. Assessment of the effective width of riparian buffer strips to reduce suspended sediment in an agricultural landscape using ANFIS and SWAT models. *Catena* **2020**, *195*, 104762. [CrossRef]
83. Johnson, S.R.; Burchell, M.R., II; Evans, R.O.; Osmond, D.L.; Gilliam, J.W. Riparian buffer located in an upland landscape position does not enhance nitrate-nitrogen removal. *Ecol. Eng.* **2013**, *52*, 252–261. [CrossRef]

84. Mayer, P.M.; Reynolds, S.K.; McCutchen, M.D.; Canfield, T.J. Meta-Analysis of Nitrogen Removal in Riparian Buffers. *J. Environ. Qual.* **2007**, *36*, 1172–1180. [CrossRef] [PubMed]
85. Lv, J.; Wu, Y. Nitrogen removal by different riparian vegetation buffer strips with different stand densities and widths. *Water Supply* **2021**, *21*, 3541–3556. [CrossRef]
86. Valkama, E.; Usva, K.; Saarinen, M.; Uusi-Kämppä, J. A Meta-Analysis on Nitrogen Retention by Buffer Zones. *J. Environ. Qual.* **2019**, *48*, 270–279. [CrossRef] [PubMed]
87. Wanyama, J.; Kristoff, H.; Isabirye, M.; Kahimba, F.; Maetens, W.; Kimaro, D.; Deckers, S. Evaluation of runoff and sediment trapping effectiveness of vegetative filter strips in the riparian zone of Lake Victoria. *Geophys. Res. Abstr.* **2011**, *13*, 48–63.
88. Schmitt, T.; Dosskey, M.G.; Hoagland, K.D. Filter strip performance and processes for different vegetation, widths, and contaminants. *J. Environ. Qual.* **1999**, *28*, 1479–1489. [CrossRef]
89. Dindaroğlu, T.; Reis, M.; Akay, A.E.; Tonguç, F. Hydroecological approach for determining the width of riparian buffer zones for providing soil conservation and water quality. *Int. J. Environ. Sci. Technol.* **2014**, *12*, 275–284. [CrossRef]
90. Kiffney, P.M.; Richardson, J.S.; Bull, J.P. Responses of periphyton and insects to experimental manipulation of riparian buffer width along forest streams. *J. Appl. Ecol.* **2003**, *40*, 1060–1076. [CrossRef]
91. Hickey, M.B.C.; Doran, B. A review of the efficiency of buffer strips for the maintenance and enhancement of riparian ecosystems. *Water Qual. Res. J. Can.* **2004**, *39*, 311–317. [CrossRef]
92. Kuglerová, L.; Ågren, A.; Jansson, R.; Laudon, H. Towards optimizing riparian buffer zones: Ecological and biogeochemical implications for forest management. *For. Ecol. Manag.* **2014**, *334*, 74–84. [CrossRef]
93. Malcher, J.; Critchell, K.; Matthews, T.G.; Lester, R.E. How wide, how much? A framework for quantifying the economic and ecological outcomes of altering riparian width on agricultural land. *Sci. Total Environ.* **2023**, *897*, 165342. [CrossRef] [PubMed]
94. Saleh, I.; Kavian, A.; Habibnezhad Roushan, M.; Jafarian, Z. The efficiency of vegetative buffer strips in runoff quality and quantity control. *Int. J. Environ. Sci. Technol.* **2017**, *15*, 811–820. [CrossRef]
95. Nóbrega, R.L.B.; Ziembowicz, T.; Torres, G.N.; Guzha, A.C.; Amorim, R.S.S.; Cardoso, D.; Johnson, M.S.; Santos, T.G.; Couto, E.; Gerold, G. Ecosystem services of a functionally diverse riparian zone in the Amazon–Cerrado agricultural frontier. *Glob. Ecol. Conserv.* **2020**, *21*, e00819. [CrossRef]
96. Rai, S.; Srinivas, R.; Magner, J. Using fuzzy logic-based hybrid modeling to guide riparian best management practices selection in tributaries of the Minnesota River Basin. *J. Hydrol.* **2022**, *608*, 127628. [CrossRef]
97. Duan, Y.; Tang, J.; Li, Z.; Yang, B.; Yan, Y.; Yang, Y. Vegetated Buffer Zone Restoration Planning in Small Urban Watersheds. *Water* **2021**, *13*, 3000. [CrossRef]
98. Chellaiah, D.; Yule, C.M. Effect of riparian management on stream morphometry and water quality in oil palm plantations in Borneo. *Limnologica* **2018**, *69*, 72–80. [CrossRef]
99. Singh, G.; Schoonover, J.E.; Williard, K.W.J.; Sweet, A.L.; Stewart, J. Giant Cane Vegetative Buffer for Improving Soil and Surface Water Quality. *J. Environ. Qual.* **2019**, *48*, 330–339. [CrossRef]
100. Dunn, R.M.; Hawkins, J.M.B.; Blackwell, M.S.A.; Zhang, Y.; Collins, A.L. Impacts of different vegetation in riparian buffer strips on runoff and sediment loss. *Hydrol. Process.* **2022**, *36*, e14733. [CrossRef] [PubMed]
101. Akter, S.; Li, X.; Dzakpasu, M.; Ifon, B.E.; Manirakiza, B.; Muyembe, D.K.; Zhang, Y.; Huang, J.; Guadie, A. Nutrient and sediment retention by riparian vegetated buffer strips: Impacts of buffer length, vegetation type, and season. *Agric. Ecosyst. Environ.* **2024**, *369*, 109050.
102. Haukos, D.A.; Johnson, L.A.; Smith, L.M.; McMurry, S.T. Effectiveness of vegetation buffers surrounding playa wetlands at contaminant and sediment amelioration. *J. Environ. Manag.* **2016**, *181*, 552–562. [CrossRef] [PubMed]
103. Hu, Y.; Gao, L.; Ma, C.; Wang, H.; Zhou, C. The Comprehensive Reduction Capacity of Five Riparian Vegetation Buffer Strips for Primary Pollutants in Surface Runoff. *Appl. Sci.* **2023**, *13*, 3898. [CrossRef]
104. Uusi-Kämppä, J.; Braskerud, B.; Jansson, H.; Syversen, N.; Uusitalo, R. Buffer zones and constructed wetlands as filters for agricultural phosphorus. *J. Environ. Qual.* **2000**, *29*, 151–158. [CrossRef]
105. Wang, Q.; Li, C.; Chen, C.; Chen, J.; Zheng, R.; Que, X. Effectiveness of narrow grass hedges in reducing atrazine runoff under different slope gradient conditions. *Environ. Sci. Pollut. Res.* **2017**, *25*, 7672–7680. [CrossRef] [PubMed]
106. Yang, F.; Yang, Y.; Li, H.; Cao, M. Removal efficiencies of vegetation-specific filter strips on nonpoint source pollutants. *Ecol. Eng.* **2015**, *82*, 145–158. [CrossRef]
107. Liang, Z.; Liu, H.; Zhao, Y.; Wang, Q.; Wu, Z.; Deng, L.; Gao, H. Effects of rainfall intensity, slope angle, and vegetation coverage on the erosion characteristics of Pisha sandstone slopes under simulated rainfall conditions. *Environ. Sci. Pollut. Res.* **2019**, *27*, 17458–17467. [CrossRef] [PubMed]
108. Salazar, O.; Rojas, C.; Avendaño, F.; Realini, P.; Nájera, F.; Tapia, Y. Inorganic nitrogen losses from irrigated maize fields with narrow buffer strips. *Nutr. Cycl. Agroecosyst.* **2015**, *102*, 359–370. [CrossRef]
109. Li, X.; Hou, L.; Liu, M.; Zheng, Y.; Yin, G.; Lin, X.; Cheng, L.; Li, Y.; Hu, X. Evidence of Nitrogen Loss from Anaerobic Ammonium Oxidation Coupled with Ferric Iron Reduction in an Intertidal Wetland. *Environ. Sci. Technol.* **2015**, *49*, 11560–11568. [CrossRef] [PubMed]
110. Kieta, K.A.; Owens, P.N.; Vanrobaeys, J.A.; Lobb, D.A. Seasonal Changes in Phosphorus in Soils and Vegetation of Vegetated Filter Strips in Cold Climate Agricultural Systems. *Agriculture* **2022**, *12*, 233. [CrossRef]

111. Thomas, S.M.; Griffiths, S.W.; Ormerod, S.J. Beyond cool: Adapting upland streams for climate change using riparian woodlands. *Glob. Chang. Biol.* **2015**, *22*, 310–324. [CrossRef]
112. Fu, D.; Xu, Z.; Wu, X.; Zhao, L.; Zhu, A.; Duan, C.; Chadwick, D.R.; Jones, D.L. Land use effects on soil phosphorus behavior characteristics in the eutrophic aquatic-terrestrial ecotone of Dianchi Lake, China. *Soil Tillage Res.* **2021**, *205*, 104793. [CrossRef]
113. Jin, B.; Liu, X.; Tan, J.; Shao, X.; Cheng, J. Effect of Plant Buffer Zone–Antifouling Curtain Wall on Reducing Non-Point Source Pollution in Paddy Fields, China. *Sustainability* **2022**, *14*, 6044. [CrossRef]
114. Habibiandehkordi, R.; Quinton, J.N.; Surridge, B.W.J. Enhancing soluble phosphorus removal within buffer strips using industrial by-products. *Environ. Sci. Pollut. Res.* **2014**, *21*, 12257–12269. [CrossRef] [PubMed]
115. Penn, C.J.; Bryant, R.B.; Callahan, M.P.; McGrath, J.M. Use of Industrial By-products to Sorb and Retain Phosphorus. *Commun. Soil Sci. Plant Anal.* **2011**, *42*, 633–644. [CrossRef]
116. Tang, C.; Liu, Y.; Li, Z.; Guo, L.; Xu, A.; Zhao, J. Effectiveness of vegetation cover pattern on regulating soil erosion and runoff generation in red soil environment, southern China. *Ecol. Indic.* **2021**, *129*, 107956. [CrossRef]
117. Yang, K.; Ma, Z.; Wang, K.; Li, G.; Zhang, X.; Zhen, Q.; Zheng, J. Combining infiltration hole and mulching techniques with fish-scale pits effectively improved soil water storage in semiarid areas with shallow buried bedrock (Pisha Sandstone) in China. *Catena* **2023**, *230*, 107249. [CrossRef]
118. Jin, C.X.; Römkens, M.J.M. Experimental studies of factors in determining sediment trapping in vegetative filter strips. *Trans. ASAE* **2001**, *44*, 277.
119. Wang, T.; Xiang, K.; Zeng, Y.; Gu, H.; Guan, Y.; Chen, S. Polycyclic aromatic hydrocarbons (PAHs) in air, foliage, and litter in a subtropical forest: Spatioseasonal variations, partitioning, and litter-PAH degradation. *Environ. Pollut.* **2023**, *328*, 121587. [CrossRef] [PubMed]
120. Hénault-Ethier, L.; Lucotte, M.; Smedbol, É.; Gomes, M.P.; Maccario, S.; Laprise, M.E.L.; Perron, R.; Larocque, M.; Lepage, L.; Juneau, P.; et al. Potential Efficiency of Grassy or Shrub Willow Buffer Strips against Nutrient Runoff from Soybean and Corn Fields in Southern Quebec, Canada. *J. Environ. Qual.* **2019**, *48*, 352–361. [CrossRef] [PubMed]
121. Tan, S.; Zhao, G.; Peng, C.; Ye, W.; Xie, D.; Chen, F.; Ni, C.; Shao, J.; Zheng, L.; Ni, J. Multi-scale effects of landscape on nitrogen (N) and phosphorus (P) in a subtropical agricultural watershed: A case of Qi river basin (QRB), China. *Ecol. Indic.* **2023**, *147*, 110017. [CrossRef]
122. Cole, L.J.; Stockan, J.; Helliwell, R. Managing riparian buffer strips to optimise ecosystem services: A review. *Agric. Ecosyst. Environ.* **2020**, *296*, 106891. [CrossRef]
123. Chen, Z.; Liu, H.Y.; Rao, C.; Jiang, C.; Tan, Y.C.; Yu, T.; Yu, E.J.; Wu, P. Soil microbial community dynamics indicate disruption of nitrogen cycling by pollution in vegetation buffer zones. *Pedobiologia* **2021**, *85*, 150722. [CrossRef]
124. Orzetti, L.L.; Jones, R.C.; Murphy, R.F. Stream condition in Piedmont streams with restored riparian buffers in the Chesapeake Bay Watershed. *J. Am. Water Resour. Assoc.* **2010**, *46*, 473–485. [CrossRef]
125. Wahl, C.M.; Neils, A.; Hooper, D. Impacts of land use at the catchment scale constrain the habitat benefits of stream riparian buffers. *Freshw. Biol.* **2013**, *58*, 2310–2324. [CrossRef]

water

MDPI

Article

Comparative Assessment of Wastewater Treatment Technologies for Pollutant Removal in High-Altitude Andean Sites

Rubén Jerves-Cobo [1,2,*], Edwin Maldonado [2], Juan Fernando Hidalgo-Cordero [3], Hernán García-Herazo [3] and Diego Mora-Serrano [2]

1 Environmental Engineering Program, Faculty of Chemical Sciences, Eco Campus Balzay, University of Cuenca, Av. Víctor Albornoz and Calle de los Cerezos, Cuenca 010207, Ecuador
2 PROMAS, Program for Water and Soil Management, Department of Civil Engineering, University of Cuenca, Av. 12 de Abril s/n and Agustín Cueva, Cuenca 010103, Ecuador; eandres.eamc@gmail.com (E.M.); diego.mora@ucuenca.edu.ec (D.M.-S.)
3 Faculty of Architecture and Urban Planning, Central Campus, University of Cuenca, 12 de Abril s/n and Agustín Cueva, Cuenca 010103, Ecuador; juan.hidalgoc@ucuenca.edu.ec (J.F.H.-C.); hernan.garcia@ucuenca.edu.ec (H.G.-H.)
* Correspondence: rubenf.jervesc@ucuenca.edu.ec

Abstract: This study evaluated the pollutant removal efficiency of two decentralized wastewater treatment plants (WWTPs) located in the high-altitude southern Andes of Ecuador, Acchayacu and Churuguzo, from 2015 to 2024. Acchayacu previously operated using an upflow anaerobic filter (UAF), and from 2021, it transitioned to using vertical-subsurface-flow constructed wetlands (VSSF-CWs). In contrast, Churuguzo employs surface-flow constructed wetlands (SF-CWs). These systems were assessed based on parameters such as the five-day biochemical oxygen demand (BOD_5), chemical oxygen demand (COD), total phosphorus, organic nitrogen, ammonia nitrogen, total solids, fecal coliforms (TTCs), and total coliforms (TCs). The data were divided into two subperiods to account for the change in technology in Acchayacu. Statistical analysis was conducted to determine whether significant differences existed between the treatment efficiencies of these technologies, and the SF-CW was found to consistently outperform both the UAF and VSSF-CW in removing organic matter and microbial pollutants. This difference is likely attributed to the longer hydraulic retention time, lower hydraulic loading rate, and vegetation type. The findings highlight the environmental implications of treatment technology selection in WWTPs, particularly regarding the quality of receiving water bodies and their potential applications for public health, proper water resource management, and the design of decentralized systems in high-altitude regions, especially in developing countries.

Keywords: constructed wetlands; wastewater treatment; high-altitude regions; efficient removal; sustainable water management

1. Introduction

Water is a vital natural resource for life, and the growing demand for water due to continuous human development makes it a valuable resource with intrinsic cultural, social, and environmental value [1].

Wastewater is a combination of liquid waste from different origins, including residential, industrial, and institutional. It presents both an environmental risk and reuse opportunities, especially in water-scarce regions [2,3].

Wastewater composition varies depending on the local socioeconomic situation and local customs [4]. Determining the chemical composition of domestic wastewater is essential for evaluating existing treatment methods and selecting the most appropriate facilities [5], as well as determining its impact on the receiving water bodies [6,7]. Water quality analysis also provides information on its potential for reuse after treatment, including applications in agricultural and landscape irrigation, urban uses, and limited residential purposes [8].

Wastewater treatment plays an important role in ensuring the sustainability of water resources, particularly in high-altitude regions where climatic, geographical, and social factors can significantly influence wastewater treatment systems' efficiency. In this context, applying decentralized systems is important since they are a viable alternative, especially in small and dispersed rural communities [9].

Constructed wetlands are a cost-effective, adaptable, and sustainable solution that have been documented since 1912 according to early records [10]. These systems use methods that harness the natural capacity of substrates, microorganisms, and aquatic macrophytes, which use their roots as a barrier to retain solids and support bacterial growth [11,12]. Pollutant removal occurs through multiple mechanisms, including sedimentation, microbial degradation, absorption, and plant uptake. It integrates microbial activity, oxygen fixation by plants, and a bed composed of gravel, sand, or other inert materials, which function as both a filter and structural support for roots [13].

Two main types of constructed wetlands are recognized: surface-flow constructed wetlands (SF-CWs) and subsurface-flow constructed wetlands (SSF-CWs) [14]. Their performance can vary depending on factors such as the design configuration, vegetation type, climatic conditions, and the operational conditions of wastewater treatment plants (WWTPs).

These nature-based solutions have proven effective at reducing energy consumption while efficiently removing organic matter and microbiological contaminants, making them both economically and energetically sustainable [15]. In the case presented by Diaz and Paredes [16], cultivating specific plant species in constructed wetlands created opportunities for production and commercialization, such as biofuel generation.

According to Córdova et al. [17], in Ecuador, although 84.85% of the population has access to safe drinking water and 90.7% has access to sewer systems or septic tanks, proper wastewater management is not always guaranteed, particularly in rural areas where treatment occurs either on-site or is entirely lacking. The state of wastewater treatment in Ecuador reveals that only 26.3% of the total water distributed nationwide undergoes treatment, with the Ecuadorian highlands accounting for 22.8% of this treated volume [18].

Since 1984, the municipal company responsible for water supply and wastewater management, ETAPA-EP, has been working to optimize environmental sanitation within its jurisdiction and to improve the water quality of the rivers flowing through Cuenca City. As part of this effort, several centralized and decentralized wastewater treatment projects have been implemented [19].

Despite their potential, technologies such as UAFs, SF-CWs, and VSSF-CWs are rarely evaluated under comparable conditions, particularly in high-altitude regions. This represents a significant research gap, as performance data may be influenced by climatic characteristics or other external factors. Therefore, evidence-based comparisons are crucial to support better-informed technology selection, particularly in small communities where implementing decentralized systems is essential.

To address these gaps, this study compares pollutant removal efficiencies in two wastewater treatment plants (WWTPs) with similar configurations. Both facilities carry out a primary treatment using a sedimentation tank, a secondary treatment consisting of a septic tank, and a tertiary treatment using constructed wetlands. The main difference between them lies in the type of technology used in the tertiary treatment. Acchayacu operated with a UAF system until 2021, after which it transitioned to a VSSF-CW planted with paramo straw (*Calamagrostis intermedia*). In contrast, the Churuguzo WWTP uses an SF-CW planted with totora (*Scirpus californicus*). This natural experiment setup provides a valuable opportunity to compare these treatment technologies under consistent climatic and geographic conditions.

Therefore, a hypothesis is proposed suggesting that high-altitude conditions characterized by low temperatures and significant daily climatic variability may significantly affect the pollutant removal performance of wastewater treatment systems. These environmental factors could impact key biological processes such as microbial activity, nitrification, and plant growth under such conditions.

This study evaluates the performances of the Acchayacu and Churuguzo WWTPs by analyzing their efficiency in removing various pollutants, using parameters such as the five-day biochemical oxygen demand (BOD_5), chemical oxygen demand (COD), total phosphorus (TP), ammonia nitrogen (N_amo), organic nitrogen (N_org), suspended solids (SSs), total solids (TSs), total coliforms (TCs), and fecal coliforms (TTCs). Additionally, it assesses the impact of technology changes and differences in system configuration. The findings offer valuable insights into the effectiveness of constructed wetlands in high-altitude environments, promoting the adequate and sustainable management of water resources in decentralized wastewater treatment systems.

2. Materials and Methods

2.1. Ubications of WWTPs

This study was conducted at two decentralized wastewater treatment plants (WWTPs), Acchayacu and Churuguzo, which are located in the parishes of Tarqui and Victoria of Portete, respectively, in the Azuay province of the southern Ecuadorian Andes (Figure 1). These WWTPs are approximately 14 km apart. Acchayacu is located at 2689 m above sea level (m.a.s.l.), and Churuguzo is at 2628 m.a.s.l.; they experience similar climatic conditions. The effluents from both systems discharge into brooks that flow into the Irquis river, a tributary of the Tarqui river. This water body plays a crucial role in supporting the agricultural and livestock activities of the surrounding communities.

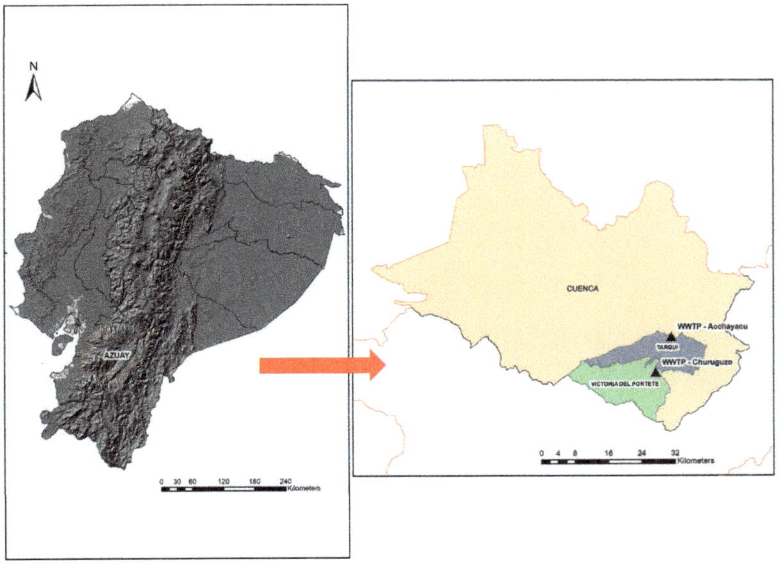

Figure 1. The locations of the Acchayacu (Tarqui parish) and Churuguzo (Victoria del Portete parish) wastewater treatment plants (WWTPs), Azuay, Ecuador.

2.2. Climate Analysis of Study Areas

Studies such as that of Operacz et al. [20] have demonstrated how climatic variability significantly affects the efficiency of constructed wetlands, as an increase in temperature can enhance the removal of parameters such as BOD_5 and COD by promoting microbial activity. To account for these climatic influences, temperature and precipitation data were obtained from the "Morascalle" station located in the Tarqui parish and precipitation data from the "Portete" station located in Victoria Portete parish. These stations are part of the meteorological network of the Water and Soil Management Program (PROMAS) at the University of Cuenca. The aforementioned stations are illustrated in Figure A12.

To compare the precipitation data from the Morascalle and Portete stations, a statistical analysis based on hypothesis tests was conducted. At first, descriptive and graphical explorations were performed using box diagrams and a precipitation value graph for each station to assess the differences between the two locations.

Subsequently, the Shapiro–Wilk test was applied to assess the normality of the data. If the data followed a normal distribution, a *t*-test was applied; otherwise, the Wilcoxon non-parametric test was used. This approach allowed for determining whether there was a significant difference in precipitation between the two sectors.

2.3. Technical Characteristics of the WWTPs

In these communities, the wastewater management system consists solely of a collection network. As previously described, the treatment process includes an inlet sedimentation tank, a septic tank, and the treatment stage, which is the focus of this study. In Acchayacu, this final stage transitioned from an upflow anaerobic filter (UAF) to vertical subsurface flow constructed wetlands (VSSF-CWs).

In the case of Churuguzo, the WWTP includes three constructed wetlands, through which the treated water flows before it is collected by a shared outlet pipe. These wetlands are planted with aquatic macrophyte species, specifically totora (*Scirpus californicus*). In both systems, the wetlands are filled with gravel of varying sizes to prevent clogging at the wastewater inlet.

These systems are decentralized, meaning that domestic wastewater is treated close to its point of origin, without a long sewer network that transports it to a centralized treatment plant. Such systems are particularly effective in small communities where geographic constraints hinder access to large-scale sewer infrastructure. In both WWTPs, the constructed wetlands receive effluent from a pretreatment stage, which comprises a sedimentation tank followed by a septic tank. At Acchayacu, the system includes three vertical-subsurface-flow constructed wetlands (VSSF-CWs) planted with paramo straw (*Calamagrostis intermedia*), each measuring 5.5 m in length, 5.1 m in width, and 1 m in depth, designed to serve a population of 720 inhabitants. In contrast, Churuguzo features two larger surface-flow constructed wetlands (SF-CWs), each measuring 34.5 m in length, 25 m in width, and 1.5 m in depth, designed to serve a population of 4560 inhabitants.

2.4. Analysis of Operating Parameters in Constructed Wetlands

Hydraulics are an important factor in the performance of a constructed wetland, as they directly influence its treatment capacity. In particular, the hydraulic retention time (HRT) and the hydraulic loading rate (H_{LR}) play fundamental roles in optimizing treatment efficiency.

The H_{LR} is a key factor in designing and operating constructed wetlands. It represents the relationship between the influent flow rate and the available treatment surface area, enabling assessments of the system's hydraulic capacity and treatment efficiency. The H_{LR} can be calculated using the following formula (Equation (1)):

$$H_{LR} = \frac{Q}{A} \tag{1}$$

where the symbols have the following meanings:

H_{LR} = Hydraulic loading rate ($m^3/(m^2 \times day)$);

Q = Design flow or average daily flow (m^3/day);

A = Effective surface area of the system (m^2).

An adequate *HRT* (Equation (2)) is crucial to preventing system overloading and ensuring optimal treatment performance. Different types of constructed wetlands have characteristic *HRT* values. For instance, horizontal-subsurface-flow constructed wetlands operate within a range of 4 to 5 days, while vertical-subsurface-flow constructed wetlands have shorter retention times, ranging from 2 to 6 h [21].

$$HRT = \frac{V}{Q} \tag{2}$$

where the symbols have the following meanings:

HRT = Hydraulic retention time;

Q = Design flow or average daily flow (m^3/day);

V = Wetland volume (m^3).

Although VSSF-CWs are typically operated under unsaturated conditions, those at the Acchayacu WWTP were designed to function under saturated conditions. Consequently, the entire volume of the wetland was considered when calculating the *HRT*.

The organic loading rate (*OLR*) (Equation (3)) was calculated to quantify the concentration of organic pollutants entering the system, which is essential for determining its treatment capacity. This parameter provided valuables insights into the performances of constructed wetlands and their efficiency in removing organic contaminants.

$$OLR = \frac{Q \times C}{A} \tag{3}$$

where the symbols have the following meanings:

Q = Influent flow rate (m^3/day);

C = Pollutant concentration (BOD or COD in kg/m^3);

A = Total surface area of the system (m^2).

2.5. Sampling and Analyzed Parameters

At the Acchayacu and Churuguzo WWTPs, sampling and analyses were conducted by personnel from the public water and sanitation company ETAPA on different dates and under varying climatic conditions. This approach enhanced the representativeness of the analyses, aligning with the requirements of the Ecuadorian Technical Standard NTE INEN 2169:2013 [22]. All analyses were performed in ETAPA's accredited laboratory, ensuring the reliability and validity of the results obtained.

The wastewater samples were analyzed at the ETAPA-EP's accredited water laboratory to determine the following parameters: electrical conductivity (Cond), five-day biochemical oxygen demand (BOD_5), chemical oxygen demand (COD), total phosphorus (TP), nitrogen ammonia (N_amo), organic nitrogen (N_org), dissolved oxygen (OD), hydrogen potential (pH), suspended solids (SSs), total solids (STs), total coliforms (TCs), and fecal coliforms (TTCs).

2.6. Comparison of Pollutant Removals Between WWTPs

The data were analyzed using R-Studio version 2024.12.1 software (Kousa Dogwood) [23], beginning with data cleaning, which involved an initial review to eliminate outliers (extreme values) and dates with missing records. These entries were excluded as they were not representative and could distort the results when compared to the other data. To determine statistical differences, *t*-tests were applied for normally distributed data, while Wilcoxon tests were used for non-normal distributions. This analysis aimed to evaluate which of the wastewater treatment technologies implemented at the Acchayacu and Churuguzo WWTPs demonstrated a better pollutant removal efficiency. No correction for multiple comparisons was applied due to the exploratory nature of this study.

Efficiency was determined for each analyzed parameter (Equation (4)) by comparing the influent and effluent concentrations at the WWTPs and between the technologies evaluated:

$$E = \frac{C_E - C_S}{C_E} \times 100 \quad (4)$$

where the symbols have the following meanings:

E = Parameter removal efficiency in the system (mg/L);

C_E = Parameter influent concentration analyzed;

C_S = Parameter effluent concentration.

In WWTP studies, data often deviate from a normal distribution due to several factors. According to Cantelmo and Ferreira [24], climatic variations in temperature and precipitation significantly affect the measurements. Additionally, biases may arise from outliers, fluctuations in pollutant loads, and the timing of sample collection. For this reason, the Shapiro–Wilk test was applied, which evaluates whether a dataset follows a normal distribution or not [25]. When normality was not found, the Wilcoxon test was used, particularly for comparing related samples to evaluate differences between data groups [26].

Finally, line graphs were used to illustrate the distribution and trends in the removal efficiency for both WWTPs and the analyzed technologies. These visualizations allowed for a comparative assessment of the results, identifying patterns and evaluating the effectiveness between the VSSF-CW with paramo straw, the SF-CW with totora, and the upflow anaerobic filter (UAF).

3. Results

This study compared the efficiency of the depuration of the Acchayacu and Churuguzo WWTPs, which are both located in high-altitude regions of the Ecuadorian Andes. The analysis included different wastewater treatment technologies: two types of constructed wetlands and an upflow anaerobic filter (UAF). The results of this analysis are presented in the following sections.

A comparison of these results is presented in Tables 1–3, which show the removal efficiency as a percentage, indicating the extent of pollutant reduction achieved by each wastewater treatment technology.

Table 1. A general comparison of treatment efficiency between the Acchayacu and Churuguzo WWTPs during 2015–2020.

Parameter	Acchayacu UAF (Efficiency %)	Standard Deviation for Acchayaccu	Churuguzo SF-CW (Efficiency %)	Standard Deviation for Churuguzo
SS	71.79	28.8	83.46	21.1
ST	47.73	32.4	63.71	22
BOD_5	60.20	25.4	74.30	23.3
COD	48.91	31	68.44	23.4
TP	37.23	27	45.04	21.2
N_amo	20.43	13	33.17	20
N_org	57.16	22	68.86	18.6
TC	58.65	28	85.24	20.7
TTC	53.18	29.3	92.85	9.47

Table 2. Comparison of treatment efficiency between WWTP after technology change (vertical sub-superficial flow wetlands and superficial flow wetlands) for 2021–2024.

Parameter	Acchayacu VSSF-CW (Efficiency %)	Standard Deviation for Acchayaccu	Churuguzo SF-CW (Efficiency %)	Standard Deviation for Churuguzo
SS	95.66	8.93	95.88	12
ST	75.20	21	83.57	13.3
BOD_5	83.90	23.5	95.56	5.09
COD	82.80	16.6	89.40	11.1
TP	53.45	19.7	57.82	19.2
N_amo	30.74	19.1	40.77	9.48
N_org	65.87	31.7	71.82	31.5
TC	69.18	29.4	94.70	5.33
TTC	75.05	18.8	96.32	4.3

Table 3. A comparison of treatment efficiency at the Acchayacu WWTP before and after the transition from UAF to VSSF-CW technology.

Parameter	Acchayacu	Standard	Acchayacu	Standard
	UAF	Deviation UAF	VSSF-CW	Deviation VSSF-CW
SS	71.79	28.8	95.66	8.93
ST	47.73	32.4	75.2	21
BOD_5	60.2	25.4	83.9	23.5
COD	48.91	31	82.8	16.6
TP	37.23	27	53.45	19.7
N_amo	20.43	13	30.74	19.1
N_org	57.16	22	65.87	31.7
TC	58.65	28	69.18	29.4
TTC	53.18	29.3	75.05	18.8

3.1. Analysis of Meteorological Parameters

The average precipitation recorded at the Morascalle (737.98 mm/year) and Portete (532.18 mm/year) stations does not show a statistically significant difference, according to the results of the t-test ($p = 0.085$) and the non-parametric Mann–Whitney test ($p = 0.296$). Although the mean precipitation at Portete is higher, the confidence intervals for the difference in means include zero values, indicating that the null hypothesis of equality between the stations cannot be rejected. Furthermore, the data distributions of the stations are similar (Figures 2 and 3), suggesting that the observed variations could occur by chance. However, it is important to note that the presence of missing data (10 from Morascalle and 50 from Portete) may have influenced the results.

The climatic conditions in the study area were evaluated using data from the Morascalle station, which is the only meteorological station in the region. The average daily temperature was 12.6 °C, with maximum and minimum averages of 22.83 °C and 3.16 °C, respectively. During the analyzed period (2015 to 2024), extreme temperatures of 26.7 °C (maximum) and −4.1 °C (minimum) were recorded. The lowest temperatures typically occurred in June, July, and August, fluctuating between −1.4 and 8.2 °C, with July typically being the coldest month. Conversely, the highest temperatures were observed from January to March, fluctuating between 19.3 and 26.7 °C. In 2024, the highest recorded temperature (26.7 °C) occurred in October.

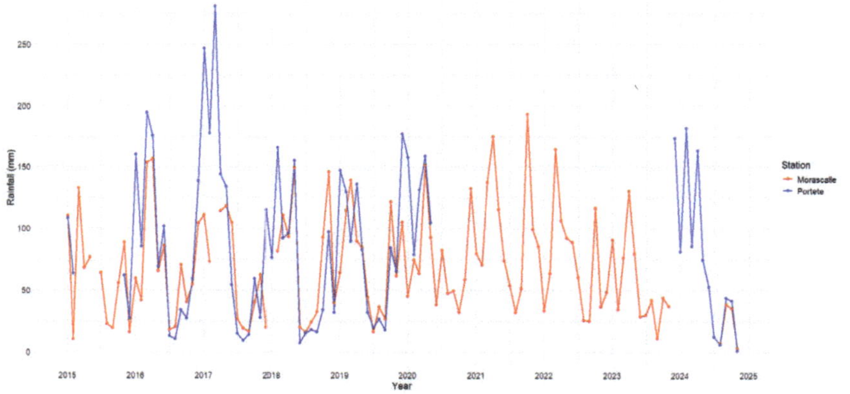

Figure 2. Comparison of rainfall data from Acchayacu and Churuguzo stations.

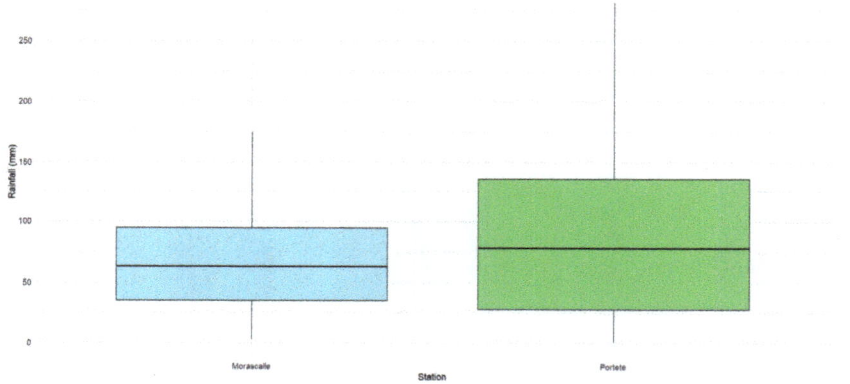

Figure 3. Precipitation distribution analysis.

3.2. Comparison of Removal Efficiency Between Treatment Technologies

3.2.1. Efficiency Comparison of the Upflow Anaerobic Filter (UAF) at the Acchayacu WWTP and the Surface Flow Constructed Wetland (SF-CW) at the Churuguzo WWTP During 2015–2020

The statistical analysis of the WWTPs from 2015 to 2020 enabled a comparison of the removal efficiency of different parameters by the UAF reactor of the Acchayacu WWTP and the SF-CW at the WWTP Churuguzo.

The results indicate that the SF-CW at Churuguzo exhibited higher efficiency, achieving superior removal rates across most of the analyzed parameters (Figure A14). These removal efficiencies were calculated based on the influent and effluent concentrations of the different parameters, as presented in Figures 4 and 5. The hydraulic retention time (*HRT*) of the UAF at Acchayacu ranged from 2.13 to 6.4 h, with an average of 3.19 h, whereas at the SF-CW at Churuguzo, it was 1.13 days. Additionally, the calculated hydraulic loading rate (H_{LR}) was 31.75 cm/day for the SF-CW, while the UAF reactor recorded an H_{LR} of 525 cm/day.

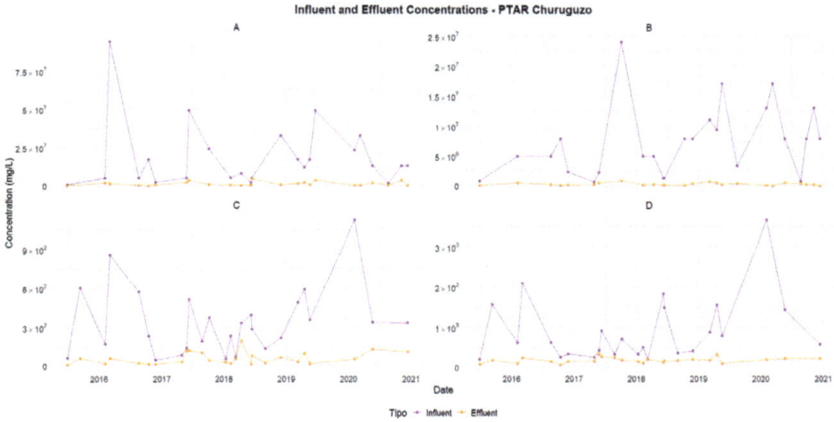

Figure 4. Churuguzo—Comparison of influent and effluent concentrations (2015–2020): (**A**) TC, (**B**) TTC, (**C**) BOD_5, and (**D**) COD.

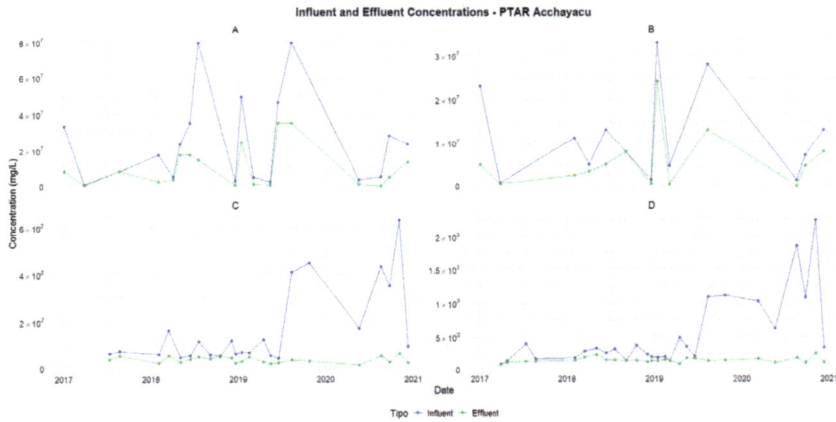

Figure 5. Acchayacu—Comparison of influent and effluent concentrations (2015–2020): (**A**) TC, (**B**) TTC, (**C**) BOD_5, and (**D**) COD.

Regarding organic matter removal, the BOD_5 shows that the Churuguzo WWTP achieved an average efficiency of 74.30%, surpassing that of Acchayacu, which recorded an average of 60.20%. Similarly, the COD removal efficiencies were 68.44% and 48.91% for Churuguzo and Acchayacu, respectively. For SSs, the UAF reactors at Acchayacu showed a removal efficiency of 71.79%, compared with 83.46% obtained for the Churuguzo WWTP. For microbiological pollutants, the SF-CW at Churuguzo exhibited a higher efficiency in removing total coliforms (TCs) (85.24%) and fecal coliforms (TTCs) (92.85%), compared to the UAF reactor at Acchayacu, for which values of 58.65% and 53.18%, respectively, were recorded. Statistical analyses comparing the UAF reactor and the SF-CW for the BOD_5, COD, TC, and TTC parameters revealed a significant difference, as indicated by p-values for the Wilcoxon test of 0.024, 0.028, 0.001, and 5.92×10^{-5}, respectively.

In contrast to the previously mentioned parameters, smaller differences were found for suspended solids (SSs) (p-value = 0.12) and total phosphorus (TP) (p-value = 0.74) between the two WWTPs. However, the SF-CW at Churuguzo still demonstrated a better overall performance, as shown in Table 1. According to the Wilcoxon test results, there was insufficient statistical evidence to confirm that the removal efficiencies for these parameters were significantly different.

At the Acchayacu WWTP, the UAF reactor presented an ammonia nitrogen (p-value = 0.049) removal efficiency of 20.43%, whereas the SF-CW at Churuguzo demonstrated a higher average efficiency of 33.17%. A similar trend was observed for organic nitrogen (N_org), (p-value = 0.82), with Acchayacu attaining an average removal efficiency of 57.16%, while Churuguzo achieved 68.86%.

The pH results indicate a slight decrease in the influent values compared to the effluent values in Acchayacu, with an average of 6.98, remaining within the neutral range of 6.5 to 8.5 (Figure A1). In contrast, Churuguzo shows a more pronounced pH decline, reaching a minimum of 5.41 (Figure A2). Regarding dissolved oxygen (DO) (p-value = 0.26), the Acchayacu WWTP records an average influent DO concentration of 4.32 mg/L, with peaks of 7.27 mg/L. However, the effluent data show a slight decrease, with an average of 4.94 mg/L, reflecting oxygen consumption due to the aerobic degradation of organic matter (Figure A3). Conversely, Churuguzo exhibits a notable increase in DO levels, with an average influent value of 2.26 mg/L and an effluent average of 4.52 mg/L, reaching a maximum of 6.1 mg/L. This suggests that greater oxygenation occurred in the horizontal subsurface flow wetland (Figure A4).

Figure A13 shows the variability in removal efficiency across different parameters, which may be attributed to fluctuating influent pollutant concentrations that occasionally exceed the treatment capacity of the UAF systems. Moreover, seasonal climate variations and maintenance practices could contribute to this variability.

The organic loading rate (OLR) calculations, based on the biochemical oxygen demand (BOD_5) and chemical oxygen demand (COD) data from Acchayacu and Churuguzo, indicate that both treatment systems operate under comparable BOD_5 loading conditions (0.202 kg/m^2·day for Acchayacu and 0.197 kg/m^2·day for Churuguzo). However, notable differences were observed in COD loading, with Acchayacu registering 0.762 kg/m^2·day compared to 0.620 kg/m^2·day in Churuguzo.

3.2.2. Efficiency Comparison of Two Constructed Wetlands Types at the Acchayacu and Churuguzo WWTPs During the Period 2021–2024

Before analyzing the removal efficiency of the wetland technologies implemented in the two WWTPs, the corresponding operational parameters, namely the hydraulic loading rate (H_{LR}) and hydraulic retention time (HRT), are presented. Thus, the calculated H_{LR} of the VSSF-CW at Acchayacu was 88.29 cm/day compared to 31.75 cm/day for the SF-CW at Churuguzo. Regarding HRT, each wetland unit in Acchayacu exhibited an HRT of 0.38 days, totaling 1.13 days across the three wetlands in series, compared to 4.72 days and 2.36 per wetland in Churuguzo.

However, the SF-CW in Churuguzo consistently demonstrated better performance in pollutant removal (Figure 6). The removal efficiencies were calculated based on the concentrations of the different parameters, as shown in Figures 6 and 7. Both wetland systems achieved a high removal efficiency for parameters such as SSs, with values exceeding 95% (Table 2).

Although both systems are effective in treating wastewater, the results suggest that the Churuguzo SF-CW is more efficient in removing organic matter and microorganisms.

The SF-CW in Churuguzo showed a higher efficiency in key parameters such as TC and TTC, with removal rates of 94.71% and 96.33%, respectively, compared to 69.18% and 75.06% for the VSSF-CW in Acchayacu. In terms of organic matter, Churuguzo also outperformed Acchayacu, reaching removal efficiencies of 95.56% for BOD_5 (p-value = 0.001) and 89.41% in COD (p-value = 0.12), exceeding the figures of 83.90% and 82.81%, respectively, recorded in Acchayacu. These differences were confirmed using the Wilcoxon test, which showed that the p-values for BOD_5 and COD indicated statistically significant differences between the two WWTPs, suggesting a variation in the removal efficiency. Likewise, microbiological parameters, such as TC and TTC, also revealed significant differences between both treatment plants, with p-values of 0.0013 and 3.19×10^{-50}, respectively. In contrast, for suspended solids (SSs), both plants have a p-value of 0.064, indicating no statistically significant difference between the plants for this parameter.

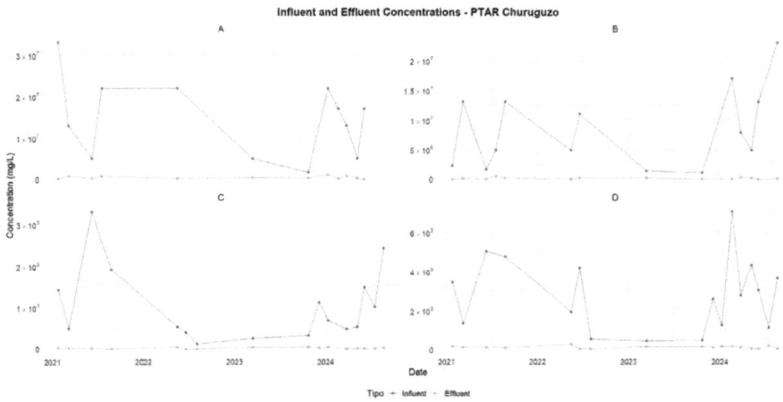

Figure 6. Churuguzo—Comparison of influent and effluent concentrations (2021–2024): (**A**) TC, (**B**) TTC, (**C**) BOD_5, and (**D**) COD.

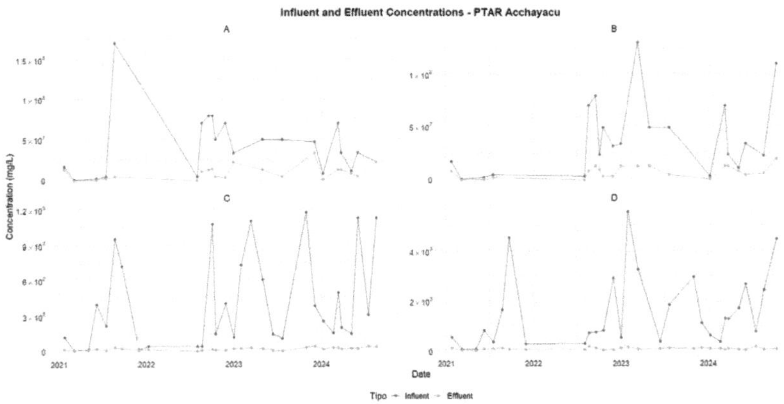

Figure 7. Acchayacu—Comparison of influent and effluent concentrations (2021–2024): (**A**) TC, (**B**) TTC, (**C**) BOD_5, and (**D**) COD.

Regarding SSs and TSs, although the differences were minor, the SF-CW in Churuguzo showed a slightly better performance. However, the differences were not statistically significant, with p-values of 0.063 and 0.015, respectively. In terms of nutrient removal, specifically organic nitrogen and TP, both systems showed moderate efficiencies, but the SF-CW in Churuguzo achieved better results, as shown in Table 2. The statistical analysis of the removal efficiencies for organic nitrogen (N_org), ammoniacal nitrogen (N_amo), and total phosphorus (TP) between the two WWTPs indicated non-statistically significant differences, with p-values of 0.566, 0.061, and 0.482, respectively, suggesting a comparable performance between the wetland technologies studied.

The pH in Acchayacu remains within a normal range, with no significant difference between influent and effluent values (7.1–7.4) and only a slight decrease in the effluent (Figure A5). In contrast, Churuguzo exhibits a more noticeable decrease in pH, reaching a minimum effluent value of 5.73; however, this still falls within acceptable limits (Figure A6).

For dissolved oxygen, DO (p-value = 0.86), Acchayacu shows an increase from 3.00 mg/L at the inlet to 4.74 mg/L at the effluent, suggesting moderate oxygenation (Figure A7). In comparison, in Churuguzo, the influent oxygen levels were low, reaching a maximum of 6.9 mg/L, indicating a higher degree of oxygenation within the wetland system (Figure A8).

3.2.3. Comparison Between Different Treatment Technologies of of Acchayacu WWTP

The change in the Acchayacu WWTP treatment system from an upflow anaerobic filter (UAF) to vertical-subsurface-flow constructed wetlands (VSSF-CWs) in 2021 resulted in a notable improvement in effluent quality. The removal efficiency of parameters such as BOD5 increased from an average of 60.2% with the UAF reactor to 83.9% after implementing the three VSSF-CWs. Similarly, DO removal improved from 48.91% with the UAF to 82.8% with the VSSF-CW system (Table 3). Moreover, the removal of microbiological parameters such as TTCs and TCs showed significant improvements, reducing the microbiological contamination of the treated wastewater. However, parameters such as TP, N_amo, and N_org continued to show only moderate removal efficiencies (Figure A15). Notably, N_amo exceeded the permissible discharge limits on one occasion, attributed to increased influent concentration recorded at that time (Figure 8). The statistical comparison between the two treatment technologies used in Acchayacu showed significant differences for BOD_5 ($p = 6.07 \times 10^{-5}$), COD ($p = 7.41 \times 10^{-6}$), SSs ($p = 5.38 \times 10^{-7}$), and TTCs ($p = 0.026$). In contrast, no statistically significant differences were observed for TC ($p = 0.111$), N_amo ($p = 0.063$), N_org ($p = 0.091$), DO ($p = 0.23$), and TP ($p = 0.093$).

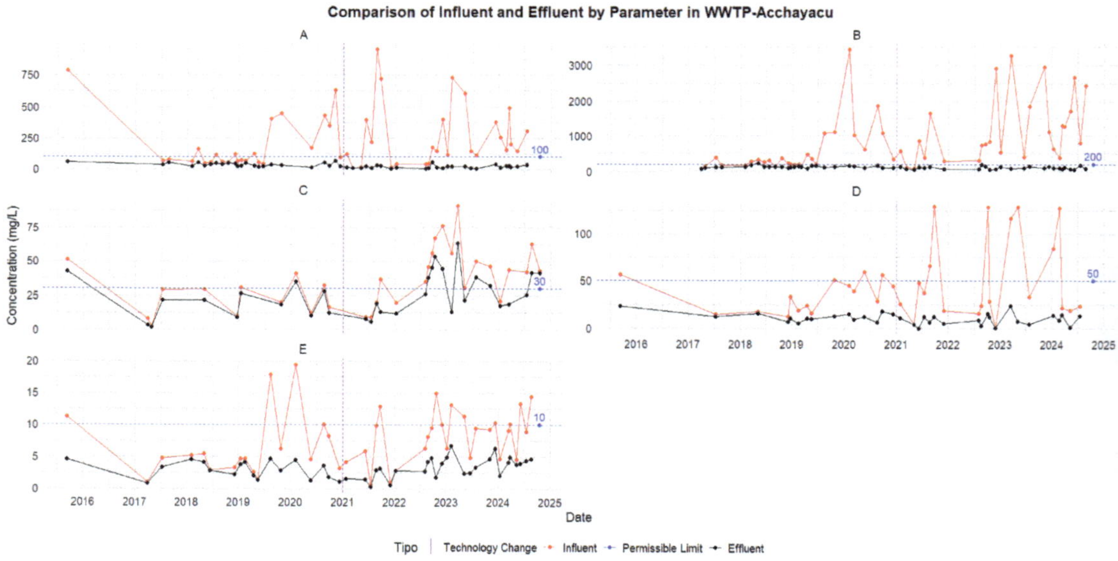

Figure 8. Concentrations in influent vs. effluent at Acchayacu WWTP: (**A**) BOD_5, (**B**) COD, (**C**) N ammonia, (**D**) N organic, and (**E**) TP.

Following the change from the UAF reactor to a VSSF-CW, no significant differences in DO and pH levels were initially observed (Figures A9 and A10) between the influent and effluent, likely due to the startup phase of the new system. However, once the VSSF-CW stabilized, the improvement in treatment efficiency was evident. Despite this, pH levels exhibited only minimal variations, which may be attributed to the adaptation of microorganisms and the planted macrophytes ("*Calamagrostis intermedia*").

The data show a high removal efficiency values for BOD_5 and COD, with significantly lower concentrations observed at the effluent, indicating a superior performance compared to Acchayacu (Figure A11). In contrast, the removal efficiencies for TP, N_amo, and N_org are more variable, with some instances showing lower removal rates. Nonetheless, it is important to highlight that, overall, these parameters demonstrate greater effectiveness in comparison to Acchayacu (Figure A12).

An analysis was conducted to compare the overall removal efficiencies of both WWTPs in the period in which all data were obtained. This comparison focused on the treatment performance of each plant as a whole, without distinguishing between the specific technologies used. However, as illustrated in Figure A15, a noticeable improvement in removal efficiency was observed following the technology upgrade in Acchayacu.

The effluent concentrations in Acchayacu generally complied with permissible limits, except for ammoniacal nitrogen, which, in some cases, exceeded the limit both before and after the technological upgrade implemented in 2021. It is important to highlight that the effluent concentrations of ammoniacal nitrogen in Churuguzo consistently remained below the permissible limit of 30 mg/L (Figure 9).

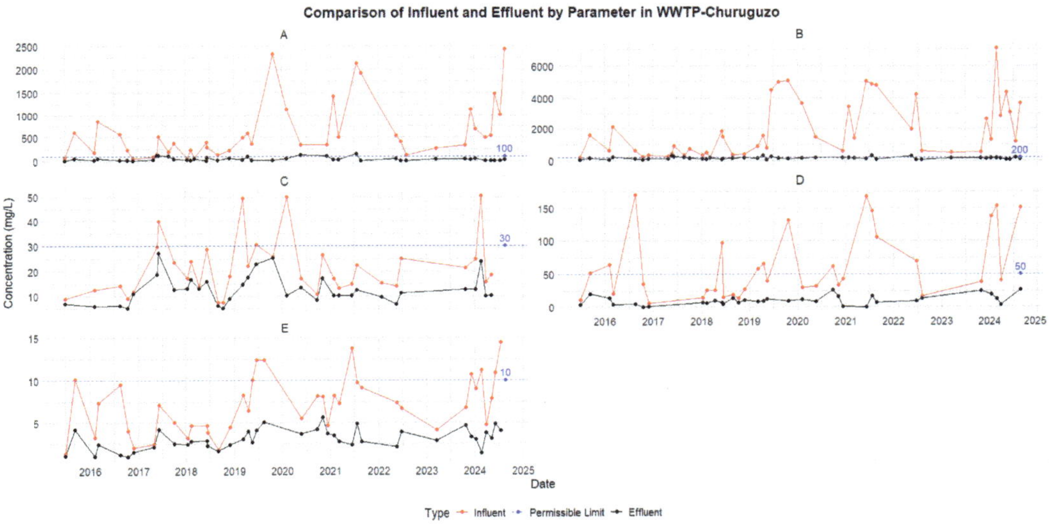

Figure 9. Concentrations in influent vs. effluent of WWTP Churuguzo: (**A**) BOD_5, (**B**) COD, (**C**) N ammonia, (**D**) N organic, and (**E**) TP.

For total suspended solids (TSSs), influent levels reached up to 180 mg/L; however, both the Acchayacu and Churuguzo systems effectively reduced TSS levels to below 100 mg/L.

A different trend was observed for total phosphorus (TP). In Acchayacu, influent concentrations ranged from 12 to 18 mg/L (Figure 8), whereas in Churuguzo, they remained below 15 mg/L (Figure 9). Despite these variations, both treatment systems successfully reduced TP levels to below the permissible limit of 10 mg/L.

Regarding fecal coliforms (TTCs), the effluent from Acchayacu consistently exceeded the permissible limit of 1000 MPN/100 mL, with peak values reaching up to 3000 MPN/100 mL. In contrast, the Churuguzo plant demonstrated more effective microbiological control, with effluent TTC concentrations falling below the regulatory threshold on several sampling dates.

4. Discussion

The results of this study highlight both the effectiveness and the limitations of the technologies implemented in the Acchayacu and Churuguzo WWTPs, specifically for treating wastewater from small populations using decentralized systems under conditions with a high pollutant load. The comparison of these WWTPs reveals significant differences in several key parameters, underscoring the influence of wetland design and the characteristics of the influent water on treatment performance.

- Analysis of meteorological parameters

The temperature data used in this study were obtained exclusively from the Morascalle weather station, as the Portete station could not be considered due to unavailable records. According to the geographical location data of the two WWTPs, both are situated in the same temperate Andean region, which is characterized by temperate-to-cold temperatures and marked seasonal variability. The Morascalle station is located approximately 3.3 km from the Acchayacu WWTP and 9.2 km from the Churuguzo WWTP, making it possible to determine that both plants are in relatively close proximity. As such, no significant climatic variability between the two sites is expected.

Temperature is a key factor influencing the efficiency of biological processes used in wastewater treatment. In particular, low temperatures can directly affect biological activity. In this case study, the average temperature ranges between 11 and 14 °C, which may impact treatment performance. In addition, in studies such as the one carried out by De La Mora-Orozco et al. [27], it has been shown that the capacity for pollutant removal, in the specific case of ammoniacal nitrogen (NH_4^+-N), decreases at temperatures below 16 °C. In contrast, water temperatures above 17 °C promote more vigorous and accelerated plant growth, enhancing treatment efficiency. These findings support the conclusion that lower temperatures can lead to reduce nutrient removal efficiency, as observed in the Acchayacu and Churuguzo WWTPs. However, during warmer months such as January to March, when temperatures exceed 19 °C, improved pollutant removal efficiency can be expected due to enhanced biological activity.

Precipitation data were analyzed using records from the Morascalle meteorological station and the Portete rainwater station. Although both stations are located within the same watershed and at similar altitudes, differences were observed in the recorded precipitation values. As noted by Buytaert et al. [28], such variations are common in mountainous regions due to factors such as exposure to prevailing winds, slope orientation, and topographic barriers. However, the differences between the stations studied were not statistically significant, suggesting that the observed variation may be primarily attributed to gaps in the available data from both stations. Therefore, this similarity in precipitation distributions reflects very similar precipitation conditions in both studied areas, which are located only 17 km away from each other.

Although the statistical analysis did not reveal significant differences in precipitation between the Morascalle and Portete stations, it is important to acknowledge that atypical events, such as extreme droughts or heavy rainfall, can introduce variability into the data. For instance, heavy rainfall causes dilution effects, temporarily enhancing treatment efficiency. This effect could be especially relevant in surface flow systems, as observed in the Churuguzo system.

- Analysis of hydraulic parameters

The hydraulic retention time (*HRT*) plays an important role in the removal efficiency of various wastewater treatment technologies. In general, a longer *HRT* enhances contaminant removal due to greater interaction between wastewater and microbial communities, thereby improving the system's ability to eliminate organic matter and nutrients. This relationship is supported by the findings of Navarro et al. [29], who show that higher efficiency is associated with increased *HRT* values.

The results obtained in this study show that the Churuguzo WWTP has a higher removal efficiency, which could be attributable to its longer *HRT*, while also acknowledging the influence of vegetation, system configuration, and wetland technology. According to the study conducted by the US EPA [30], an optimal *HRT* for an SF-CW is approximately three days for preventing algae blooms, a value exceeding that recorded in the SF-CW of Churuguzo. Conversely, for VSSF-CWs, an extensive *HRT* (greater than 1–3 days) is recommended for nutrient removal [31], a value much higher than that determined in the Acchayacu VSSF-CWs, potentially explaining their comparatively lower performances in nutrient reduction.

For the upflow anaerobic filter (UAF), the *HRT* suitable for operation is estimated to be between 4 and 10 h. When operating at low temperatures (10–20 °C), the efficiency of a UAF system tends to decrease due to slower reaction rates, which would require a longer *HRT* to maintain performance [32]. In a previous study conducted by González and Narváez [33], the UAF reactor at Acchayacu would operate at an average *HRT* below the recommended range, which could have influenced the removal efficiency of this technology.

- Comparative analysis of removal efficiency between SF-CW and VSSF-CW

The following section presents an analysis of the performances of and a comparison between two wetland technologies: the SF-CWs used at the Churuguzo WWTP and the VSSF-CWs implemented at the Acchayacu WWTP.

According to Quintero García et al. [34], a constructed wetland can be considered to function adequately for treating domestic wastewater when the elimination of bioindicators of fecal contamination and the presence of nitrifying and denitrifying bacteria responsible for the nitrification and denitrification processes required for nitrogen removal occur.

In general, the analysis of data obtained from the two wetland systems confirms that the Churuguzo WWTP demonstrated a superior performance for removing most of the parameters analyzed, particularly for eliminating organic matter (BOD_5 and COD) and microbiological contaminants (total and fecal coliforms). This higher efficiency may be attributed to the enhanced pollutant removal capacity of SF-CWs for pollutants, as reported in studies by Bedoya et al. [35]. The average BOD_5 removal efficiency in the SF-CWs at Churuguzo was higher than that of the VSSF-CWs at Acchayacu, supporting the hypothesis that wetlands with a longer *HRT* and denser vegetation promote more effective wastewater treatment. The difference in hydraulic loading rates (H_{LR}s) between the WWTPs suggests that although Acchayacu has a lower flow rate, its proportionally smaller wetland area limits the interaction time between the wastewater and the treatment system. Furthermore, it should be noted that Churuguzo's SF-CW technology allows for greater interaction with the atmosphere, enhancing oxygenation and plant development, which in turn contribute to more efficient pollutant removal. These findings highlight Churuguzo's greater capacity to treat domestic wastewater with high organic and microbiological loads, possibly attributable to its longer *HRT*. In addition, microbiological parameters such as TCs and TTCs exhibit significant differences between the two WWTPs, underscoring the effectiveness of the Churuguzo SF-CW in mitigating microbiological risks, crucial for protecting both public health and environmental quality.

When comparing the capacity for the removal of nutrients, such as nitrogen, phosphorus, and suspended solids (SSs), between the VSSF-CWs at Acchayacu, which uses the plant species paramo straw (*Calamagrostis intermedia*), and the SF-CWs in Churuguzo, the latter demonstrates a higher efficiency. However, these differences were not statistically significant. The variation in performance may be attributed to several factors, including differences in the wetland technology, H_{LR}, *HRT*, and type of vegetation. This observation is corroborated by Romero-Aguilar et al. [36], who found that a higher *HRT* in surface-flow constructed wetlands enhances the sedimentation of suspended solids and promotes nutrient uptake by vegetation. Additionally, Zahraeifard and Deng [37] reported that prolonged *HRT* in subsurface flow wetlands, such as the VSSF-CW at Acchayacu, favors processes such as nitrification and organic matter removal by allowing more time for microbial activity within the filter media. The extended contact between the wastewater, filter media, and microorganisms optimizes key biological processes, including nitrification and denitrification. Conversely, in systems such as the Acchayacu VSSF-CW, where the *HRT* is shorter, lower removal efficiencies for organic matter and nutrients are observed due to the limited contact time for the treatment process. This reduced retention time increases the risk of diminished performance in biological treatment processes [38]. Regarding vegetation, constructed wetlands may have a greater capacity to remove pollutants such as nutrients (N and P) and organic matter when larger macrophytes such as reeds (*Typha*) are used. This species improves system oxygenation by transporting oxygen through its roots, in turn stimulating microbial activity within the wetland substrate [39,40].

Considering that paramo straw is a smaller macrophyte species compared to totora, the latter requires more nutrients to sustain its growth. Additionally, the hydraulic loading rate (H_{LR}) of the SF-CW at the Churuguzo WWTP is lower, resulting in a larger treatment area. This expanded surface area promotes greater microbiological development, enhancing the breakdown and removal of organic matter.

This effect could be enhanced with a longer hydraulic retention time (*HRT*), as observed in the SF-CWs at the Churuguzo WWTP. Considering that paramo straw is a smaller macrophyte species compared to totora, the latter requires a higher nutrient input to sustain its growth. Moreover, the lower hydraulic loading rate (H_{LR}) in the SF-CW at the Churuguzo WWTP results in a larger treatment area, supporting the more extensive microbial development responsible for degrading organic matter. However, neither of the two wetland systems analyzed achieved the complete removal of ammoniacal nitrogen, suggesting the potential need to complement the treatment process with additional systems specifically targeting nutrient removal. These findings are consistent with the previous study conducted by Abdelhakeem et al. [41], which emphasizes the efficiency of subsurface wetlands in enhancing aerobic processes and improving nutrient removal from wastewater. The design of the VSSF-CW, characterized by a smaller effective surface area and higher H_{LR}, may have adversely impacted its treatment performance.

- Influence of technological change

Upflow anaerobic filters (UAFs) have been widely used in wastewater treatment systems as they use a biofilm fixed on a substrate for the removal (primarily) of organic matter under anaerobic conditions [42]. However, these systems have limitations in the removal of nutrients and pathogenic microorganisms, which may require further treatment.

In this study, the UAF exhibits a lower removal efficiency compared to the other technologies evaluated (SF-CW and VSSF-CW), particularly for eliminating ammoniacal nitrogen and fecal coliforms. Furthermore, the variability in treatment performance may be influenced by several factors, including the hydraulic load, stability of anaerobic biomass, and site-specific temperature conditions.

In this comparative analysis, it is necessary to evaluate the period prior to the change in technology at the Acchayacu WWTP—specifically, from 2015 to 2020—when the plant operated using a UAF reactor. The transition in 2021 from the UAF to vertical-subsurface-flow constructed wetlands (VSSF-CWs) led to notable improvements, as reflected in the increased removal efficiency for parameters such as BOD_5, COD, SSs, and TTCs compared to the previous system. However, the performance of the Acchayacu VSSF-CW still falls short of the removal efficiencies achieved by the SF-CWs in Churuguzo, particularly for the key parameters such as BOD_5, COD, TTC, and TC. This highlights the critical role of wetland design, vegetation selection, adaptation to local environmental conditions, and, in particular, the hydraulic retention time (*HRT*). A longer *HRT* not only enhances the removal of organic matter and nutrients but also contributes to eliminating pathogens by allowing the wastewater to have extended contact with the environment and vegetation, thereby facilitating microbial action in the removal process.

A notable decrease in pollutant concentrations, along with an increase in the removal efficiency for several parameters, can be observed following the transition from using UAF reactor technology to VSSF-CW systems. For instance, in the removal of SSs, BOD_5, and COD, a peak can be observed, which could be interpreted as an initial decline following the transition from UAF reactor technology to VSSF-CW systems. However, it is important to consider the influence of the wetland's startup or commissioning phase. As stated by Mosquera [43], for the removal of BOD_5, which typically occurs rapidly, treatment efficiency tends to be lower during the first months of the startup process. This phenomenon is closely linked to the development and stabilization of microbial consortia within the system.

Constructed wetlands are typically known for their stability in treatment efficiency once they reach operational equilibrium. However, as shown in Figures 5–7, in this case, they exhibit stable behavior in terms of contaminant removal efficiency, even under variations in the load entering the WWTP. This stability can be attributed to their good capacity to absorb peak loads despite the challenges that high-altitude conditions may present. However, it is important to consider factors such as the quality of the effluent and the startup period.

In this study, it was observed that during the first three to six months following the implementation of the new VSSF-CW system, the removal efficiencies for some parameters, particularly those dependent on microbial activity, were relatively low. This trend corresponds to the typical microbial and vegetative stabilization period. Such a startup phase is common in newly constructed wetlands and may influence the results recorded during the initial monitoring period.

The removal of ammoniacal nitrogen (N_amo) across the two evaluated periods improved following the change in technology at the Acchayacu WWTP. However, its efficiency remains lower than that of the Churuguzo WWTP. In contrast, the removal of organic nitrogen (N_org) and N_amo do not exhibit significant variability across different technologies and time periods. Nonetheless, both parameters consistently demonstrate low removal efficiencies. This limited performance may be attributed to several factors, including temperature—an aspect analyzed in previous studies such as that by Zhang et al. [44]. Their study indicates that the removal of these contaminants can fluctuate significantly depending on temperature and seasonal variations, a pattern also observed in both Acchayacu and Churuguzo.

These temperature variations can also significantly impact the efficiency of both constructed wetlands and the UAF reactor, as higher temperatures tend to enhance microbial activity, while lower temperatures can hinder biological processes, thereby reducing the overall treatment performance.

Although the change in technology has enhanced the performance of the Acchayacu WWTP compared to the former UAF system, the *HRT* may act as a limiting factor, restricting the system from reaching its maximum potential efficiency.

Although the change in technology at the Acchayacu WWTP led to improvements in wastewater treatment, the persistent presence of nutrients such as ammoniacal nitrogen and phosphorus in the effluents from both treatment plants suggests the need to implement complementary treatment strategies.

Considering the case of Acchayacu, additional treatments such as absorbent material filters or anaerobic bioreactors could offer potential solutions for enhanced nutrient removal. Various studies have demonstrated that these systems can effectively reduce elevated nutrient concentrations in effluents by combining physical, chemical, and biological processes across multiple stages. Consequently, future research should explore implementing integrated or hybrid systems to evaluate their performances under high-altitude conditions.

- Implications and limitations

This analysis highlights the positive impacts of integrating nature-based wastewater treatment systems, which offer sustainable and effective solutions, particularly in high-altitude regions. These systems also present strong potential for implementation in areas where the construction of sewer networks and centralized treatment facilities is limited by geographic, economic, or technical constraints. However, several critical factors must be considered when designing and operating constructed wetland-based treatment systems, including the configuration of the WWTP. It is important to note that no wetland treatment system is entirely maintenance-free. One of the most significant operational challenges in horizontal subsurface flow wetlands (HSSF-CWs) is clogging, which occurrs when solids accumulate and block the pore spaces in the media [45]. This reduces treatment efficiency and compromises system performance. Regular maintenance is, therefore, essential to ensure the proper functioning of constructed wetlands. For example, vegetation, such as totora (*Scirpus californicus*), must be pruned periodically, possibly annually or every two years, depending on the management objectives [46].

Vertical-subsurface-flow constructed wetlands (VSSF-CWs) have been widely studied for their effectiveness in contaminant removal, and they are comparable to other wetland types. However, their distinct structural design presents both advantages and limitations. While their primary advantage lies in restoring aerobic conditions during dry periods, they are limited by their dependence on substrate aeration and their susceptibility to clogging. To mitigate these challenges, VSSF-CWs are typically operated with intermittent loading and controlled organic matter input to prevent system overload [47].

Free-water-surface constructed wetlands (SF-CWs) offer a viable alternative for wastewater treatment, particularly in decentralized systems, as demonstrated in this study. However, their treatment efficiency is closely linked to the hydraulic retention time (*HRT*). In this study, the Acchayacu system, with an *HRT* of 1.13 days, exhibited lower removal rates compared to the Churuguzo system, which operated with a longer *HRT* of 4.72 days. This observation is consistent with findings from previous studies, such as that by Guerra et al. [48], indicating that a longer *HRT* can enhance nitrification and sedimentation processes.

A notable limitation of SF-CWs is their requirement for more frequent maintenance compared to horizontal flow wetlands (HSSF-CWs). This includes regularly pruning vegetation and periodically removing accumulated substrate material, which can result in higher operational and maintenance costs [49]. In general, surface-flow constructed wetlands (SF-CWs) tend to have lower construction costs due to their simple design and continuous operation, which eliminates the need for intermittent pumping. In contrast, vertical-subsurface-flow constructed wetlands (VSSF-CWs) typically require a higher initial investment. However, operational considerations must also be taken into account. SF-CWs often require more maintenance, particularly for vegetation management, and may present a greater risk of vector proliferation due to the exposed water surface.

Although no visible obstructions were observed in this case and no issues related to reduced flow or wetland overflow were reported, the potential for clogging due to sediment accumulation cannot be ruled out. This underscores the importance of regular and adequate maintenance.

Maintenance activities are managed by ETAPA-EP, which generally prunes vegetation every six months. However, this schedule may vary or adherence may be inconsistent. As such, it is essential to further strengthen the analysis presented in this study by incorporating data related to the frequency and timing of pruning. This would allow for the identification of potential variations in wastewater treatment performance associated with maintenance practices.

An important aspect to consider is that both WWTPs discharge into the Irquis river, which serves as the receiving body for the treated effluent. While the preliminary results indicate improvements in effluent quality, further adjustments may be required to fully comply with discharge standards, particularly in terms of nutrient concentrations. The water quality of the Irquis river has not only environmental significance but also social implications, as local communities rely on this resource for agriculture, domestic use, and other essential activities.

This study did not include an analysis that normalized pollutant loads based on flow rates, which is an acknowledged limitation. However, it is important to emphasize that the comparison was based on removal efficiencies derived from influent and effluent concentrations in the constructed wetlands using actual data provided by the ETAPA-EP, which has been monitoring these decentralized WWTPs. Nevertheless, it is recommended that future studies incorporate flow-based normalization to enable a more comprehensive evaluation of treatment system performance.

It is important to emphasize that the comparisons of treatment technologies were based on influent and effluent concentrations at the wastewater treatment plants (WWTPs), expressed in mg/L. Although this study did not include an analysis normalizing pollutant loads based on flow rates—a factor that could be considered a limitation—the primary objective was to compare the efficiency of each technology by assessing the removal capacity of the WWTPs under similar conditions, with the main distinction being the type and characteristics of the treatment technology employed for domestic wastewater. In future research, it is recommended to calculate influent loads to enable a more comprehensive and robust evaluation of system performance.

Although precipitation (P) data were available during the study period, no data on potential evapotranspiration (ETP) were collected. This constitutes a limitation, as it hindered a more comprehensive analysis of the system's performance. It is recommended that future studies incorporate ETP data to enable a more complete evaluation of the constructed wetlands.

5. Conclusions

The results obtained in this study demonstrate the importance of design, management, and continued research on decentralized wastewater treatment systems in rural and high-altitude regions. In this context, it is essential to consider key operational parameters such as the hydraulic retention time (*HRT*) and the hydraulic loading rate (H_{LR}). Additionally, selecting endemic vegetation with a high capacity for contaminant removal is crucial to ensuring the effectiveness and sustainability of such systems within the specific environmental conditions of the implementation area.

Regarding public policies, there is evidently a pressing need to strengthen regulatory frameworks and support initiatives that promote the adoption of nature-based sanitation technologies, particularly in areas where access to centralized systems or significant investment in infrastructure is limited or unfeasible. These alternatives offer viable solutions for satisfying water discharge regulations and safeguarding environmental quality, while simultaneously promoting equitable access to essential sanitation services. It is worth noting that constructed wetland systems, such as those analyzed in this study, have demonstrated both efficiency and viability as alternative wastewater treatment solutions, even under the challenging conditions of high-altitude environments and significant daily temperature fluctuations. These characteristics make them particularly well -suited to implementation in such regions as part of a broader strategy for decentralized wastewater management

The technological shift at the Acchayacu WWTP from an upflow anaerobic filter (UAF) to vertical-subsurface-flow constructed wetlands (VSSF-CWs) significantly enhanced pollutant removal, especially for organic matter (BOD_5, COD) and microbiological parameters (TCs and TTCs). Nonetheless, the removal efficiency for nutrients such as phosphorus and ammonia nitrogen remains an area requiring improvement.

The comparative analysis between the surface flow constructed wetland (SF-CW) at the Churuguzo WWTP and the VSSF-CW at the Acchayacu WWTP reveals significant differences in the removal efficiency of various contaminants. These differences are likely due to the distinct design characteristics and configurations of each system.

The SF-CW at Churuguzo, with a longer hydraulic retention time (*HRT*) and a lower hydraulic loading rate (H_{LR}), demonstrated greater efficiency in removing organic matter (BOD_5, COD), as well as in microbiological parameters (fecal and total coliforms). This improved performance could be attributed to the greater interaction between the water, filter media, and planted vegetation, with totora (*Scirpus californicus*) in the SF-CW system and paramo grass (*Calamagrostis intermedia*) in the VSSF-CW system. Additionally, the SF-CW design enhances oxygenation and microbial activity, further contributing to its higher treatment efficiency.

The findings of this study highlight the significance of the design and configuration of wastewater treatment systems that utilize nature-based technologies in high-altitude regions. The results demonstrated that factors such as the hydraulic retention time (*HRT*), hydraulic loading rate (H_{LR}), and vegetation type play crucial roles in contaminant removal efficiency. These insights are essential for planning and optimizing more effective and sustainable sanitation strategies. Improving effluent quality has direct implications for enhancing the environmental quality of receiving water bodies, particularly the Tarqui and Irquis rivers and their tributaries, and for protecting public health. These insights are valuable for guiding future designs for sustainable wastewater treatment strategies in high-altitude and decentralized systems.

One limitation of this study is the absence of normalization for the influent contaminant loads; future research should address this issue by conducting analyses based on pollutant loads rather than solely concentrations.

Future research should aim to enhance the existing treatment systems, with a particular focus on nutrient removal. Emphasis should be placed on promoting hybrid technologies, evaluating new endemic vegetation species suitable for constructed wetlands, and incorporating variables such as seasonal fluctuations and potential changes associated with extreme climatic events.

Author Contributions: R.J.-C. contributed to sampling preparation, assisted with field sampling, analyzed the data, and wrote the manuscript. E.M. coordinated and conducted the sampling campaign, contributed to sampling preparation and fieldwork, and also participated in data analysis and manuscript writing. J.F.H.-C., D.M.-S. and H.G.-H. contributed to writing the manuscript. All authors have read and agreed to the published version of the manuscript.

Funding: This study was funded by the Vice-Rectorate for Research at the University of Cuenca, research projects 2023.

Data Availability Statement: The raw data supporting the conclusions of this article will be made available by the authors on request.

Acknowledgments: The authors would like to express their sincere gratitude to the municipal water supply and wastewater management company, ETAPA-EP, for providing access to field data collected between 2015 and 2024 at both the Acchayacu and Churuguzo wastewater treatment plants (WWTPs).

Conflicts of Interest: The authors declare no conflict of interest.

Appendix A

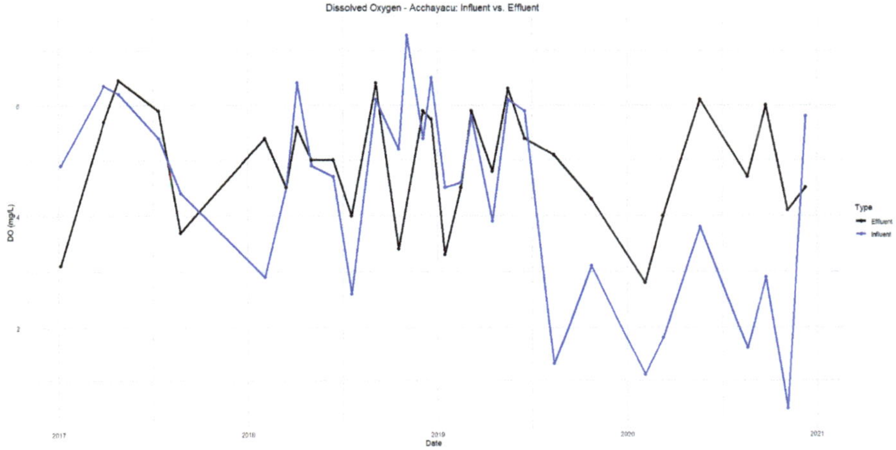

Figure A1. Dissolved oxygen in Acchayacu influent and effluent in 2015–2020.

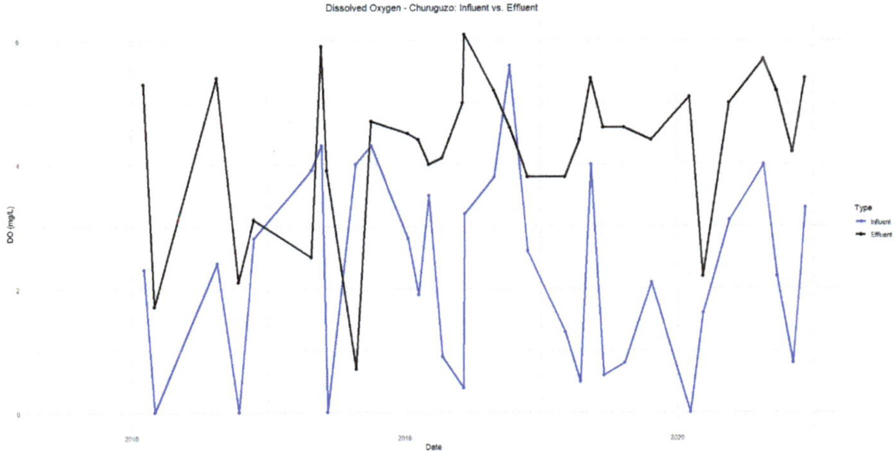

Figure A2. Dissolved oxygen in Churuguzo influent and effluent in 2015–2020.

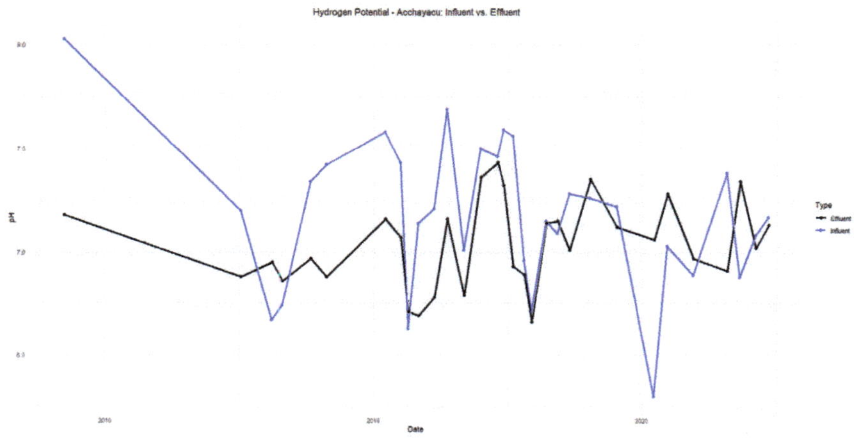

Figure A3. pH of Acchayacu influent and effluent in 2015–2020.

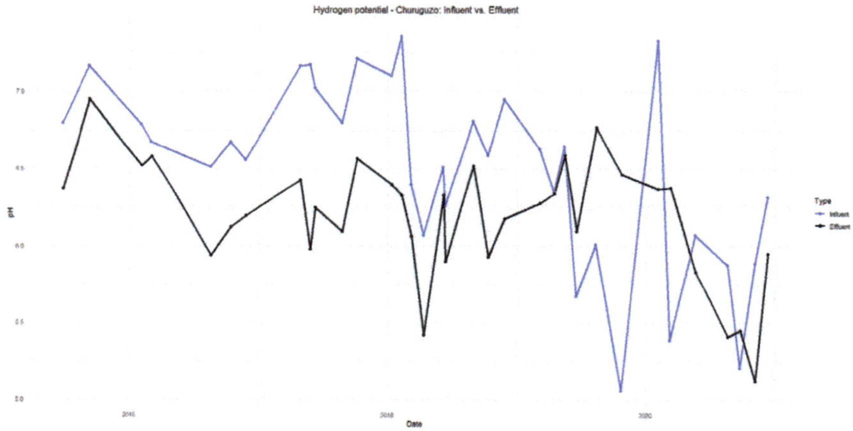

Figure A4. pH of Churuguzo influent and effluent in 2015–2020.

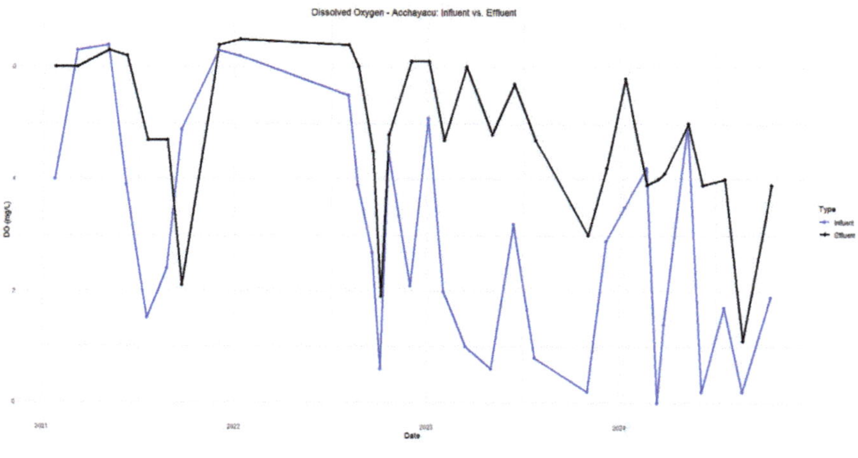

Figure A5. Dissolved oxygen in Acchayacu influent and effluent in 2021–2024.

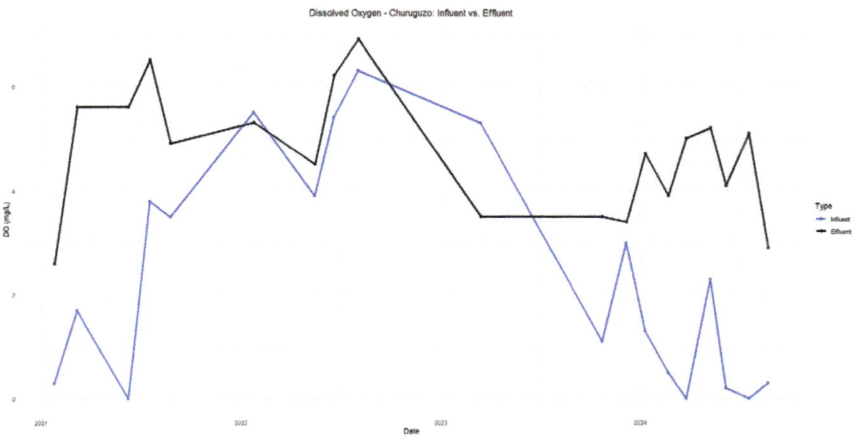

Figure A6. Dissolved oxygen in Churuguzo influent and effluent in 2021–2024.

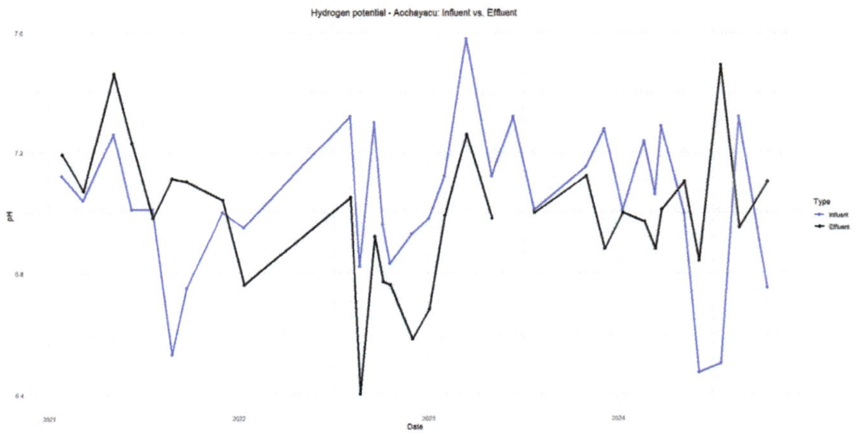

Figure A7. pH of Acchayacu influent and effluent in 2021–2024.

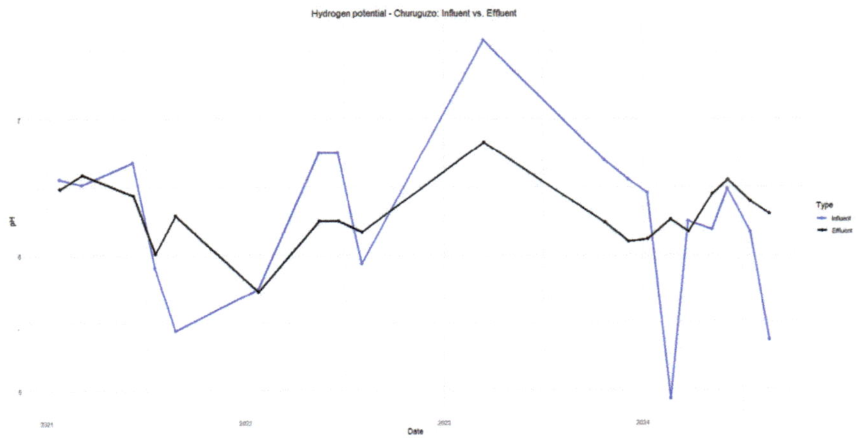

Figure A8. pH of Churuguzo influent and effluent in 2021–2024.

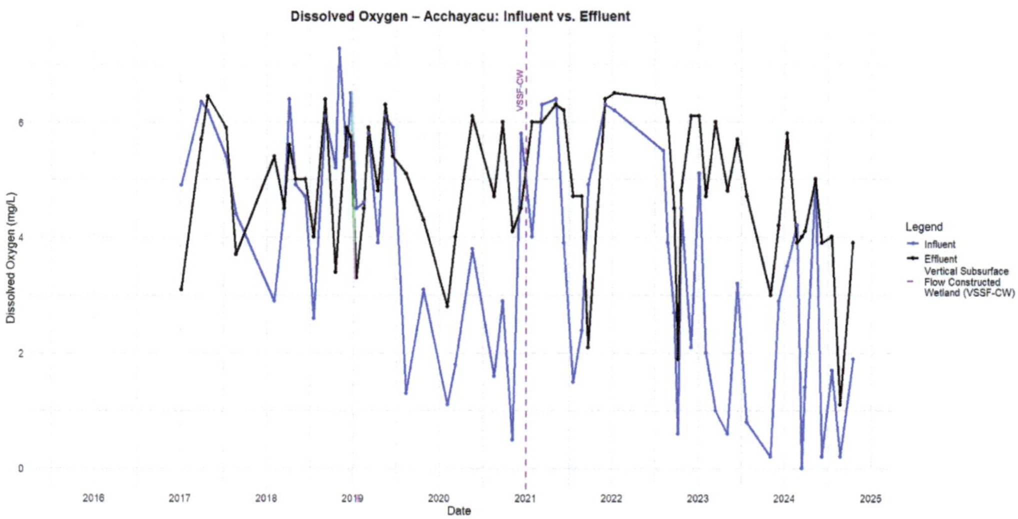

Figure A9. Dissolved oxygen in Acchayacu.

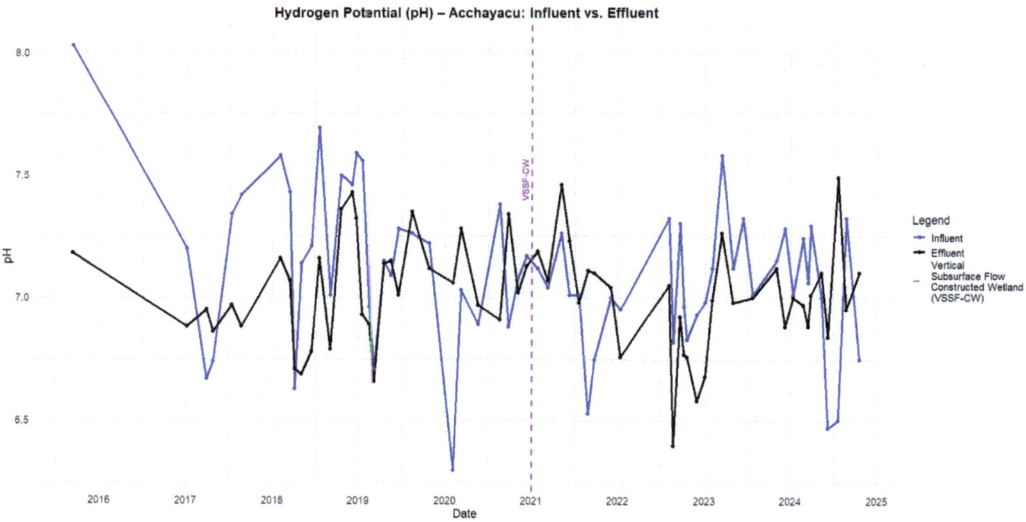

Figure A10. Hydrogen potential (pH) in Acchayacu.

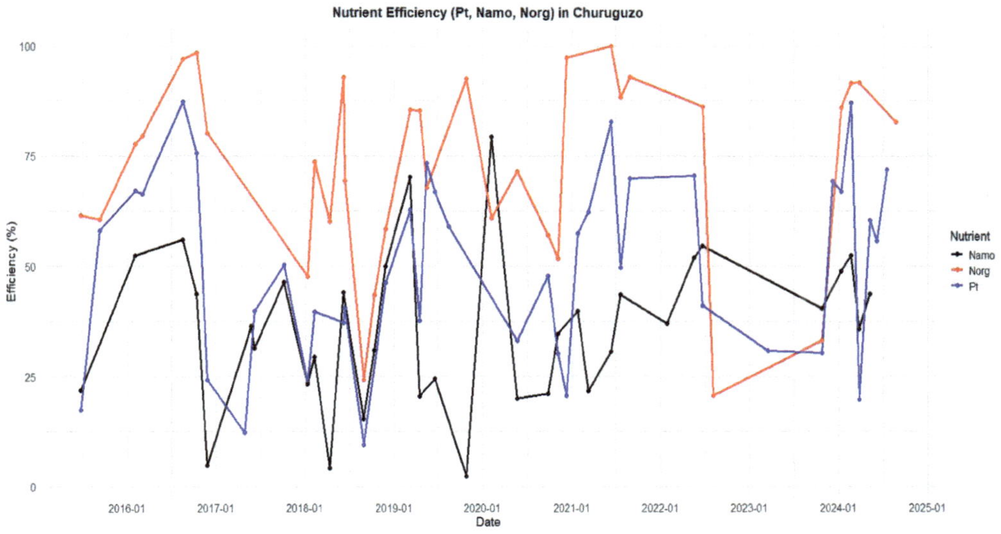

Figure A11. Removal efficiency—parameters: ammonia nitrogen (N_amo), organic nitrogen (N_org), and total phosphorus (TP) in WWTP Churuguzo.

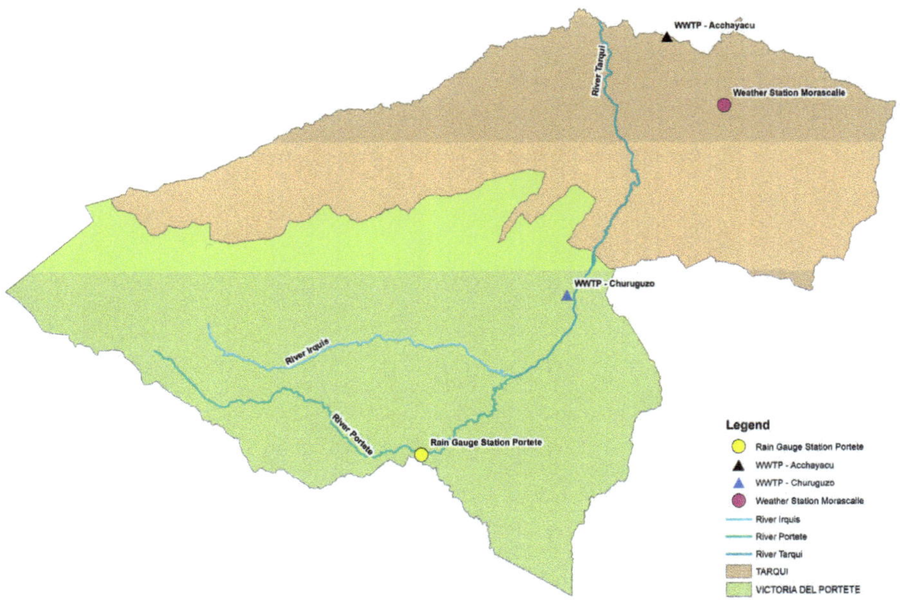

Figure A12. Locations of Morascalle weather station and Portete rainfall station.

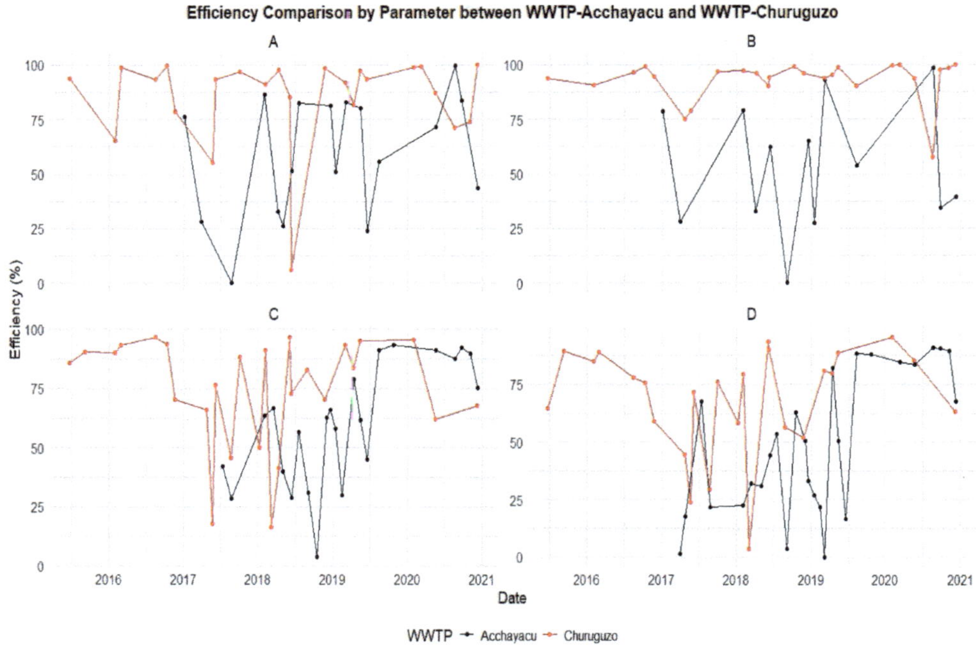

Figure A13. An efficiency comparison between Acchayacu and Churuguzo WWTPs during the period 2015–2020: (**A**) total coliforms (TCs), (**B**) fecal coliforms (TTCs), (**C**) biochemical oxygen demand (BOD_5), and (**D**) chemical oxygen demand (COD).

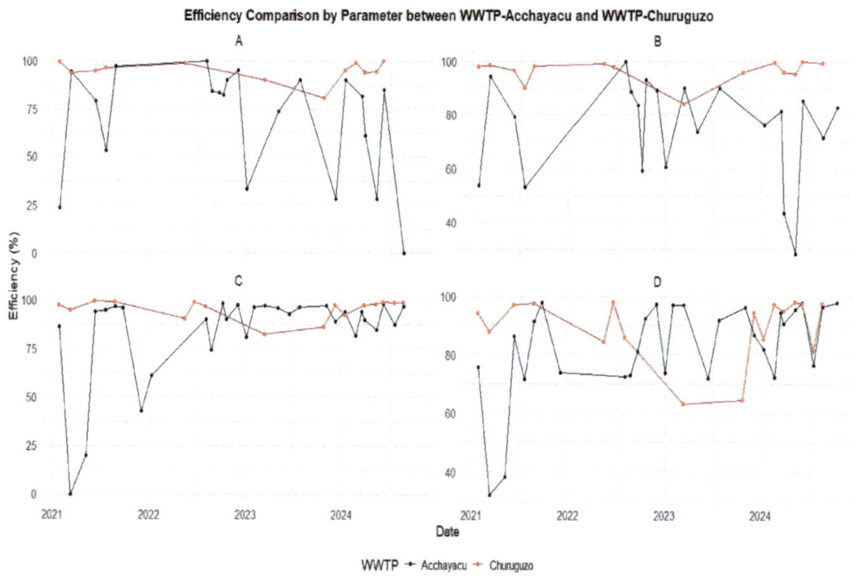

Figure A14. A comparison of removal efficiency between the VSSF-CW at the Acchayacu WWTP and the SF-CW at the Churuguzo WWTP (2021–2024): (**A**) TC, (**B**) TTC, (**C**) BOD_5, and (**D**) COD.

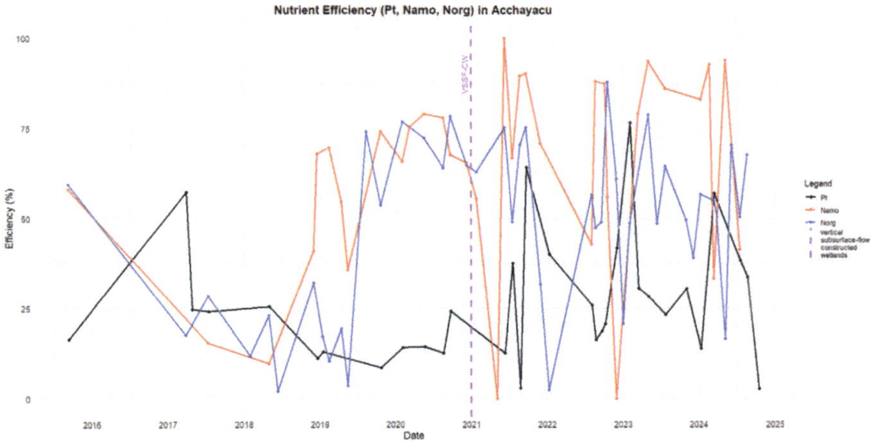

Figure A15. Removal efficiency of ammonia nitrogen (N_amo), organic nitrogen (N_org), and total phosphorus (TP) at the Acchayacu WWTP.

References

1. Coello, E.; Burbano, I.; Molina, A. Polluting factors of the San Pablo River in the city of Babahoyo, Ecuador. *J. Soc. Dev. Stud. Cuba. Lat. Am.* **2023**, *11*, 20–26.
2. Tchobanoglous, G.; Burton, F.L. *Wastewater Engineering: Treatment, Disposal, and Reuse*, 3rd ed.; Impresos y Revistas, S.A.: Madrid, Spain, 1995; Volume 1.
3. Hanjra, M.A.; Blackwell, J.; Carr, G.; Zhang, F.; Jackson, T.M. Wastewater irrigation and environmental health: Implications for water governance and public policy. *Int. J. Hyg. Environ. Health* **2012**, *215*, 255–269. [CrossRef]
4. Humanante, J.; Moreno, L.; Grijalva, A.; Saldoya, R.; Suárez, J. Removal Efficiency and impact of wastewater treatment system in the urban and rural sector of the Santa Elena Province. *Manglar* **2022**, *19*, 177–187. [CrossRef]
5. Wijaya, I.M.W.; Soedjono, E.S. Physicochemical Characteristic of Municipal Wastewater in Tropical Area: Case Study of Surabaya City, Indonesia. *IOP Conf. Ser. Earth Environ. Sci.* **2018**, *135*, 012018. [CrossRef]
6. Rojas, J.J.A.R. *Wastewater Treatment: Theory and Design Principles*, 3rd ed.; Nuevas Ediciones S.A.: Bogotá, Colombia, 2000; Volume 1.
7. Jerves-Cobo, R.; Forio, M.A.E.; Lock, K.; Van Butsel, J.; Pauta, G.; Cisneros, F.; Nopens, I.; Goethals, P.L. Biological water quality in tropical rivers during dry and rainy seasons: A model-based analysis. *Ecol. Indic.* **2020**, *108*, 105769. [CrossRef]
8. Isaza, C.A.A. Constructed wetlands for wastewater treatment. *Gen. José María Córdova Sci. J.* **2005**, *3*, 40–44. Available online: https://www.redalyc.org/articulo.oa?id=476259066011 (accessed on 31 December 2024).
9. Chung, G.; Lansey, K.; Blowers, P.; Brooks, P.; Ela, W.; Stewart, S.; Wilson, P. A general water supply planning model: Evaluation of decentralized treatment. *Environ. Model. Softw.* **2008**, *23*, 893–905. [CrossRef]
10. Llagas, W.; Gómez, E.G. Design of constructed wetlands for wastewater treatment at the National University of San Marcos. *J. Inst. Res. Fac. Geol. Min. Metall. Geogr. Eng.* **2006**, *15*, 85–96.
11. Anil, A.; Sumavalli, K.; Charan, S.; Mounika, M.; Praveen, P.; Gayathri, A.; Ganesh, V.M. Constructed Wetland for Low-Cost Waste Water Treatment. *Int. J. Sci. Res. Sci. Eng. Technol.* **2023**, *10*, 328–334. [CrossRef]
12. Rizaldi, M.A.; Limantara, L.M. Wetland as revitalization pond at urban area based on the eco hydrology concept. *Int. J. Eng. Technol.* **2018**, *7*, 143. [CrossRef]
13. Delgadillo, O.; Camacho, A.; Pérez, L.F.; Andrade, M. *Wastewater Treatment Through Constructed Wetlands*; Antequera, N., Ed.; Andean Center for Water Management and Use (Centro AGUA): Bolivia, Cochabamba, 2010.
14. United States Environmental Protection Agency (EPA). *Subsurface Flow Constructed Wetlands for Wastewater Treatment: A Technology Assessment*; EPA: Washington, DC, USA, 1993.
15. Aguado, R.; Parra, O.; García, L.; Manso, M.; Urkijo, L.; Mijangos, F. Modelling and simulation of subsurface horizontal flow constructed wetlands. *J. Water Process Eng.* **2022**, *47*, 102676. [CrossRef]
16. Díaz, A.J.D.-G.; Paredes, P.M.P.-R. Use of treated domestic wastewater through a constructed wetland with Jatropha and sunflower. *Knowl. Hub J.* **2022**, *7*, 1478–1495. Available online: https://polodelconocimiento.com/ojs/index.php/es/article/view/3556 (accessed on 31 December 2024).

17. Córdova, M.; Paz, D.; Santelices, M. *Governance for Monitoring Access to Sanitation in Ecuador*; Latin American Faculty of Social Sciences—FLACSO: Quito, Ecuador, 2023. [CrossRef]
18. Terán, C.; Argüello, J.; Cando, C. *Technical Bulletin of Environmental Economic Statistics in Municipal Decentralized Autonomous Governments: Drinking Water and Sanitation Management 2022*; Agricultural and Environmental Statistics Directorate—DEAGA: Quito, Ecuador, 2023.
19. Jerves-Cobo, R.; Benedetti, L.; Amerlinck, Y.; Lock, K.; De Mulder, C.; Van Butsel, J.; Cisneros, F.; Goethals, P.; Nopens, I. Integrated ecological modelling for evidence-based determination of water management interventions in urbanized river basins: Case study in the Cuenca River basin (Ecuador). *Sci. Total Environ.* **2020**, *709*, 136067. [CrossRef]
20. Operacz, A.; Jóźwiakowski, K.; Rodziewicz, J.; Janczukowicz, W.; Bugajski, P. Impact of Climate Conditions on Pollutant Concentrations in the Effluent from a One-Stage Constructed Wetland: A Case Study. *Sustainability* **2023**, *15*, 13173. [CrossRef]
21. Ramírez, J.S.; Loaiza, D.C.R.; Asprilla, W.J. Subsurface Constructed Wetlands: Comparison of Design Methodologies for Calculating Surface Area Based on Organic Matter Removal. *Ing. USBMed* **2022**, *11*, 65–73. [CrossRef]
22. *NTE INEN 2169:2013—Water*; Water Quality. Sampling. Handling and Preservation of Samples (First Revision). Ecuadorian Institute for Standardization (INEN): Quito, Ecuador, 2013.
23. R Core Team. *R: A Language and Environment for Statistical Computing*; R Foundation for Statistical Computing: Vienna, Austria, 2024.
24. Cantelmo, N.F.; Ferreira, D.F. Performance of Multivariate Normality Tests Evaluated by Monte Carlo Simulation. *Ciência Agrotecnologia* **2007**, *31*, 1630–1636. [CrossRef]
25. Khan, R.A.; Ahmad, F. Power Comparison of Various Normality Tests. *Pak. J. Stat. Oper. Res.* **2017**, *11*, 331–345. [CrossRef]
26. Saculinggan, M.; Balase, E.A. Empirical Power Comparison of Goodness of Fit Tests for Normality in the Presence of Outliers. *J. Phys. Conf. Ser.* **2013**, *435*, 012041. [CrossRef]
27. De La Mora-Orozco, C.; Saucedo-Terán, R.A.; González-Acuña, I.J.; Gómez-Rosales, S.; Flores-López, H.E. Effect of Water Temperature on the Reaction Rate Constant of Pollutants in a Constructed Wetland for Swine Wastewater Treatment. *Rev. Mex. Cienc. Pecu.* **2020**, *11*, 1–17. [CrossRef]
28. Buytaert, W.; Celleri, R.; Willems, P.; De Bièvre, B.; Wyseure, G. Spatial and temporal rainfall variability in mountainous areas: A case study from the south Ecuadorian Andes. *J. Hydrol.* **2006**, *329*, 413–421. [CrossRef]
29. Navarro, L.; Torres, J.; Carlos, R. Review of Groundwater Quality Impacts Generated by Constructed Wetlands in Rural Areas. *Desarro. Innovación Ingeniería* **2021**, *2*, 10–24.
30. EPA. *Wastewater Technology Fact Sheet: Free Water Surface Wetlands*; EPA 832-F-00-024; U.S. Environmental Protection Agency: Washington, DC, USA, 2000.
31. EPA. *Wastewater Technology Fact Sheet: Subsurface Flow Wetlands*; EPA 832-R-00-023; U.S. Environmental Protection Agency: Washington, DC, USA, 2000.
32. CONAGUA. *Design of Municipal Wastewater Treatment Plants: Constructed Wetlands. Manual of Drinking Water, Sewerage and Sanitation (MAPAS)*; Book 30; National Water Commission (CONAGUA): Mexico City, Mexico, 2015.
33. González, T.C.R.; Narváez, T.A.C. *Evaluation and Redesign of the Acchayacu Wastewater Treatment Plant*; Universidad de Cuenca: Cuenca, Ecuador, 2018.
34. García, K.L.Q.; Zúñiga, D.P.R.; Duque, M.E.G.; Rojas, J.A.A. Evaluation of Nitrogen and Organic Matter Removal through Subsurface Flow Constructed Wetlands Coupled with Fixed-Bed Reactors with Microalgae at the Institución Universitaria Colegio Mayor de Antioquia. *Ing. Región* **2021**, *25*, 82–94. [CrossRef]
35. Bedoya, J.C.; Ardila, A.N.; Reyes, J. Evaluation of a Subsurface Flow Constructed Wetland for the Treatment of Wastewater Generated at the Institución Universitaria Colegio Mayor de Antioquia, Colombia. *Rev. Int. Contam. Ambient.* **2014**, *3*, 275–283.
36. Romero-Aguilar, M.; Colin-Cruz, A.; Sanchez-Salinas, E.; Ortiz-Hernandez, M.L. Wastewater treatment by a pilot constructed wetland system: Evaluation of organic load removal. *Rev. Int. Contam. Ambient.* **2009**, *25*, 157–167.
37. Zahraeifard, V.; Deng, Z. Hydraulic residence time computation for constructed wetland design. *Ecol. Eng.* **2011**, *37*, 2087–2091. [CrossRef]
38. Rodríguez-Miranda, J.P.; Gómez, E.; Garavito, L.; López, F. Comparative study of domestic wastewater treatment using duckweed and water hyacinth in constructed wetlands. *Tecnol. Cienc. Agua* **2010**, *1*, 59–68. [CrossRef]
39. Brix, H. Do macrophytes play a role in constructed treatment wetlands? *Water Sci. Technol.* **1997**, *35*, 11–17. [CrossRef]
40. Wang, Q.; Hu, Y.; Xie, H.; Yang, Z. Constructed Wetlands: A Review on the Role of Radial Oxygen Loss in the Rhizosphere by Macrophytes. *Water* **2018**, *10*, 678. [CrossRef]
41. Abdelhakeem, S.G.; Aboulroos, S.A.; Kamel, M.M. Performance of a vertical subsurface flow constructed wetland under different operational conditions. *J. Adv. Res.* **2016**, *7*, 803–814. [CrossRef]
42. Carrillo-Anchundia, B.; Vera-Loor, J.; Loor-Vélez, J. Design of an upflow anaerobic filter for wastewater treatment. *INGENIAR Eng. Technol. Res.* **2022**, *5*, 2–16.
43. Mosquera, Y. Leachate treatment using constructed wetlands. *Tumbaga* **2012**, *1*, 73–99.

44. Zhang, D.Q.; Jinadasa, K.B.S.N.; Gersberg, R.M.; Liu, Y.; Ng, W.J.; Tan, S.K. Application of constructed wetlands for wastewater treatment in developing countries—A review of recent developments (2000–2013). *J. Environ. Manag.* **2014**, *141*, 116–131. [CrossRef]
45. Dotro, G.; et al. *Constructed Wetlands for Pollution Control*; IWA Publishing, Alliance House: London, UK, 2017; Volume 7, SW1H 0QS.
46. Keyport, S.; Carson, B.D.; Johnson, O.; Lawrence, B.A.; Lishawa, S.C.; Tuchman, N.C.; Kelly, J.J. Effects of experimental harvesting of an invasive hybrid cattail on wetland structure and function. *Restor. Ecol.* **2019**, *27*, 389–398. [CrossRef]
47. Arteaga-Cortez, V.M.; Quevedo-Nolasco, A.; del Valle-Paniagua, D.H.; Castro-Popoca, M.; Bravo-Vinaja, Á.; Ramírez-Zierold, J.A. State of the art: A current review on the mechanisms used by constructed wetlands for nitrogen and phosphorus removal. *Tecnol. Cienc. Agua* **2019**, *10*, 319–342. [CrossRef]
48. Guerra, J.D.T.; Vargas, J.S.M.; Aguirre, R.R.P.; Huaranga, M.A.C. Evaluation of the efficiency in wastewater treatment for irrigation using Free Water Surface (FWS) Constructed Wetlands with Cyperus papyrus and Phragmites australis in Carapongo-Lurigancho. *Rev. Investig. Cienc. Tecnol. Desarro.* **2018**, *3*, 48–64. [CrossRef]
49. Lahora, A. Wastewater purification through constructed wetlands: The Edar de los Gallardos (Almería). *Ecol. Manag. Conserv. Wetl.* **2003**, *49*, 99–112.

water

Article

Jinluo Low-Density Lotus Pond Wetland Water Purification Practice Experiment—A Case of Limited Efficacy

Bo Liu [1], Yuan Gao [2,3,*], Jing Zhou [4], Yun Wang [2] and Junxia He [1]

1 Shandong Linyi Station of Positioning Observation and Research for Forest Ecosystem, College of Resources and Environment, Linyi University, Linyi 276005, China; liubo@lyu.edu.cn (B.L.); hejunxia@lyu.edu.cn (J.H.)
2 Shandong Provincial Key Laboratory of Water and Soil Conservation and Environmental Protection, College of Resources and Environment, Linyi University, Linyi 276005, China; wangyun@lyu.edu.cn
3 Linyi Scientific Exploration Laboratory, Linyi 276037, China
4 School of Life Sciences, Qufu Normal University, Jining 273165, China; jingzhou-2004@163.com
* Correspondence: 13589652495@163.com

Abstract: (1) Although lotus ponds exhibit ecological benefits in wetland restoration, their efficacy in water purification and eutrophication mitigation remains unclear. (2) This study utilized Jinluo lotus pond as the experimental group and the adjacent river as the control. Five sampling points were established in each area, with water samples collected in June 2022, April 2025, and May 2025. (3) The pH, BOD, COD, TN, and NH_3-N concentrations in Jinluo lotus pond water are higher than those in rivers, while the TP, NO_3-N, Chl-a, and algal cell density in rivers are higher. However, there was no significant difference in the nine parameters ($p > 0.05$) in June 2022. The pH, DO, algal cell density, and algal biomass of the Jinluo lotus pond were significantly higher ($p < 0.05$ for DO); the concentrations of BOD, COD, TN, TP, NH_3-N, NO_3-N, PI, and Chl-a in rivers are higher, with significant differences in Chl-a ($p < 0.05$) in April 2025. The BOD, COD, TP, NO_3-N, and PI of the Jinluo lotus pond were relatively high ($p < 0.05$ for PI); the pH, TN, NH_3-N, DO, Chl-a, algal cell density, and algal biomass of rivers are higher, with significant differences in Chl-a ($p < 0.05$) in May 2025. The results showed that there was no significant difference in the four diversity indicators in June 2022, April 2025, and May 2025. There was no significant difference in the algal diversity indices, including species richness (S), Shannon–Wiener diversity index (H), Simpson diversity index (P), and Pielou evenness index (E) between Jinluo lotus pond and rivers. (4) Conclusions and Recommendations: The Jinluo lotus pond and adjacent rivers suffer from severe nutrient overload, especially with BOD, COD, and TN all being classified as Class 5 water. Expanding natural and constructed reed communities is recommended to enhance nutrient removal. However, given the limited purification capacity of lotus ponds, maintaining or increasing their area may not be justified.

Keywords: lotus ponds; constructed wetland; water purification effect; water quality assessment; phytoplankton

1. Introduction

With the rapid increase in global water demand, a series of water ecological issues caused by eutrophication, such as water quality deterioration and biodiversity loss, have become a global focus of concern [1]. Water eutrophication stands as one of the greatest challenges facing the global water environment, as it undermines the stability and functionality of aquatic ecosystems [2,3]. Water scarcity has emerged as a critical issue threatening sustainability in many regions worldwide [4,5]. Water environment governance constitutes

https://doi.org/10.3390/w17131985

an essential component of the United Nations Sustainable Development Goals (SDGs), with the United Nations Environment Programme (UNEP) emphasizing the pivotal role of water governance in addressing sustainable development challenges [6,7].

Selecting appropriate restoration methods is essential for the recovery of aquatic ecosystems [1]. Among these, ecological approaches—known for their cost-effectiveness and efficiency—have emerged as a preferred solution for water quality improvement in diverse aquatic systems [8]. Specifically, aquatic macrophyte-based remediation is widely regarded as one of the most economical, efficient, and environmentally sustainable methods, offering distinct advantages [8]. For instance, constructed wetlands are increasingly utilized for sustainable wastewater treatment, effectively removing organic matter and nutrients while delivering ecosystem services and recreational benefits. Notably, hybrid constructed wetlands represent the most efficient approach for enhancing water quality and mitigating greenhouse gas emissions. Their performance depends on multiple factors, including plant species, substrate selection, and environmental/hydraulic conditions, with pollutant removal efficiency largely influenced by temperature, hydraulic retention time, and pollutant loading rates [9].

Constructed wetlands composed of diverse aquatic macrophyte species, acknowledged as a convenient, environmentally friendly, low-cost, and efficient phytoremediation technology, have been widely used globally for polluted water treatment [10–12]. Artificial water bodies are increasingly becoming prominent features in urban landscapes [13], serving as temporary sanctuaries and phased wetland reserves [14], though most landscape water bodies face risks of pollution and eutrophication.

As a rapidly developing economic powerhouse, China has confronted escalating water pollution challenges resulting from decades of intensive industrialization and urbanization. In response to these pressing environmental concerns, the Chinese government initiated the groundbreaking "Sponge City" program in 2015 as a comprehensive national infrastructure strategy. This innovative approach systematically addresses multiple urban water management challenges, including the following: (1) urban surface water flooding mitigation, (2) stormwater runoff purification, (3) peak flow regulation, and (4) sustainable water resource utilization [15]. Through coordinated efforts led by the State Council and implemented by various governmental departments, sponge city development has been vigorously promoted nationwide. By 2024, the cumulative investment in this initiative had surpassed 60 billion yuan (approximately 8.3 billion USD), facilitating the creation of urban ecosystems with enhanced natural hydrological functions including water retention, infiltration, and purification capacity. However, the effectiveness of natural wetlands in pollution control remains constrained by two fundamental limitations: (1) restricted spatial expansion potential within urban environments and (2) overburdened biogeochemical processing capacities [16]. In this context, constructed wetlands have emerged as strategically important engineered ecosystems, demonstrating a proven efficacy in water quality remediation through controlled biological and physical–chemical processes [17]. Nevertheless, their operational sustainability faces challenges from two primary environmental impacts: (1) greenhouse gas emissions associated with microbial metabolic processes and (2) potential secondary pollution from accumulated contaminants [18].

Constructed wetland systems consistently demonstrate high removal efficiencies (typically 70–90%) for organic pollutants including biochemical oxygen demand (BOD), chemical oxygen demand (COD), and suspended solids, although nitrogen removal performance shows a greater variability (30–80%) depending on system configuration [19]. Typical designs incorporate extensive reed beds utilizing either single or mixed plant species to achieve multiple treatment objectives: the physical filtration of suspended solids, biochemical transformation of nutrients, and enhanced sedimentation processes through rhizosphere interactions [20,21]. Selected macrophyte species such as reeds (*Phragmites australis*), sedges, and other emergent vegetation effectively assimilate nutrients and pollutants through their root systems before translocating them to aerial biomass [22,23]. Particularly high-performing species commonly employed in these systems include *Phragmites australis*, *Nelumbo nucifera*, *Typha domingensis*, *T. latifolia*, *Eichhornia crassipes*, *Pistia stratiotes*, and *Vallisneria natans*, which are specifically valued for their exceptional nutrient uptake capacities demonstrated in numerous studies [22–26].

Numerous case studies have demonstrated the remarkable treatment efficiency of constructed wetlands across various applications: in Xiantao, a constructed wetland system achieved removal rates of 94% for total nitrogen, 90% for total phosphorus, 68% for COD, and 95% for ammonia nitrogen in municipal wastewater treatment [27]; at Universiti Sains Malaysia, wetlands dominated by *Typha angustifolia* and *Eleocharis variegata* effectively reduced nitrites, nitrates, ammonia nitrogen, and phosphates [28]; reed floating beds showed removal efficiencies of 55–60% for total solids, 45–55% for NH_3-N, 33–45% for NO_3^--N, 45–50% for TKN, and 40–50% for BOD, proving particularly suitable for the in-situ treatment of shallow, slow-flowing water bodies [29]; the Lotus Lake National Wetland Park in Tieling City, featuring Phragmites and *Nelumbo nucifera*, significantly reduced total phosphorus and ammonia nitrogen concentrations [30]; while lotus pond wetlands demonstrated effectiveness in treating garlic processing wastewater through the substantial removal of organic pollutants and the reduction of COD_{60} and BOD_5 levels [31]. Comparative research has revealed distinct species-specific treatment efficiencies, with *Eichhornia crassipes* and *Phragmites australis* exhibiting superior nitrogen removal capabilities, whereas *Pistia stratiotes* and *Nelumbo nucifera* show an enhanced phosphorus removal performance [24,32]. Further studies conducted at Wuliangsuhai Lake and Baiyangdian Lake have elucidated *Nelumbo nucifera*'s dual effects on algal dynamics, demonstrating that low-density plantings can promote algal growth while high-density configurations effectively suppress it, underscoring the critical importance of optimal density management in wetland design [33–35].

The Yi River, a major watercourse in the Huai River Basin [36], is located in southern Shandong and northern Jiangsu, with geographical coordinates 34°23′–36°20′ N and 117°25′–118°42′ E. Spanning approximately 574 km, it originates from Yiyuan County in Shandong and flows into the Yellow Sea at Yanwei Port through the Xin-Yi River (Yi River Diversion Channel) from Wu Lou Village in Pi County, Jiangsu [37–39]. The Yi River has been listed as a key control and monitoring river in the Huai River Basin Water Pollution Prevention Plan. The Liuqing River, a tributary of the Yi River, suffers from severe excessive nutrient loads in its upper reaches.

Extensive research confirms that aquatic plants play a beneficial role in mitigating water eutrophication [4,10,40]. Common species employed in urban wetlands include reeds and reed ponds [4,41], as well as lotus and lotus ponds [10,32,33,42], all demonstrating ecological benefits for wetland restoration. The study site, Jinluo lotus pond, is situated on the north bank of the Liuqing River in Linyi City, Shandong Province, covering a total area of 7.109 km^2 with lotus ponds accounting for 5.5667 km^2. This project was designed to utilize lotus roots for regulating nutrient concentrations in wetland waters, alleviating eutrophication, and restoring polluted water bodies. Nevertheless, the actual efficacy of such systems in water quality purification remains questionable, particularly regarding their ability to achieve sustainable eutrophication control without causing secondary ecological impacts.

2. Materials and Methods

2.1. Site Description and Sampling Procedure

In this study, Jinluo lotus pond served as the experimental group, while the adjacent river outside the pond was designated as the control group, with five sampling areas established in each location. Sampling was performed in June 2022, April 2025, and May 2025 using sterilized 4 L sampling buckets to collect water samples for laboratory analysis in Figure 1.

Figure 1. The sampling point layout of Jinluo lotus pond and the river outside the pond. (**A**,**B**) show the exterior views of the lotus pond, (**C**) displays the distribution of sampling points in this study (The red line in the diagram represents the boundary of the lotus pond, yellow triangles indicate the lotus pond area, and pink triangles represent the river outside the lotus pond).

2.2. Analytical Methods

pH, Oxygen Demand (BOD), Chemical Oxygen Demand (COD), Total Nitrogen (TN), Total Phosphorus (TP), Ammonia Nitrogen (NH_3-N), Nitrate Nitrogen (NO_3^--N), Dissolved Oxygen (DO), Planktonic Index (PI), Chlorophyll-a (Chl-a), and algal cell density were studied. These selected parameters served dual purposes: (1) assessing the water purification efficacy of the lotus pond system, and (2) providing scientific basis for developing effective environmental management strategies for the Liuqing River watershed.

pH: Quantify using electrode method (HJ 1147-2020) and ST2100 pH meter (HLJC-243-2) [43]. BOD: Quantified using the standard Dilution and Inoculation Method (HJ 505-2009) with a 25 mL acid burette (Model B193) [44]. COD: Analyzed by the Potassium Dichromate Method (HJ 828-2017) employing a 50 mL acid burette (Model B192) [45]. TN: Determined through Ultraviolet Spectrophotometry (HJ 636-2012) using a UV-1750 UV-Vis Spectrophotometer (Model A11605031003CS) [46]. TP and NH_3-N: Measured, respectively, by Ammonium Molybdate Spectrophotometry (GB 11893-1989) and Nessler's Reagent Spectrophotometry (HJ 535-2009), with measurements conducted on a DR2008 Visible Spectrophotometer (Serial No. 1429121) [47,48]. NO_3^--N: Determined via ion chromatography (HJ 84-2016) using a DIONEX AQUION ion chromatograph (HLJC-231) [49]. DO: Detection using electrochemical probe method (HJ 506-2009) and JPB-607A portable dissolved oxygen analyzer (HLJC-285) [50]. PI: Determination of permanganate index (GB/T 11892-1989) and detection using a 25mL acid burette (B-S-25-2) [51]. Phytoplankton analysis: Species identification and quantification were performed under an optical microscope using standardized counting chambers, with results expressed as cell density (cells/L) and species diversity indices [52,53]. Chl-a: Quantified following Acetone Spectrophotometry (HJ 897-2017) [54].

2.3. Diversity Indices Calculation

Species Richness (S): Total number of identified phytoplankton species per sample.

$$\text{Shannon–Wiener Index } (H){:}\ H = -\sum(P_i \times \ln P_i)$$

$$\text{Simpson's Diversity Index } (P){:}\ D = 1 - \sum(P_i^2)$$

$$\text{Pielou's Evenness Index } (E){:}\ E = H/\ln S$$

where P_i denotes the proportion of individuals of the i-th species relative to the total phytoplankton count [55–57].

2.4. Statistical and Spatial Analysis

All data were processed using SPSS 19.0 for statistical analysis.

3. Results

3.1. Differences in Water Quality Factors

This study conducted a systematic comparison of water quality parameters between the Jinluo lotus pond and its adjacent river system in Figures 2 and 3. Results indicated that while the lotus pond exhibited elevated levels of water pH, BOD, COD, TN, and NH_3-N compared to the river, the adjacent river conversely showed higher concentrations of TP, NO_3^--N, Chl-a, and algal cell density. Statistical analysis revealed that none of the nine measured parameters demonstrated statistically significant differences ($p > 0.05$) in June 2022.

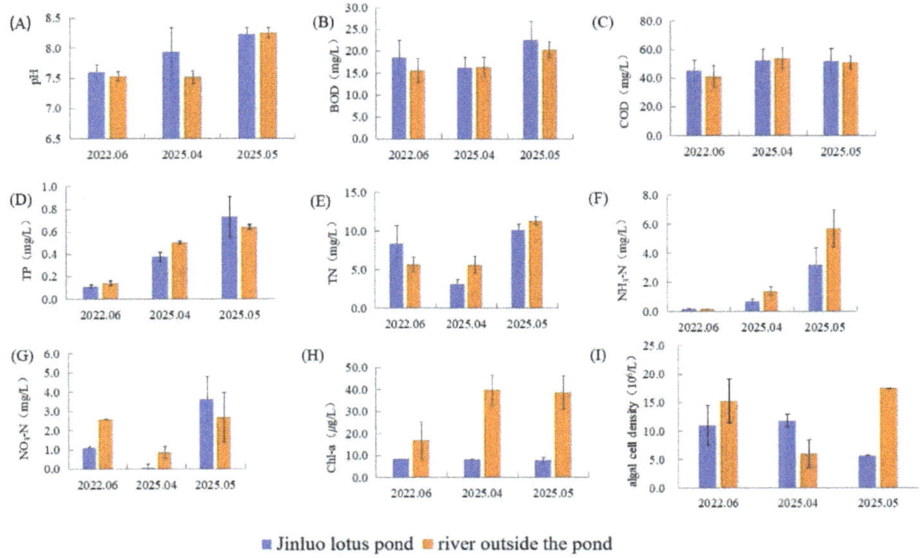

Figure 2. Differences of **(A)** pH, **(B)** BOD, **(C)** COD, **(D)** TP, **(E)** TN, **(F)** NH_3-N, **(G)** $NO_3{}^-$-N, **(H)** Chl-a, and **(I)** algal cell density in the water quality factors of Jinluo lotus pond and the river outside the pond.

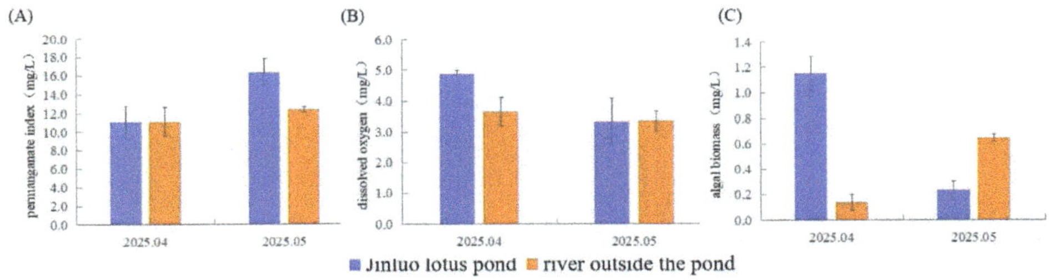

Figure 3. Differences of **(A)** PI, **(B)** DO, and **(C)** algal biomass in the water quality factors of Jinluo lotus pond and the river outside the pond.

Relative to adjacent rivers, the lotus pond exhibited a significantly higher pH, dissolved oxygen (DO), algal cell density, and algal biomass ($p < 0.05$ for DO). In contrast, river water showed elevated levels of BOD, COD, TN, TP, NH_3-N, $NO_3{}^-$-N, planktonic index (PI), and Chl-a, with a statistically significant difference in Chl-a ($p < 0.05$) in April 2025.

Compared to rivers, the lotus pond had a higher BOD, COD, TP, $NO_3{}^-$-N, and PI, with a significant difference in PI ($p < 0.05$). Conversely, river samples displayed a greater pH, TN, NH_3-N, DO, Chl-a, algal cell density, and algal biomass, with Chl-a showing a significant difference ($p < 0.05$) in May 2025.

3.2. Water Quality Assessment

In accordance with China's Surface Water Environmental Quality Standards (GB 3838-2002) [52], we evaluated the water quality of Jinluo lotus pond and its adjacent river using seven key parameters: BOD, COD, TN, TP, NH_3-N, PI, and DO (Table 1). The results demonstrated severe exceedances of national standards for BOD, COD, and TN in both systems, reflecting significant organic pollution and nutrient loading. All sampling sites consistently exceeded Class V water quality thresholds—the most polluted classification under GB 3838-2002—indicating critically degraded water conditions throughout the study area.

Table 1. The water quality assessment of Jinluo lotus pond and the river outside the pond.

	Jinluo Lotus Pond	River Outside the Pond
BOD	Super Class V water (2022) Super Class V water (2025)	Super Class V water (2022) Super Class V water (2025)
COD	Super Class V water (2022) Super Class V water (2025)	Super Class V water (2022) Super Class V water (2025)
TN	Super Class V water (2022) Super Class V water (2025)	Super Class V water (2022) Super Class V water (2025)
TP	Class III water (2022) Super Class V water (2025)	Class III water (2022) Super Class V water (2025)
NH_3-N	Class II water (2022) Class III water (2025)	Class II water (2022) Class IV water (2025)
PI	Class V water (2025)	Class V water (2025)
DO	Class IV water (2025)	Class IV water (2025)

3.3. The Correlation of Water Quality Factors

As shown in Table 2, BOD and COD exhibited a strong positive correlation ($p < 0.01$), demonstrating a close relationship between these key organic pollution indicators. A significant positive correlation was also observed between COD and chlorophyll-a ($p < 0.05$). Among nitrogen components, TN and NH_3-N showed a particularly strong positive correlation ($p < 0.01$), implying shared sources or transformation processes.

Table 2. The correlation of water quality factors of Jinluo lotus pond and the river outside the pond (2022).

	pH	BOD	COD	TP	TN	NH_3-N	NO_3^--N	Chl-a	ACD
pH	1.000								
BOD	−0.240	1.000							
COD	−0.123	0.927 **	1.000						
TP	−0.452	−0.027	−0.104	1.000					
TN	−0.030	0.318	0.169	−0.105	1.000				
NH_3-N	−0.160	0.381	0.331	−0.206	0.775 **	1.000			
NO_3^--N	0.230	−0.431	−0.452	0.331	0.103	−0.292	1.000		
Chl-a	0.023	0.252	0.547 *	−0.066	−0.119	0.341	−0.293	1.000	
ACD	−0.035	0.176	0.245	0.231	0.224	0.584 *	−0.106	0.536 *	1.000

Note: ** $p < 0.01$; * $p < 0.05$; ACD, Algae cell density.

The analysis revealed significant linkages between algal dynamics and multiple water quality parameters. Notably, algal cell density showed positive correlations with both NH_3-N and Chl-a (all $p < 0.05$), suggesting a potential coupled biogeochemical cycling of these nutrients within the lotus pond ecosystem.

In Table 3, pH is significantly negatively correlated with COD ($p < 0.05$). BOD is significantly positively correlated with COD, TN, TP, NH_3-N, Permanganate index, and dissolved oxygen ($p < 0.01$). COD is significantly positively correlated with TP, NH_3-N, NO_3-N, Permanganate index, dissolved oxygen, Algae cell density, and Algal biomass ($p < 0.05$). TP is significantly positively correlated with TN and NH_3-N ($p < 0.01$), Permanganate index, and dissolved oxygen ($p < 0.05$). TN is significantly positively correlated with NH_3-N, Permanganate index, and dissolved oxygen ($p < 0.01$), NO_3-N ($p < 0.05$). NH_3-N is significantly positively correlated with dissolved oxygen ($p < 0.01$). Permanganate index is significantly positively correlated with dissolved oxygen ($p < 0.01$). Dissolved oxygen is significantly positively correlated with Chl-a ($p < 0.05$). Algae cell density is significantly positively correlated with Algal biomass ($p < 0.01$).

Table 3. The correlation of water quality factors of Jinluo lotus pond and the river outside the pond (2025).

	pH	BOD	COD	TP	TN	NH_3-N	NO_3^--N	PI	DO	Chl-a	ACD	AB
pH	1.000											
BOD	−0.089	1.000										
COD	−0.428 *	0.835 **	1.000									
TP	0.181	0.642 **	0.452 *	1.000								
TN	0.179	0.536 **	0.179	0.526 **	1.000							
NH_3-N	0.235	0.647 **	0.384 *	0.518 **	0.685 **	1.000						
NO_3^--N	0.136	−0.181	−0.392 *	0.004	0.461 *	−0.243	1.000					
PI	−0.093	0.745 **	0.536 **	0.414 *	0.589 **	0.298	0.245	1.000				
DO	0.098	−0.603 **	−0.463 *	−0.492 *	−0.656 **	−0.622 **	−0.028	−0.529 **	1.000			
Chl-a	−0.366	0.124	0.266	0.052	0.290	0.124	0.227	0.040	−0.392 *	1.000		
ACD	−0.215	0.378 *	0.454 *	0.190	0.186	0.355	−0.273	0.342	−0.267	−0.094	1.000	
AB	−0.310	0.301	0.460 *	0.037	−0.004	0.076	−0.213	0.340	−0.033	−0.159	0.902 **	1.000

Note: ** $p < 0.01$; * $p < 0.05$; PI, Permanganate index; DO, dissolved oxygen; ACD, Algae cell density; AB, Algal biomass.

3.4. Phytoplankton Diversity

Phytoplankton diversity indices—including species richness (*S*), Shannon–Wiener diversity index (*H*), Simpson diversity index (*P*), and Pielou evenness index (*E*)—were analyzed for the lotus pond. The results showed that there was no significant difference in the four diversity indicators in June 2022, April 2025, and May 2025 in Figure 4.

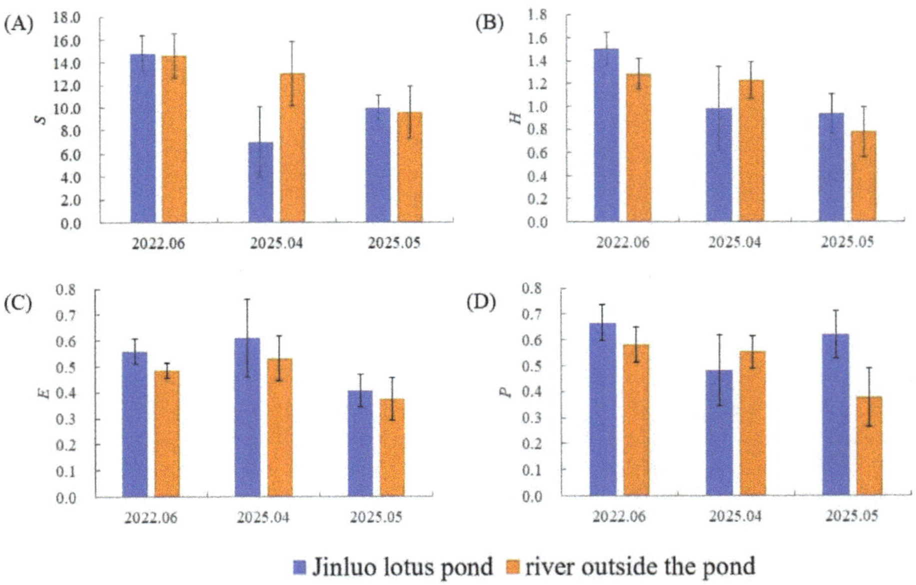

Figure 4. (**A**) The Phytoplankton richness index (*S*), (**B**) Shannon–Wiener diversity index (*H*), (**C**) Simpson diversity index (*P*), and (**D**) Pielou evenness index (*E*) of Jinluo lotus pond and the river outside the pond.

There is no significant difference between the Phytoplankton cell density of Jinluo lotus pond and the river outside the pond. In order of cell density, the phytoplankton in the Jinluo lotus pond water are *Cyclotella meneghiniana*, *Pseudanabaena* sp., and *Scenedesmus quadricauda*, but in the river outside the pond, the water has *P.* sp., *C. meneghiniana*, and *S. quadricauda* in 2022 in Table 4. The phytoplankton in the Jinluo lotus pond water is

Tetrastrum staurogeniaforme, but in the river outside the pond, the water has *Tetrastrum staurogeniaforme*, *Coelastrum microporum*, *Scenedesmus quadricauda*, and *Scenedesmus bicaudatus* in 2025 in Table 5.

Table 4. The Phytoplankton cell density of Jinluo lotus pond and the river outside the pond (2022).

	Jinluo Lotus Pond	River Outside the Pond
1	*Cyclotella meneghiniana*, 3,986,800, 36.30%	*Pseudanabaena* sp., 6,576,000, 43.13%
2	*Pseudanabaena* sp., 3,336,000, 30.37%	*Cyclotella meneghiniana*, 3,718,000, 24.39%
3	*Scenedesmus quadricauda*, 1,155,200, 10.52%	*Scenedesmus quadricauda*, 1,476,800, 9.69%
4	*Aphanocapsa delicatissima*, 560,000, 5.10%	*Trachelomonas superba*, 592,000, 3.88%
5	*Crucigenia tetrapedia*, 497,600, 4.53%	*Oscillatoria chlorine*, 560,000, 3.67%
6	*Coelastrum microporum*, 320,000, 2.91%	*Crucigenia tetrapedia*, 544,000, 3.57%
7	*Oscillatoria chlorine*, 280,000, 2.55%	*Coelastrum microporum*, 358,400, 2.35%
8		*Acanthosphaera* sp., 204,000, 1.34%
9		*Dictyosphaerium pulchellum*, 192,000, 1.26%
total	10,135,600/10,984,400, 92.27%	14,221,200/15,246,400, 93.28%

Note: This table only lists phytoplankton that account for over 1% of the total.

Table 5. The Phytoplankton cell density of Jinluo lotus pond and the river outside the pond (2025).

	Jinluo Lotus Pond	River Outside the Pond
1	*Tetrastrum staurogeniaforme*, 1,492,800, 17.04%	*Tetrastrum staurogeniaforme*, 2,964,000, 25.27%
2	*Anabaena* sp., 567,300, 6.47%	*Coelastrum microporum* 1,900,200, 16.20%
3	*Scenedesmus quadricauda*, 499,500, 5.70%	*Scenedesmus quadricauda*, 1,694,400, 14.45%
4	*Coelastrum microporum*, 377,700, 4.30%	*Scenedesmus bicaudatus*, 1,689,000, 14.40%
5	*Scenedesmus bicaudatus*, 176,100, 2.01%	*Scenedesmus dimorphus*, 641,100, 5.47%
6		*Scenedesmus denticulatus*, 175,800, 1.50%
7		*Actinastrum hantzschii*,168,600, 1.44%
8		*Pediastrum tetras*, 163,800, 1.40%
9		*Pediastrum biradiatum*, 157,300, 1.34%
10		*Scenedesmus acuminatus*, 150,300, 1.28%
11		*Microcystis* sp., 146,700, 1.25%
total	3,113,400/8,762,100, 35.53%	9,851,200/11,729,100, 83.99%

Note: This table only lists phytoplankton that account for over 1% of the total.

4. Discussion

Ecological restoration strategies frequently incorporate three principal phytoremediation approaches: riparian vegetation buffer zones, ecological floating beds, and constructed wetlands, each offering distinct advantages for water quality improvement [58–60]. The effectiveness of these phytoremediation systems is primarily determined by two critical design parameters: appropriate plant species selection and optimal planting density, which collectively govern nutrient uptake capacity and treatment performance [61–63]. As fundamental elements of ecological engineering solutions, aquatic plants demonstrate marked variations in removal efficiencies depending on pollutant composition, with species-specific responses to different contaminant mixtures [64,65]. Research indicates that under eutrophic conditions, superior nutrient removal performance correlates strongly with three key plant traits: (1) high biomass production capacity, (2) elevated leaf dry matter content (LDMC), and (3) reduced specific leaf area (SLA), suggesting that wetland species exhibiting this combination of characteristics—particularly those with high biomass and LDMC coupled with low SLA values—may represent optimal candidates for nutrient-rich wastewater treatment applications [66]. Notably, engineered sequential wetland systems have proven particularly effective for purifying polluted urban waterways even in challenging cold climate conditions, demonstrating robust year-round treatment capabilities [67].

This study investigates the wastewater treatment capacity of constructed wetlands in the context of livestock effluent, particularly given the upstream location of a globally significant swine slaughtering center that generates substantial volumes of nutrient-rich wastewater containing elevated concentrations of carbon, nitrogen, and phosphorus compounds. Aquaculture effluents, similarly characterized by high nutrient loads, present significant environmental challenges with nitrogen and phosphorus being primary contributors to ecological degradation [68]. Research demonstrates that constructed wetland systems can reliably achieve total nutrient removal efficiencies exceeding 60% when treating bullfrog aquaculture wastewater [69], while Euryale ferox Salisb-based ecological ponds exhibit exceptional performance in both the in-situ and ex-situ treatment of shrimp aquaculture effluent, showing particularly high removal rates for total nitrogen (TN) and total phosphorus (TP) [70]. Advanced multi-stage treatment systems combining lotus ponds with surface flow wetlands have proven particularly effective for swine wastewater remediation, capable of transforming heavily contaminated influent into lightly polluted effluent through efficient nitrogen and phosphorus removal mechanisms [71], though these systems require careful operational management including seasonal lotus root harvesting during winter–spring periods and periodic sediment dredging to maintain treatment efficacy.

Constructed wetlands have emerged as a highly efficient, cost-effective, and environmentally sustainable solution for wastewater treatment, offering distinct advantages over conventional treatment systems in terms of operational simplicity, minimal maintenance requirements, and ecological compatibility [72]. The field has witnessed an evolutionary shift from basic treatment wetlands to sophisticated, multi-functional integrated systems. This transition is exemplified by the Integrated Constructed Wetland (ICW) concept, which adopts a comprehensive design philosophy that systematically incorporates four key dimensions—economic viability, social acceptance, environmental performance, and landscape integration—throughout all project phases from initial planning to long-term operation. Beyond their core treatment functions, these integrated wetland systems provide substantial secondary benefits, particularly in terms of biodiversity enhancement and habitat creation [73,74]. In this context, while performance assessments reveal that the Jinluo lotus pond demonstrates relatively modest water purification capabilities as a standalone treatment system, it nonetheless makes valuable contributions as a constructed wetland through its significant ecological services and exceptional aesthetic qualities that enrich the surrounding landscape.

Derived from Liebig's Law of the Minimum, this paradigm suggests algal growth in aquatic ecosystems becomes N-limited when the aqueous N:P ratio falls below 16 (molar basis), while P-limitation occurs when ratios exceed 16 [75,76]. Alternative formulations using the TN:TP mass ratio propose N-limitation thresholds at TN:TP < 9 and P-limitation thresholds at TN:TP > 23 [42], with some studies establishing a nitrogen limitation threshold at TN:TP < 25 [77]. Our research shows that from 2022 to 2025, the nitrogen phosphorus ratio in this water area will decrease in Jinluo lotus pond and the river outside the pond in Table 6.

Table 6. The N:P of Jinluo lotus pond and the river outside the pond.

	N:P	Restrictive	Sample Time
Jinluo lotus pond	39.67	P restrictive	2022
	12.00		2025
River outside the pond	98.97	P restrictive	2022
	14.71		2025

5. Conclusions

The main findings indicate that Jinluo lotus pond, as a low-density lotus pond artificial wetland system, has not shown significant water quality improvement effects. Based on these findings, we propose the following management recommendations: (1) a strategic expansion of both natural and constructed reed (*Phragmites australis*) communities along riparian zones, as these demonstrate superior nitrogen removal capabilities, a and (2) discontinuation of current lotus pond maintenance practices and an avoidance of further lotus cultivation area expansion, given their demonstrated tendency to exacerbate nitrogen accumulation in aquatic systems. This integrated approach would optimize nutrient removal efficiency while enhancing the overall ecological functionality of the watershed.

Author Contributions: Conceptualization, B.L., Y.G., J.Z., Y.W. and J.H.; Methodology, B.L., Y.G., J.Z. and Y.W.; Software, Y.G., J.Z. and J.H.; Validation, Y.G.; Formal analysis, Y.G. and Y.W.; Investigation, Y.G.; Resources, B.L. and Y.G.; Data curation, Y.G. and J.H.; Writing—original draft, B.L., Y.G. and J.Z.; Writing—review & editing, Y.G., J.Z., Y.W. and J.H.; Visualization, Y.G.; Supervision, Y.G.; Project administration, B.L. and Y.G.; Funding acquisition, B.L. All authors have read and agreed to the published version of the manuscript.

Funding: This research was funded by Shandong Provincial Key Laboratory of Water and Soil Conservation and Environmental Protection (No. STKF202307), and Shandong Province Youth Innovation and Technology Support Program for Higher Education Institutions (2020KJE009).

Data Availability Statement: The original contributions presented in this study are included in the article. Further inquiries can be directed to the corresponding author.

Conflicts of Interest: The authors declare no conflicts of interest.

References

1. Wang, D.; Gan, X.; Wang, Z.; Jiang, S.; Zheng, X.; Zhao, M.; Zhang, Y.; Fan, C.; Wu, S.; Du, L. Research status on remediation of eutrophic water by submerged macrophytes: A review. *Process. Saf. Environ. Prot.* **2023**, *169*, 671–684. [CrossRef]
2. Nimma, D.; Devi, O.R.; Laishram, B.; Ramesh, J.V.N.; Boddupalli, S.; Ayyasamy, R.; Tirth, V.; Arabil, A. Implications of climate change on freshwater ecosystems and their biodiversity. *Desalination Water Treat.* **2025**, *321*, 100889. [CrossRef]
3. Li-Kun, Y.; Sen, P.; Xin-Hua, Z.; Xia, L. Development of a two-dimensional eutrophication model in an urban lake (China) and the application of uncertainty analysis. *Ecol. Model.* **2017**, *345*, 63–74. [CrossRef]
4. Asad, A.A.; El-Hawary, A.M.; Abbas, M.H.H.; Mohamed, I.; Abdelhafez, A.A.; Bassouny, M.A. Reclamation of wastewater in wetlands using reed plants and biochar. *Sci. Rep.* **2022**, *12*, 19516. [CrossRef]
5. Greve, P.; Kahil, T.; Mochizuki, J.; Schinko, T.; Satoh, Y.; Burek, P.; Fischer, G.; Tramberend, S.; Burtscher, R.; Langan, S.; et al. Global assessment of water challenges under uncertainty in water scarcity projections. *Nat. Sustain.* **2018**, *1*, 486–494. [CrossRef]
6. Pahl-Wostl, C.; Knieper, C.; Lukat, E.; Meergans, F.; Schoderer, M.; Schütze, N.; Schweigatz, D.; Dombrowsky, I.; Lenschow, A.; Stein, U.; et al. Enhancing the capacity of water governance to deal with complex management challenges: A framework of analysis. *Environ. Sci. Policy* **2020**, *107*, 23–35. [CrossRef]
7. Zhao, X.; Yang, D. Research on Regional Disparities, Dynamic Evolution, and Influencing Factors of Water Environment Governance Efficiency in China. *Water* **2025**, *17*, 515. [CrossRef]
8. Bakker, E.S.; Van Donk, E.; Declerck, S.A.J.; Helmsing, N.R.; Hidding, B.; Nolet, B.A. Effect of macrophyte community composition and nutrient enrichment on plant biomass and algal blooms. *Basic Appl. Ecol.* **2010**, *11*, 432–439. [CrossRef]
9. Wu, H.; Wang, R.; Yan, P.; Wu, S.; Chen, Z.; Zhao, Y.; Cheng, C.; Hu, Z.; Zhuang, L.; Guo, Z.; et al. Constructed wetlands for pollution control. *Nat. Rev. Earth Environ.* **2023**, *4*, 218–234. [CrossRef]
10. Abd Rasid, N.S.; Naim, M.N.; Che Man, H.; Abu Bakar, N.F.; Mokhtar, M.N. Evaluation of surface water treated with lotus plant; *Nelumbo nucifera*. *J. Environ. Chem. Eng.* **2019**, *7*, 103048. [CrossRef]
11. Mustafa, H.M.; Hayder, G. Recent studies on applications of aquatic weed plants in phytoremediation of wastewater: A review article. *Ain Shams Eng. J.* **2021**, *12*, 355–365. [CrossRef]
12. Xiao, H.; Peng, S.; Liu, X.; Jia, J.; Wang, H. Phytoremediation of nutrients and organic carbon from contaminated water by aquatic macrophytes and the physiological response. *Environ. Technol. Innov.* **2021**, *21*, 101295. [CrossRef]
13. Kou, S.; Yang, J.; Li, Y.; Li, S.; Shen, Y.; Gao, J. Design and Application of an Integrated Landscape Water Purification Device: Long-Term Performance in Nitrogen and Phosphorus Removal. *Water* **2025**, *17*, 556. [CrossRef]
14. Rillig, M.C. Global change refugia could shelter species from multiple threats. *Nat. Rev. Biodivers.* **2025**, *1*, 10–11. [CrossRef]
15. Chan, F.K.S.; Griffiths, J.A.; Higgitt, D.; Xu, S.; Zhu, F.; Tang, Y.-T.; Xu, Y.; Thorne, C.R. "Sponge City" in China—A breakthrough of planning and flood risk management in the urban context. *Land Use Policy* **2018**, *76*, 772–778. [CrossRef]
16. Ferreira, C.S.S.; Kašanin-Grubin, M.; Solomun, M.K.; Sushkova, S.; Minkina, T.; Zhao, W.; Kalantari, Z. Wetlands as nature-based solutions for water management in different environments. *Curr. Opin. Environ. Sci. Health* **2023**, *33*, 100476. [CrossRef]

17. Wilkins, N.R.; Fallowfield, H.; Baring, R. Spatial performance assessment of reed bed filtration in a constructed wetland. *Sci. Total Environ.* **2022**, *820*, 153060. [CrossRef]
18. Chen, J.; Guo, F.; Wu, F.; Bryan, B.A. Costs and benefits of constructed wetlands for meeting new water quality standards from China's wastewater treatment plants. *Resour. Conserv. Recycl.* **2023**, *199*, 107248. [CrossRef]
19. Zhao, Y.; Ji, B.; Liu, R.; Ren, B.; Wei, T. Constructed treatment wetland: Glance of development and future perspectives. *Water Cycle* **2020**, *1*, 104–112. [CrossRef]
20. Gebeyehu, A.; Shebeshe, N.; Kloos, H.; Belay, S. Suitability of nutrients removal from brewery wastewater using a hydroponic technology with *Typha latifolia*. *BMC Biotechnol.* **2018**, *18*, 74. [CrossRef]
21. Hadad, H.R.; Mufarrege, M.M.; Pinciroli, M.; Di Luca, G.A.; Maine, M.A. Morphological Response of *Typha domingensis* to an Industrial Effluent Containing Heavy Metals in a Constructed Wetland. *Arch. Environ. Contam. Toxicol.* **2010**, *58*, 666–675. [CrossRef] [PubMed]
22. Grisey, E.; Laffray, X.; Contoz, O.; Cavalli, E.; Mudry, J.; Aleya, L. The Bioaccumulation Performance of Reeds and Cattails in a Constructed Treatment Wetland for Removal of Heavy Metals in Landfill Leachate Treatment (Etueffont, France). *Water Air Soil Pollut.* **2012**, *223*, 1723–1741. [CrossRef]
23. Licata, M.; Gennaro, M.C.; Tuttolomondo, T.; Leto, C.; La Bella, S.; Singer, A.C. Research focusing on plant performance in constructed wetlands and agronomic application of treated wastewater—A set of experimental studies in Sicily (Italy). *PLoS ONE* **2019**, *14*, e0219445. [CrossRef] [PubMed]
24. Qin, H.; Zhang, Z.; Liu, M.; Liu, H.; Wang, Y.; Wen, X.; Zhang, Y.; Yan, S. Site test of phytoremediation of an open pond contaminated with domestic sewage using water hyacinth and water lettuce. *Ecol. Eng.* **2016**, *95*, 753–762. [CrossRef]
25. Rezania, S.; Ponraj, M.; Talaiekhozani, A.; Mohamad, S.E.; Din, M.F.M.; Taib, S.M.; Sabbagh, F.; Sairan, F.M. Perspectives of phytoremediation using water hyacinth for removal of heavy metals, organic and inorganic pollutants in wastewater. *J. Environ. Manag.* **2015**, *163*, 125–133. [CrossRef]
26. Wang, J.; Chen, G.; Liu, F.; Song, X.; Zou, G. Combined ozonation and aquatic macrophyte (*Vallisneria natans*) treatment of piggery effluent: Water matrix and antioxidant responses. *Ecol. Eng.* **2017**, *102*, 39–45. [CrossRef]
27. Wu, H.; Gao, X.; Wu, M.; Zhu, Y.; Xiong, R.; Ye, S. The efficiency and risk to groundwater of constructed wetland system for domestic sewage treatment—A case study in Xiantao, China. *J. Clean. Prod.* **2020**, *277*, 123384. [CrossRef]
28. Shaharuddin, S.; Chan, N.W.; Zakaria, N.A.; Ab Ghani, A.; Chang, C.K.; Roy, R. Constructed Wetlands as a Natural Resource for Water Quality Improvement in Malaysia. *Nat. Resour.* **2014**, *5*, 292–298. [CrossRef]
29. Billore, S.K.; Prashant; Sharma, J.K. Treatment performance of artificial floating reed beds in an experimental mesocosm to improve the water quality of river Kshipra. *Water Sci. Technol.* **2009**, *60*, 2851–2859. [CrossRef]
30. Yang, L.; Zhang, Z.; Zhang, W.; Zhang, T.; Meng, H.; Yan, H.; Shen, Y.; Li, Z.; Ma, X. Wetland Park Planning and Management Based on the Valuation of Ecosystem Services: A Case Study of the Tieling Lotus Lake National Wetland Park (LLNWP), China. *Int. J. Environ. Res. Public Health* **2023**, *20*, 2939. [CrossRef]
31. Pang, J.; Feng, X.; Wang, X. Purification and utilization of garlic processing wastewater in lotus pond wetlands. *Water Sci. Eng.* **2014**, *7*, 395–402. [CrossRef]
32. Xu, S.Z.; Wang, Y.X.; Wang, Y.D.; Zhao, Y.J.; Gao, Y. Seasonal influence of reed (*Phragmites australis*) and lotus (*Nelumbo nucifera*) on urban wetland of Yi river. *Appl. Ecol. Environ. Res.* **2019**, *17*, 7891–7900. [CrossRef]
33. He, L.S.; Meng, F.L.; Meng, R.; Huang, C.H.; Li, Y.W.; Xi, B.D.; Shu, J.M. In Situ Enclosure Experiment on *Nelumbo nucifera* for Eutrophication Control in Baiyangdian Lake. *Wetl. Sci.* **2013**, *11*, 282–285. [CrossRef]
34. Li, L.; Hou, W.-H. Inhibitory effects of liquor cultured with *Nelumbo nucifera* and *Nymphaea tetragona* on the growth of *Microcystis aeruginosa*. *Environ. Sci.* **2007**, *28*, 2180–2186.
35. Li, X.; Xu, X.Q.; Gou, M.M. Study on the influence of lotus planting in Wuliangsu of Inner Mongolia on water environment. *J. Environ. Health* **2018**, *35*, 457–459. [CrossRef]
36. Li, X.; Gao, Y. Influence of the island with grass and the island with trees to water quality in Yihe River, China. *Desalination Water Treat.* **2018**, *121*, 186–190. [CrossRef]
37. Gao, Y.; Ci, H.X.; Qi, S.C.; Su, Y.X. Seasonal changes of phytoplankton diversity and assessment of water quality in four tributaries of Yi River. *Res. Environ. Sci.* **2009**, *22*, 176–180.
38. Gao, Y.; Qi, S.C.; Su, Y.X. Seasonal changes of phytoplankton diversity and water quality in Yi River and Beng River. *Trans. Oceanol. Limnol.* **2010**, *32*, 109–113. [CrossRef]
39. Gao, Y.; Su, Y.X.; Qi, S.C. Phytoplankton and evaluation of water quality in Yi River watershed. *J. Lake Sci.* **2008**, *20*, 544–548.
40. Zhang, W.-H.; Gao, Y.; Wang, Y.; Zhou, J. Water Quality Assessment and Management Strategies for Nishan Reservoir, Sihe River, and Yihe River Based on Scientific Evaluation. *Water* **2024**, *16*, 1958. [CrossRef]
41. Ge, Z.W.; Fang, S.Y.; Li, C.; Li, Q.; Bu, Q.Q.; Xue, Y.H. Analysis of the plant N and P sequestration from common reed and common reed+ cattail communities in wetland soil in Qinhu Lake of northern Jiangsu, China. *J. Lake Sci.* **2017**, *29*, 585–593. [CrossRef]

42. Guildford, S.J.; Hecky, R.E. Total Nitrogen, Total Phosphorus, and Nutrient Limitation in Lakes and Oceans: Is There A Common Relationship? *Limnol. Oceanogr.* **2000**, *45*, 1213–1223. [CrossRef]
43. *HJ 1147-2020*; Water Quality-Determination of pH-Electrode Method. Ministry of Ecology and Environment of the People's Republic of China: Beijing, China, 2020.
44. *HJ 505-2009*; Water Quality-Determination of Biochemical Oxygen Demand After 5 Days (BOD5) for Dilution and Seeding Method. Ministry of Environmental Protection of the People's Republic of China: Beijing, China, 2009.
45. *HJ 828-2017*; Water Quality-Determination of the Chemical Oxygen Demand-Dichromate Method. Ministry of Environmental Protection of the People's Republic of China: Beijing, China, 2017.
46. *HJ 636-2012*; Water Quality-Determination of Total Nitrogen-Alkaline Potassium Persulfate Digestion UV Spectrophotometric Method. Ministry of Environmental Protection of the People's Republic of China: Beijing, China, 2012.
47. *GB 11893-89*; Water Quality-Determination of Total Phosphorus-Ammonium Molybdate Spectrophotometric Method. State Administration of Technical Supervision of the People's Republic of China: Beijing, China, 1989.
48. *HJ 535-2009*; Water Quality-Determination of Ammonia Nitrogen-Nessler's Reagent Spectrophotometry. Ministry of Environmental Protection of the People's Republic of China: Beijing, China, 2009.
49. *HJ 84-2016*; Water Quality-Determination of Inorganic Anions (F−, Cl−, NO2−, Br−, NO3−, PO43−, SO32−, SO42−)-Ion Chromatography. Ministry of Environmental Protection of the People's Republic of China: Beijing, China, 2016.
50. *HJ 506-2009*; Water Quality-Determination of Dissolved Oxygen-Electrochemical Probe Method. Ministry of Environmental Protection of the People's Republic of China: Beijing, China, 2009.
51. *GB 11892-89*; Water Quality-Determination of Permanganate Index. Environmental Protection Administration of the People's Republic of China: Beijing, China, 1989.
52. *GB 3838-2002*; Environmental Quality Standards for Surface Water. State Environmental Protection Administration of the People's Republic of China and General Administration of Quality Supervision, Inspection and Quarantine of the People's Republic of China: Beijing, China, 2002.
53. *HJ 1296-2023*; Technical Guidelines for Water Ecological Monitoring-Aquatic Organism Monitoring and Evaluation of Lakes and Reservoirs (on Trial). Ministry of Ecology and Environment of the People's Republic of China: Beijing, China, 2023.
54. *HJ 897-2017*; Water Quality-Determination of Chlorophyll a-Spectrophotometric Method. Ministry of Environmental Protection of the People's Republic of China: Beijing, China, 2017.
55. Borics, G.; Abonyi, A.; Salmaso, N.; Ptacnik, R. Freshwater phytoplankton diversity: Models, drivers and implications for ecosystem properties. *Hydrobiologia* **2021**, *848*, 53–75. [CrossRef] [PubMed]
56. Pielou, E.C. The measurement of diversity in different types of biological collections. *J. Theor. Biol.* **1966**, *13*, 131–144. [CrossRef]
57. Shannon, C.E. A Mathematical theory of communication. *Bell Syst. Tech. J.* **1948**, *27*, 379–423. [CrossRef]
58. Colares, G.S.; Dell'Osbel, N.; Wiesel, P.G.; Oliveira, G.A.; Lemos, P.H.Z.; da Silva, F.P.; Lutterbeck, C.A.; Kist, L.T.; Machado, Ê.L. Floating treatment wetlands: A review and bibliometric analysis. *Sci. Total. Environ.* **2020**, *714*, 136776. [CrossRef]
59. Wang, W.-H.; Wang, Y.; Sun, L.-Q.; Zheng, Y.-C.; Zhao, J.-C. Research and application status of ecological floating bed in eutrophic landscape water restoration. *Sci. Total Environ.* **2020**, *704*, 135434. [CrossRef]
60. Zhu, Z.; Wang, Y.; Han, X.-Y.; Wang, W.-H.; Li, H.-M.; Yue, Z.-Q.; Chen, W.; Xue, F.-R. Strengthen the purification of eutrophic water and improve the characteristics of sediment by functional ecological floating bed suspended calcium peroxide and sponge iron jointly. *J. Environ. Manag.* **2023**, *325*, 116610. [CrossRef]
61. Carleton, J.N.; Montas, H.J. An analysis of performance models for free water surface wetlands. *Water Res.* **2010**, *44*, 3595–3606. [CrossRef]
62. Pan, X.; Liu, S.; Li, R.; Sun, H.; Feng, J.; Cheng, X.; Yao, J. Research on the purification enhancement of ecological ponds: Integrating water cycle optimization and plants layout. *J. Environ. Manag.* **2023**, *344*, 118487. [CrossRef]
63. Sabokrouhiyeh, N.; Bottacin-Busolin, A.; Savickis, J.; Nepf, H.; Marion, A. A numerical study of the effect of wetland shape and inlet-outlet configuration on wetland performance. *Ecol. Eng.* **2017**, *105*, 170–179. [CrossRef]
64. Gong, Y.; Zhang, X.; Li, H.; Zhang, X.; He, S.; Miao, Y. A comparison of the growth status, rainfall retention and purification effects of four green roof plant species. *J. Environ. Manag.* **2021**, *278*, 111451. [CrossRef] [PubMed]
65. Ong, S.-A.; Uchiyama, K.; Inadama, D.; Ishida, Y.; Yamagiwa, K. Performance evaluation of laboratory scale up-flow constructed wetlands with different designs and emergent plants. *Bioresour. Technol.* **2010**, *101*, 7239–7244. [CrossRef] [PubMed]
66. Frenzel, P.; Brunings, O.; Esler, K.J.; le Maitre, D.; Rebelo, A.J. How well do endemic wetland plant species perform in water purification? *Wetlands* **2024**, *44*, 124. [CrossRef]
67. Bai, X.; Zhu, X.; Jiang, H.; Wang, Z.; He, C.; Sheng, L.; Zhuang, J. Purification effect of sequential constructed wetland for the polluted water in Urban River. *Water* **2020**, *12*, 1054. [CrossRef]
68. Kurniawan, S.B.; Abdullah, S.R.S.; Imron, M.F.; Ahmad, A.; Said, N.S.M.; Rahim, N.F.M.; Alnawajha, M.M.; Abu Hasan, H.; Othman, A.R.; Purwanti, I.F. Potential of valuable materials recovery from aquaculture wastewater: An introduction to resource reclamation. *Aquac. Res.* **2021**, *52*, 2954–2962. [CrossRef]

69. Chao, C.; Gong, S.; Xie, Y. The Performance of a multi-stage surface flow constructed wetland for the treatment of aquaculture wastewater and changes in epiphytic biofilm formation. *Microorganisms* **2025**, *13*, 494. [CrossRef]
70. Ni, M.; Yuan, J.; Zhang, L.; Hua, J.; Rong, H.; Gu, Z. In- situ and ex- situ purification effect of ecological ponds of *Euryale ferox* Salisb on shrimp aquaculture. *Aquaculture* **2021**, *540*, 736678. [CrossRef]
71. Wang, C.; Lei, J.; Guo, C. Application studies on purification of piggery waste water by multi-level artificial system of ecological lotus pond-surface flow wetland. *Ecol. Environ. Sci.* **2019**, *28*, 2289–2298.
72. Harrington, C.; Scholz, M. Assessment of pre-digested piggery wastewater treatment operations with surface flow integrated constructed wetland systems. *Bioresour. Technol.* **2010**, *101*, 7713–7723. [CrossRef]
73. Harrington, R.; McInnes, R. Integrated Constructed Wetlands (ICW) for livestock wastewater management. *Bioresour. Technol.* **2009**, *100*, 5498–5505. [CrossRef] [PubMed]
74. Scholz, M.; Harrington, R.; Carroll, P.; Mustafa, A. The Integrated Constructed Wetlands (ICW) concept. *Wetlands* **2007**, *27*, 337–354. [CrossRef]
75. Keck, F.; Lepori, F. Can we predict nutrient limitation in streams and rivers? *Freshw. Biol.* **2012**, *57*, 1410–1421. [CrossRef]
76. Liu, J.; Wang, Y.; Li, Y.; Liu, X.; Jiang, Y.; Fu, Y.; Jin, W.; Wu, J. Ecosystem N:P stoichiometric ratios determine the catchment surface water N:P ratio through subsurface hydrological processes. *CATENA* **2020**, *194*, 104740. [CrossRef]
77. Gal, G.; Hipsey, M.; Parparov, A.; Wagner, U.; Makler, V.; Zohary, T. Implementation of ecological modeling as an effective management and investigation tool: Lake Kinneret as a case study. *Ecol. Model.* **2009**, *220*, 1697–1718. [CrossRef]

water

MDPI

Review

Arsenic Contamination in Sludge and Sediment and Relationship with Microbial Resistance Genes: Interactions and Remediation

Menglong Xing [1,2,†], Dajiang Yan [2,3,†], Mengmeng Hai [2], Yanhao Zhang [2], Zhibin Zhang [2,3,*] and Fengmin Li [1,*]

1 Institute of Coastal Environmental Pollution Control, Ministry of Education Key Laboratory of Marine Environment and Ecology, Marine Ecology and Environmental Science Laboratory, Pilot National Laboratory for Marine Science and Technology, College of Environmental Science and Engineering, Ocean University of China, Qingdao 266100, China; m15513079866_2@163.com

2 School of Municipal and Environmental Engineering, Shandong Jianzhu University, Jinan 250101, China; y601741130@gmail.com (D.Y.); zlsktz@126.com (M.H.); sdzyh66@126.com (Y.Z.)

3 College of Safety and Environmental Engineering, Shandong University of Science and Technology, Qingdao 266590, China

* Correspondence: zbinzhang@yeah.net (Z.Z.); fengminli2024@126.com (F.L.)

† These authors contributed equally to this work.

Abstract: Arsenic contamination in sludge and sediment has emerged as a pressing environmental issue with far-reaching implications. This review delves into the multifaceted problem of arsenic contamination, focusing on its complex interactions with microbial resistance genes (MRGs). It explores the key role of microorganisms in the biogeochemical cycling of arsenic, including processes such as reduction, oxidation, methylation, and volatilization. It describes how microorganisms resist arsenic through resistance genes that encode proteins such as efflux pumps, enzymatic detoxification, and intracellular sequestration. Arsenic, a naturally occurring element, can enter sludge and sediment through various natural and anthropogenic pathways, leading to detrimental effects on environmental quality. Understanding the role of microorganisms in arsenic mobilization, transformation, and their ability to resist arsenic toxicity through MRGs is essential for effective mitigation and remediation strategies. This review discusses the sources and distribution of arsenic in sludge and sediment, the intricate mechanisms of microbial arsenic resistance, and the potential implications for environmental management and human health. It also examines current research trends and identifies areas requiring further investigation. By unraveling the interplay between arsenic contamination, microorganisms, and MRGs, this review aims to contribute to a deeper understanding of the issue and guide future research and environmental protection efforts.

Keywords: arsenic; sludge; sediment; microbial resistance genes; environmental health; remediation strategies

1. Introduction

Sludge and sediment are important components of the Earth's ecosystems. They not only serve as important media for plant growth but also form a complex ecosystem that provides habitats for various microorganisms and animals [1]. Sludge and sediment are also storage media for heavy metal pollutants. These can adsorb heavy metal ions and allow the ions to enter plant tissues through plant absorption, subsequently affecting human health through the food chain [2].

Arsenic (As) is a naturally occurring metalloid element that has garnered global attention due to its widespread presence in sludge and sediment and its adverse effects on ecosystems and human health [3]. This review aims to explore the multifaceted issue of arsenic contamination in sludge and sediment, with a particular emphasis on its intricate relationship with microbial resistance genes (MRGs) [4]. Arsenic contamination can arise from both natural processes and human activities, making it a complex environmental challenge. Microorganisms play a pivotal role in the biogeochemical cycling of arsenic and have developed various mechanisms to resist its toxicity, making them central to our understanding of this issue [5].

The objective of this review is to provide a comprehensive overview of arsenic contamination in sludge and sediment. It encompasses the sources, distribution, and fate of arsenic in these environments, as well as the genetic basis of microbial resistance to arsenic. Additionally, the review discusses the potential risks associated with MRGs and evaluates the current state of research in the field. By shedding light on the intricate interplay between arsenic contamination, microorganisms, and MRGs, this review aims to facilitate informed decision-making for environmental protection and remediation efforts.

2. The Properties and Hazards of Arsenic

2.1. Physical and Chemical Properties of Arsenic

Arsenic is a metalloid element. It is widely distributed in the Earth's crust, primarily in the form of inorganic compounds such as arsenides and oxides [6]. It can be released into the environment through natural geological processes like volcanic activity and rock weathering, as well as through human activities such as mining, smelting, agriculture, and industrial wastewater discharge, leading to serious pollution issues [7]. Arsenic exists in several chemical forms and oxidation states, with the most common being pentavalent (arsenate, As(V)) and trivalent (arsenite, As(III)). Arsenic has significant applications in industry, manufacturing, and medicine. In the semiconductor industry, gallium arsenide (GaAs) is widely used in the manufacture of high-speed electronic and optoelectronic devices [8]; in the medical field, arsenic trioxide (As_2O_3) is utilized as an effective treatment for acute promyelocytic leukemia [9]; in the metallurgical industry, arsenic can be utilized as an alloy additive to enhance the strength and corrosion resistance of alloys [10].

In water bodies, the chemical behavior of arsenic primarily depends on the redox potential and pH value [11]. In oxidizing environments, arsenic typically exists in the form of As(V), mainly dissolving in water as $H_2AsO_4^-$ or $HAsO_4^{2-}$ [12]. In reducing environments, arsenic exists in the form of As(III), usually as H_3AsO_3 or $H_2AsO_3^-$, where it has increased solubility and higher toxicity. The mobility of arsenic in water is controlled by its adsorption reactions with iron, aluminum, and manganese oxides; under oxidizing conditions, As(V) is prone to precipitate with these oxides, while As(III) is more likely to remain in solution under reducing conditions [13]. Additionally, organic matter and microorganisms in the water can also affect the transformation of arsenic forms [14]; these microorganisms can convert As(V) to As(III) through reduction, or transform inorganic arsenic into organic arsenic compounds through methylation, thereby altering arsenic's mobility and toxicity [15].

In sediments and bottom mud, the behavior of arsenic is similar to that in water bodies, but due to the more complex physicochemical structure of sediments and mud, the adsorption, precipitation, and re-release processes of arsenic are more pronounced [16]. Arsenic can bind with metal oxides such as iron and aluminum to form precipitates [17]; under anaerobic conditions, arsenic can combine with sulfides to form insoluble arsenides (e.g., As_2S_3). However, when environmental conditions change, such as changes in the redox state of the water, fluctuations in pH, or disturbances to the sediment, arsenic in the sediment may be re-released back into the water, increasing its mobility and bioavailability [18]. In oxidizing environments, arsenic primarily exists as As(V) and forms stable adsorption states by binding with hydroxyl groups on the surfaces of iron oxides, especially under neutral and acidic conditions [19]. This process significantly reduces the mobility and bioavailability of arsenic. In contrast, As(III) is more stable in neutral and reducing environments and has weaker adsorption capabilities, depending primarily on the pH of the solution and the type of iron oxides. When iron oxides are reduced or disintegrate, particularly in anaerobic environments, Fe(III) is reduced to Fe(II), leading to the release of arsenic, especially As(III), increasing its mobility and toxicity [20]. Under high pH conditions, adsorbed arsenic is likely to desorb, entering the water phase and further exacerbating the risk of arsenic contamination [21].

2.2. The Hazards of Arsenic

Arsenic is not only a metalloid element with complex physical and chemical properties but also a highly toxic substance [22]. The toxicity of arsenic depends on its chemical form, oxidation state, and exposure pathway. Inorganic arsenic compounds, including pentavalent arsenic (As(V), such as arsenates) and trivalent arsenic (As(III), such as arsenites), are the main toxic forms of arsenic, with trivalent arsenic being significantly more toxic than pentavalent arsenic [23]. The toxicity of arsenic poses threats not only to human health but also disrupts the balance of ecosystems.

For humans, long-term exposure to low doses of arsenic can lead to chronic arsenic poisoning, with common symptoms including skin lesions, hyperkeratosis, pigmentation changes, neuropathy, cardiovascular diseases, and cancer [24]. Arsenic is classified as a Group 1 carcinogen by the International Agency for Research on Cancer (IARC), with skin cancer, lung cancer, and bladder cancer being closely associated with long-term arsenic exposure [25]. Arsenic contamination in drinking water is a global public health issue, especially in South Asia, including Bangladesh, India, and Nepal, where many residents have long-term exposure to arsenic-laden groundwater, leading to large-scale poisoning events. The World Health Organization (WHO) sets the arsenic limit in drinking water at 10 μg/L, but in many developing countries, this limit is often significantly exceeded [26].

The mobility and bioaccumulation of arsenic in the environment mean that its toxicity is not limited to water bodies but can also spread through sediments and bottom mud [27]. Under certain conditions (such as changes in the redox environment and disturbances to sediments), arsenic in sediments may be re-released into water bodies, increasing the concentration and toxicity of arsenic in the water. This cyclical process makes the management of arsenic pollution complex and exacerbates ecological damage [28]. Arsenic has significant toxicity to aquatic organisms, potentially causing death or reproductive disorders in fish, shellfish, and other aquatic life, and affecting humans through the food chain. Especially in crops like rice, arsenic can be absorbed by plant roots from arsenic-contaminated irrigation water and accumulate within the plant, thus entering the human food chain and further increasing the potential health risks of arsenic to humans [29].

There are significant differences between As(V) and As(III) in terms of toxicity and environmental behavior, which mainly arise from their chemical properties, bioavailability, and effects on organisms. As(V) primarily exists in anionic forms, such as hydrogen arsenate ($H_2AsO_4^-$) and arsenate ($HAsO_4^{2-}$), and its high solubility allows it to be widely distributed in water bodies. However, due to As(V)'s larger ionic radius and hydrophilic nature, it has a weaker ability to penetrate cell membranes, making its intracellular toxicity relatively low [15]. As(V) competes with phosphate by binding to phosphate transporters in cells, thus disrupting the phosphate metabolism process. Phosphorus is a key component of cellular energy metabolism and nucleic acid synthesis, so when phosphate metabolism is disrupted, the cell's energy supply and biosynthesis are affected, leading to some level of toxicity [30].

In contrast, As(III) exists in a neutral form, such as H_3AsO_3, and its smaller molecular size and higher lipophilicity enable it to more easily penetrate cell membranes. The chemical structure of As(III) causes its toxic effects within cells to be more direct and pronounced. Studies have shown that As(III) can directly interfere with mitochondrial function, inhibit oxidative phosphorylation, and reduce ATP production, thereby affecting the cell's energy supply [31]. Additionally, As(III) can bind to sulfur-containing proteins (such as thiol enzymes) within the cell, forming arsenic–sulfur compounds that inhibit these proteins' normal functions, disrupting the cell's antioxidant defense system. As(III) can also cause oxidative stress by generating excessive reactive oxygen species (ROS), which damage cell membranes, proteins, and DNA, leading to apoptosis or necrosis. These mechanisms make As(III) tens of times more toxic than As(V) [32].

In the environment, the high bioavailability and toxicity of As(III) lead to its greater accumulation capacity in aquatic organisms. For example, aquatic plants and animals can efficiently absorb As(III) and accumulate it in their bodies, passing it up the food chain to higher trophic-level predators, including humans. This biomagnification effect increases the threat of arsenic to ecosystems and public health [33]. Additionally, As(III) is relatively stable in water and not easily broken down, with its high toxicity and mobility making its environmental impact more persistent and severe.

3. Sources and Distribution of Arsenic

3.1. Natural Sources of Arsenic

Arsenic's natural sources are extensive and complex, primarily released into the environment through geological processes, hydrological conditions, hot spring activity, marine sediments, and biological metabolism [34]. First, rock weathering is a significant pathway for the release of arsenic. Arsenic often binds with metallic minerals such as arsenopyrite (FeAsS), orpiment (As_2S_3), and realgar (As_4S_4). During the weathering of these arsenic-containing minerals, arsenic is released into the soil and water bodies, entering the ecosystem [35]. Additionally, volcanic activity is a significant natural source of arsenic entering the environment. Volcanic eruptions release arsenic and other minerals into the atmosphere in gaseous or particulate form, deposited in the soil and water bodies [36]. Particularly in volcanic and geothermal areas, groundwater often contains high levels of arsenic because the groundwater dissolves significant amounts of arsenic-containing minerals from the crust through geothermal activity [37].

Hydrogeological conditions also affect the natural concentrations of arsenic in groundwater. In some regions, groundwater passes through arsenic-rich rock layers or sediments, especially under anoxic, reducing conditions, where arsenic in iron and manganese oxides is reduced and released into the water [38]. This situation is common in many high-arsenic regions globally, such as in the groundwater of Bangladesh and India. Evaporation in arid and semi-arid areas also leads to increased concentrations of arsenic. In these regions, rapid water evaporation leads to the gradual concentration of dissolved solids, including arsenic, in salt lakes or groundwater [39]. Hot springs and geothermal springs are also significant natural sources of arsenic, especially in regions with frequent geothermal activity [40]. The water from hot springs and geothermal springs dissolves a large amount of minerals, including arsenic, as it passes through deep crustal layers [41]. This arsenic-rich geothermal water eventually flows into rivers or lakes, leading to concentration of arsenic in local water bodies [42]. Additionally, the natural concentration of arsenic in soil depends on the geological background of the area. In some regions, the parent rock contains high levels of arsenic, which gradually enters the soil through rock weathering and may enter a broader ecosystem through plant absorption or water infiltration. Natural sources of arsenic contamination may result in localized hotspots in regions with specific geological characteristics [43].

3.2. Anthropogenic Sources of Arsenic

Human activities have significantly contributed to arsenic contamination in sludge and sediment [44]. The anthropogenic sources of arsenic mainly stem from various activities in industry, agriculture, and daily life. These activities release large amounts of arsenic into the environment, leading to significantly elevated levels of arsenic in water bodies, soil, and the atmosphere, posing threats to ecosystems and human health. Firstly, mining and metallurgical industries are among the primary sources of anthropogenic arsenic emissions [45]. Arsenic often coexists with metal ores (such as copper, gold, lead, etc.), and during the mining and smelting processes, arsenic is released as a byproduct. Particularly during the smelting process, arsenic-containing minerals decompose at high temperatures, and arsenic is emitted into the atmosphere in gaseous form, which then enters water bodies and soil through precipitation. Additionally, tailings and wastewater from mining areas often contain high concentrations of arsenic. These wastes release arsenic into the surrounding environment through weathering or leaching, further exacerbating arsenic pollution. Secondly, agricultural activities are also a significant source of arsenic pollution. Historically, arsenic-based compounds (such as arsenates and lead arsenate) were widely used as insecticides, herbicides, and fungicides, particularly in cotton and fruit cultivation in the early 20th century [46]. Although these chemicals have been banned or severely restricted in many countries, their residues still persist in the soil and enter water bodies through agricultural runoff. Additionally, in some regions, the use of arsenic-contaminated groundwater for irrigation can lead to the accumulation of arsenic in soil and crops, especially in rice fields, where rice has a high capacity to absorb arsenic, thereby affecting human health through the food chain [47].

Industrial wastewater discharge is another major pathway for arsenic entering the environment. The chemical industry, especially in the production of pesticides, dyes, electronics, and glass manufacturing processes, generates arsenic-containing wastewater. Untreated or improperly treated industrial wastewater discharged directly into rivers, lakes, or groundwater increases the arsenic concentration in water bodies. Moreover, waste incineration is one of the pathways for anthropogenic arsenic emissions, especially when processing arsenic-containing waste [48]. Arsenic can volatilize at high temperatures and be released into the atmosphere during incineration. Coal combustion is also a significant source of anthropogenic arsenic release. Coal naturally contains certain amounts of arsenic, and the process of burning coal releases arsenic into the atmosphere, especially in power plants and home heating. The burning of untreated coal produces a significant amount of arsenic-containing gases and particulates [49]. Once released into the atmosphere, these arsenic-containing gases enter soil and water bodies through dry and wet deposition, causing widespread environmental pollution. Understanding the sources and pathways of anthropogenic arsenic contamination is crucial for effective management and mitigation strategies.

3.3. Global Distribution of Arsenic Contamination

Arsenic contamination is a global issue, affecting diverse regions across the world [33]. The global distribution of arsenic pollution exhibits significant regional variations, influenced by both natural geological conditions and human activities. Arsenic pollution is primarily concentrated in areas with unique geological conditions and high levels of industrialization, particularly in parts of Asia, South America, and North America [50]. The status of arsenic contamination in some areas is shown in Table 1.

Table 1. Distribution of arsenic contamination in selected countries.

Continents	Country	Region	Pollution Status	References
Asia	Bangladesh	Nationwide	More than 30% of tube wells exceed arsenic limits, affecting about 35 million people	[12]
	India	Ganges Plain, Indus River Basin	Arsenic concentrations exceed 50 μg/L in some areas, affecting 140 million people	[38]
	China	Northern and Central regions	Industrial and mining areas affected, impacting 5–10 million people	[51]
South America	Argentina	La Pampa, Mendoza, etc.	Groundwater arsenic concentrations up to 200 μg/L, affecting about 4 million people	[52]
	Chile	Northern Atacama region	Groundwater arsenic concentrations as high as 500 μg/L	[45]
	Peru	Andes mining areas	Groundwater arsenic concentrations exceeding 50 μg/L	[45]
North America	United States	Western and Midwestern regions	Groundwater arsenic concentrations up to 50 μg/L or higher in Nevada, Arizona, etc.	[53]
	Mexico	Durango, Laguna	Groundwater concentrations of 100–200 μg/L affect more than 2 million people	[54]

Table 1. *Cont.*

Continents	Country	Region	Pollution Status	References
Europe	Hungary	Mining areas	Groundwater arsenic concentrations up to 100 μg/L	[55]
	Czech Republic	Bohemia mining areas	Groundwater arsenic concentrations above 50 μg/L	[56]
	Italy	Naples volcanic region	Groundwater arsenic concentrations of 20–40 μg/L	[38]
	Spain	Galicia, Andalusia	Groundwater arsenic concentrations of 15–50 μg/L	[57]
Africa	Ethiopia, Kenya	East African Rift Valley region	Groundwater arsenic concentrations of 50–150 μg/L	[58]
	Zimbabwe	Mining areas	Groundwater arsenic concentrations of 30–80 μg/L	[51]
	Ghana	Gold mining areas	Groundwater and soil arsenic concentrations of 20–100 μg/L	[59]
Oceania	Australia	New South Wales, Queensland, Tasmania	Groundwater arsenic concentrations of 30–100 μg/L	[60]
	New Zealand	North Island agricultural areas	Soil arsenic concentrations of 20–50 mg/kg, groundwater arsenic exceeding limits	[61]

Arsenic contamination in Asia is one of the most severe global issues, affecting vast geographic areas and hundreds of millions of people [62]. The most impacted countries include Bangladesh, India, China, Pakistan, Vietnam, Nepal, and Cambodia. For instance, approximately 35 million people in Bangladesh rely on arsenic-contaminated groundwater, with more than 30% of tube well water exceeding safe arsenic levels [12]. In India, more than 140 million people in the Ganges Plain and the Indus River Basin are affected by arsenic contamination, with concentrations often exceeding 50 μg/L [38]. In parts of China, particularly in the northern and central regions, industrial activities and mining have led to significant arsenic pollution, affecting between 5 and 10 million people [51]. The primary causes of arsenic contamination include natural geological factors, such as the dissolution of arsenic-bearing minerals in young alluvial plains and delta regions. These areas typically contain abundant organic matter and reducing conditions, which facilitate arsenic mobility and release. Additionally, human activities, such as agricultural irrigation and industrial wastewater discharge, exacerbate the problem. The widespread use of arsenic-laden groundwater for crop irrigation, particularly in agriculture, has led to excessive arsenic levels in food products such as rice. Long-term consumption of arsenic-contaminated water has resulted in severe health issues, including skin diseases, cancer, neurological damage, and cardiovascular diseases. In Bangladesh and India, approximately 5 million people are affected by arsenic poisoning, with children and pregnant women being particularly vulnerable. They face heightened risks of developmental disorders and cognitive impairment.

Arsenic contamination in South America is primarily concentrated in Argentina, Chile, and Peru, posing serious health risks to millions of people. In Argentina, the provinces of La Pampa, Mendoza, and Córdoba report groundwater arsenic concentrations as high as 200 μg/L, far exceeding the World Health Organization's (WHO) safe drinking water limit of 10 μg/L, affecting approximately 4 million people [52]. This contamination is largely attributed to the dissolution of arsenic-rich minerals in the geological substrate, exacerbated by the arid climate, which facilitates arsenic release. In Chile, the northern Atacama region is one of the most severely affected areas, with groundwater arsenic levels reaching up to 500 μg/L [45]. The contamination stems from long-standing mining activities and volcanic geological features. Similarly, in Peru, arsenic pollution is prevalent in mining regions along the Andes, with groundwater arsenic concentrations exceeding 50 μg/L in certain areas, directly linked to mining operations [45]. Long-term consumption of arsenic-contaminated water has led to severe health impacts, including skin diseases, skin cancer, lung cancer, bladder cancer, and other chronic illnesses. Rural populations in parts of Chile and Argentina have already experienced these health issues. While both Argentina and Chile have implemented remediation measures such as adsorption, chemical precipitation, and membrane filtration to address arsenic contamination, economic and technical limitations, particularly in rural and remote areas, pose significant challenges to these efforts.

Arsenic contamination in North America is primarily concentrated in the United States and Mexico, particularly in areas with complex geological conditions and intensive industrial activities, posing significant health risks to millions of people. In the United States, arsenic pollution is most prevalent in the western and midwestern states, including Nevada, Arizona, New Mexico, and California, affecting more than 21 million residents. In some regions, arsenic concentrations exceed 50 μg/L [53]. Similarly, northern Mexico, particularly in Durango and the Laguna region, faces severe arsenic contamination, with an estimated 2 million people relying on groundwater containing arsenic levels as high as 100–200 μg/L [54]. The contamination in North America is driven by both natural geological factors and human activities. Geologically, the weathering and dissolution of arsenic-rich volcanic rocks and sedimentary minerals, especially in the volcanic regions of the U.S. West, contribute significantly to elevated groundwater arsenic levels. Human activities, such as mining, industrial discharges, and agricultural irrigation, further exacerbate arsenic mobility and pollution. Historic metal mining regions in the U.S. have released large quantities of arsenic-laden wastewater and mine tailings, spreading contamination. Additionally, in arid and semi-arid areas, evaporation intensifies arsenic concentration in groundwater, compounding the issue.

Arsenic contamination in Europe is primarily concentrated in Central and Eastern Europe and the Mediterranean region, particularly in countries like Hungary, the Czech Republic, Italy, and Spain, posing significant health risks to local populations. In certain mining areas of Hungary and the Czech Republic, groundwater arsenic concentrations can reach up to 100 μg/L, primarily due to historical mining activities and the leaching of mine tailings [55,56]. In Italy, particularly in volcanic regions near Naples, groundwater arsenic levels in the range of 20–40 μg/L stem from the natural weathering of volcanic rocks and water–rock interactions [38]. Spain's Galicia and Andalusia regions are also affected by historical mining activities, with groundwater arsenic concentrations typically in the range of 15–50 μg/L [57]. To address these arsenic pollution issues, Hungary and the Czech Republic have implemented water treatment facilities and alternative water source solutions, while Italy and Spain have focused on water source substitution and drinking water purification. However, the ongoing natural release of arsenic and limitations in economic and technological resources present significant challenges to remediation efforts.

Arsenic contamination in Africa is primarily concentrated in the East African Rift Valley, Southern Africa, and certain mining areas of West Africa, where elevated arsenic levels in groundwater and soil pose serious health risks to local populations. In the East African Rift Valley, including Ethiopia, Kenya, and Tanzania, volcanic activity and geothermal processes result in groundwater arsenic concentrations typically ranging from 50 to 150 μg/L [58]. Residents along the Rift Valley are at high risk due to long-term exposure to arsenic-contaminated drinking water. In Southern Africa, particularly Zimbabwe, arsenic levels in groundwater can reach 30–80 μg/L, mainly due to historical mining activities and the disposal of mine tailings [51]. In West Africa, Ghana is a major hotspot, especially in gold mining areas, where arsenic concentrations in groundwater and soil range from 20 to 100 μg/L, directly linked to cyanide processing of gold ore and wastewater discharge [59]. Although water treatment facilities and mine wastewater management measures have been implemented in some regions to mitigate arsenic pollution, the effectiveness of these efforts remains limited due to economic constraints and the complex geological conditions. This issue is particularly pronounced in rural and remote areas, where many residents continue to rely on arsenic-contaminated water sources.

Arsenic contamination in Oceania is primarily concentrated in certain historical mining and agricultural areas of Australia and New Zealand, significantly affecting local water and soil quality. In Australia, regions such as New South Wales, Queensland, and Tasmania are heavily impacted. Historical mining operations and the disposal of mine tailings have resulted in groundwater arsenic concentrations ranging from 30 to 100 μg/L, posing serious health risks to nearby communities [60]. In Tasmania, gold and tin mining areas are particularly affected, where high arsenic levels in groundwater directly threaten the safety of drinking water. In New Zealand, arsenic pollution is mainly concentrated in agricultural areas of the North Island. Historical use of arsenic-based pesticides has led to significant soil arsenic accumulation, with concentrations ranging from 20 to 50 mg/kg in some regions [61]. Additionally, groundwater near volcanic zones has been found to contain arsenic levels exceeding safety standards. The primary causes of arsenic pollution in these areas include both geological factors and human activities, such as mining and prolonged use of arsenic-containing pesticides. These contamination sources not only degrade groundwater quality but also affect soil, hindering crop growth and potentially introducing arsenic into the food chain, posing long-term health risks to the population.

4. The Role of Microorganisms in Arsenic Cycling and Remediation

4.1. Involvement of Microorganisms in Arsenic Geochemical Cycling

Microorganisms play a central role in the biogeochemical cycling of arsenic, directly influencing its forms, mobility, and toxicity through a series of oxidation, reduction, methylation, and demethylation reactions [63]. Microbial arsenic transformation processes not only regulate the distribution of arsenic in the environment but also affect its chemical behavior and ecological impact in water, soil, and sediments [64]. The role played by microorganisms in arsenic geochemical cycling is shown in Figure 1.

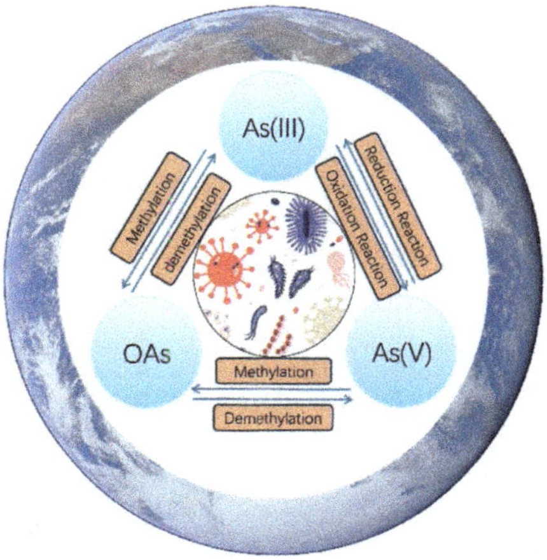

Figure 1. Role played by microorganisms in arsenic geochemical cycling.

Microorganisms can oxidize trivalent arsenic (As(III)) to pentavalent arsenic (As(V)), a process known as microbial oxidation [65]. As(III) is more toxic and mobile than As(V), so microbial oxidation is important in reducing the toxicity and bioavailability of arsenic in the environment. The microorganisms involved in arsenic oxidation include aerobic bacteria and archaea, such as As(III)-oxidizing bacteria (e.g., *Pseudomonas* and *Thiomonas* species). These microorganisms use As(III) as an electron donor, oxidizing it to As(V), which is then adsorbed or precipitated in sediments or on mineral surfaces [66]. This process typically occurs in oxygen-rich environments such as surface waters and soils.

In contrast to oxidation, certain microorganisms can reduce As(V) to As(III), a process known as microbial reduction. In anaerobic environments (such as sediments and sludge), reduction conditions dominate, and microorganisms use As(V) as an electron acceptor, reducing it to As(III) to gain energy [67]. These bacteria are often referred to as As(V)-respiring bacteria, including species like *Geobacter*, *Shewanella*, and *Desulfotomaculum* [68]. The reduction reaction significantly increases arsenic's mobility and toxicity because As(III) is more soluble in water and more toxic to organisms than As(V). Therefore, microbial reduction is a key mechanism for the release of arsenic from sediments or soil in reducing environments.

Microorganisms can convert inorganic arsenic into organic arsenic compounds (such as monomethyl arsenic acid and dimethylarsenic acid) through methylation reactions. This process is typically carried out by arsenic-methylating bacteria and archaea, including some fungi and bacteria such as *Methanobacterium* and *Methanosarcina* species [69]. Arsenic methylation transforms inorganic arsenic into volatile or water-soluble organic arsenic compounds, reducing its toxicity and bioaccumulation potential in water bodies. Organic arsenic compounds, such as the methylated forms monomethylarsenic acid (MMA) and dimethylarsenic acid (DMA), are relatively less toxic and are easily excreted by organisms [15]. Thus, microbial methylation not only plays an important role in the biodegradation of arsenic but also helps volatilize arsenic (e.g., as arsine, AsH_3) from water or soil, reducing its accumulation in the local environment.

Conversely, some microorganisms can demethylate organic arsenic compounds back into inorganic arsenic. These microorganisms utilize organic arsenic compounds as carbon or energy sources, metabolizing them into As(III) or As(V) [70]. Demethylation increases the concentration of inorganic arsenic, raising its toxicity and bioavailability in the environment. Therefore, microbial activity in the balance between methylation and demethylation directly affects the transformation and distribution of arsenic forms. The genetic basis of microbial arsenic resistance is a subject of intense research and is discussed in detail in subsequent sections [71].

4.2. *Involvement of Microorganisms in Arsenic Remediation*

Microorganisms play a crucial role in arsenic remediation, primarily due to their unique metabolic capabilities, which allow them to process arsenic through various biochemical pathways [72]. These metabolic processes, including oxidation, reduction, adsorption, and precipitation, enable microorganisms to effectively transform the chemical forms of arsenic, thereby reducing its toxicity and bioavailability [73]. The role played by microorganisms in the remediation of arsenic contamination is shown in Figure 2.

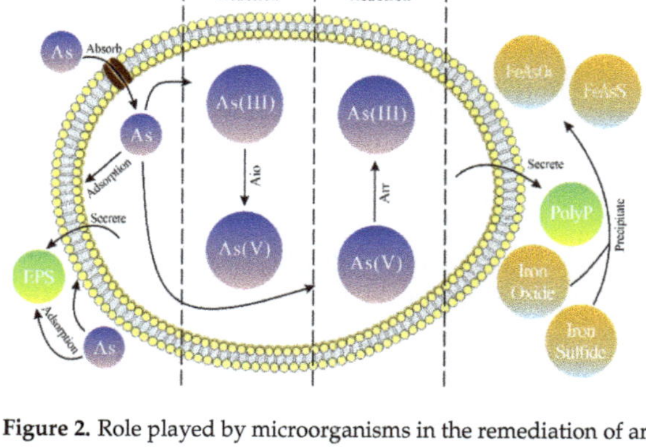

Figure 2. Role played by microorganisms in the remediation of arsenic contamination.

4.2.1. Oxidation Reaction

Some microorganisms can oxidize the more toxic As(III) to the less toxic As(V) through arsenic oxidase [69]. This transformation not only reduces arsenic toxicity but also facilitates subsequent treatment steps, such as adsorption or precipitation. Studies have shown that bacteria belonging to *Pseudomonas* and *Acidovorax* can perform this oxidation [74,75]. The reaction primarily relies on specific arsenic oxidases (Aio), which are usually located on the outer membrane of microorganisms and belong to the multicopper oxidase family. These enzymes effectively convert environmental As(III) to As(V). During this process, As(III) is first taken up by specific transport proteins in the microbial cell wall and delivered to the site containing arsenic oxidase [76]. The active center of arsenic oxidase contains multiple copper ions, which play a crucial catalytic role in transferring electrons from As(III) to the oxidase, thus completing the oxidation process. The oxidized As(V) is then released back into the environment, where it is more amenable to further biological or chemical processes. This microbial oxidation not only mitigates arsenic toxicity but also helps control its mobility and accumulation in the environment [77]. Furthermore, microbially mediated arsenic oxidation is an effective pathway for altering arsenic's chemical form and a critical step in the biogeochemical cycling of arsenic, providing a viable and environmentally friendly strategy for biotechnological applications in environmental remediation.

4.2.2. Reduction Reaction

In anaerobic environments, some microorganisms reduce As(V) to As(III) using arsenate reductase (Arr), a process that temporarily increases arsenic toxicity but creates favorable conditions for subsequent fixation and precipitation [78]. This reduction is primarily catalyzed by specific arsenate reductases in anaerobic conditions. These enzymes are typically found in bacteria such as *Shewanella* and *Geobacter*, which can use As(V) as a terminal electron acceptor in their energy metabolism [79]. During this process, the microorganisms utilize electron donors generated through their metabolic pathways (e.g., organic acids or hydrogen gas) to supply the necessary electrons. These electrons are captured by arsenate reductase and transferred to As(V), reducing it to As(III). Due to its higher chemical reactivity, As(III) readily reacts with anions such as sulfides in the environment to form insoluble precipitates [80]. This enables the effective immobilization and removal of arsenic from the environment.

4.2.3. Adsorption

The microbial remediation of arsenic pollution through adsorption mechanisms involves complex biochemical interactions, primarily relying on the functional groups within microbial cell walls and extracellular polymeric substances (EPS). For instance, *Rhodococcus* and *Mycobacterium* can secrete EPS [81]. The microbial cell wall, composed of proteins, polysaccharides, and lipids, contains abundant functional groups such as amino, carboxyl, phosphate, and hydroxyl groups, which can form coordination bonds with arsenic ions, enabling effective arsenic adsorption. Furthermore, many microorganisms secrete biofilms and EPS, which not only increase the adsorption surface but also create more stable complexes with arsenic through their internal functional groups, enhancing the adsorption efficiency. The presence of biofilms provides a protective barrier for microorganisms and establishes a reactive interface between the microbes and their environment, facilitating the capture and immobilization of arsenic ions [82]. Under specific environmental conditions, such as anaerobic or low-oxygen environments, microorganisms adjust their metabolic activities and surface charge properties to enhance arsenic adsorption. This capability enables microorganisms to efficiently remove arsenic from water bodies and stabilize it in polluted soils. This microbially mediated arsenic adsorption process is not only crucial for environmental remediation but also provides a scientific basis and practical approach for developing new bioremediation technologies [83]. It highlights the significant potential and sustainability of microbial technology in environmental management.

4.2.4. Precipitation

Some microorganisms also facilitate the formation of minerals such as iron oxides, promoting the co-precipitation or adsorption of arsenic. Microbial remediation of arsenic pollution through precipitation mechanisms involves complex biogeochemical processes [84]. During their growth and metabolic activities, microorganisms not only alter the chemical conditions of their environment, such as pH and redox potential, but also directly participate in the transformation of arsenic's chemical forms and the formation of minerals. Specifically, certain microorganisms can secrete polyphosphates (PolyP) or other inorganic compounds, inducing the precipitation of minerals like iron oxides or sulfides [85]. These minerals exhibit a high affinity for arsenic ions in the environment, forming insoluble compounds such as iron arsenate ($FeAsO_4$) or iron arsenic sulfide (FeAsS). This process significantly reduces the solubility of arsenic, effectively immobilizing it in soil or water. Additionally, microorganisms can influence arsenic's chemical state by regulating environmental pH or producing reducing or oxidizing compounds, further enhancing arsenic precipitation [86]. This microbially mediated arsenic precipitation process not only decreases arsenic's bioavailability and environmental mobility but also provides an eco-friendly and cost-effective strategy for the long-term stabilization of arsenic pollution.

4.2.5. Microbe–Adsorbent Synergy

In arsenic remediation, the synergistic interaction between microorganisms and adsorbents represents an efficient strategy, relying on the interplay of microbial biochemical capabilities and the physicochemical properties of the adsorbents [87]. Adsorbents such as biochar and metal oxides inherently possess high specific surface areas, enabling effective arsenic adsorption. When combined with microorganisms, such as iron oxide-modified biochar, the presence of microbes can enhance adsorption by forming new binding sites or altering the surface chemistry of the adsorbent through biomineralization processes. In one study, iron-magnetic biochar modified with *Bacillus* sp. K1 demonstrated significant synergistic effects in improving arsenic removal efficiency [88]. This composite material not only increased the removal rate of Cd(II) but also expanded active sites for As(III) adsorption by forming type B ternary surface complexes on the MBB surface. This microorganism-adsorbent system retains high efficiency under a wide range of environmental conditions, including varying pH and temperature. Moreover, the involvement of microorganisms can aid in the regeneration and recycling of the adsorbent, extending its lifespan and reducing treatment costs [89]. This integrated, efficient, and environmentally friendly solution offers significant potential for arsenic remediation, showcasing the promising intersection of biotechnology and materials science.

4.2.6. Microbe–Plant Synergy

In the field of arsenic bioremediation, the synergistic interaction between plants and microorganisms offers an efficient and sustainable remediation strategy. For example, the fern *Pteris vittata* has been extensively studied for its exceptional arsenic accumulation capability [90]. These plants absorb arsenic from the soil through their root systems and translocate it to their aerial parts, primarily accumulating it in the leaves, effectively removing arsenic from the soil. Rhizosphere microorganisms, on the other hand, influence the chemical state of arsenic through their biochemical activities, reducing its potential hazards to the environment and living organisms. Additionally, these microorganisms secrete organic acids and other metabolic products that alter soil pH, thereby increasing arsenic bioavailability, making it more accessible for plant uptake. Rhizosphere microorganisms also promote root health and growth, thereby enhancing the plant's arsenic absorption and accumulation efficiency. For instance, one study found that rhizosphere microorganisms increased the concentration of As(V) in the rhizosphere through their metabolic activities, enabling *Pteris vittata* to absorb arsenic more effectively [91]. This plant–microbe interaction not only boosts the plant's arsenic removal capability but also provides an effective solution for the ecological restoration of arsenic-contaminated soils.

5. Microbial Metal Resistance Genes

Microbial metal resistance genes (MRGs) refer to the genes that confer microorganisms the ability to resist, tolerate, or transform heavy metal ions present in the environment [92]. These genes are typically located either on the microbial chromosome or on mobile genetic elements such as plasmids and transposons, which can facilitate horizontal gene transfer. The presence of MRGs enables microorganisms to survive and thrive in heavy metal-contaminated environments, and these genes exhibit high adaptability and transmissibility within microbial communities [93]. The involvement of metal resistance genes in heavy metal pollution is shown in Figure 3.

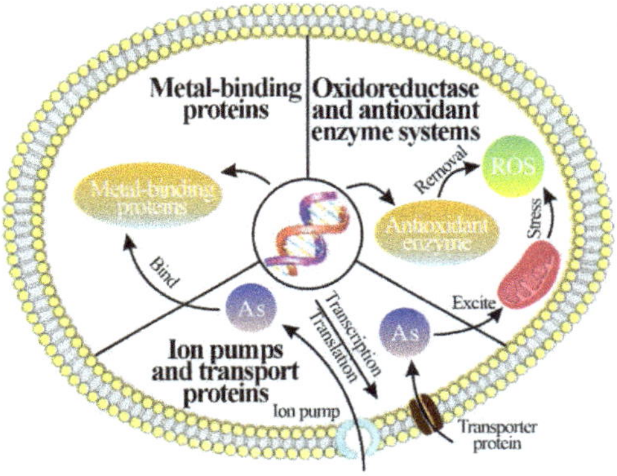

Figure 3. Involvement of metal resistance genes in heavy metal pollution.

5.1. Ion Pumps and Transport Proteins

Microbial metal resistance genes (MRGs) typically encode ion pumps and transport proteins, which play crucial roles in maintaining the balance of metal ions inside and outside the cell. These proteins enable microorganisms to survive in environments containing heavy metals. Ion pumps, such as ATPases, utilize energy released from ATP hydrolysis to drive metal ions across cell membranes [94]. Some pumps expel toxic metal ions like copper and silver from the cell to reduce toxicity, while others import essential nutrients like iron and zinc. Additionally, proton pumps are a specialized type of ion pump that expel protons to help maintain transmembrane pH gradients and electrical potential, which are vital for energy production and intracellular regulation. A study highlighted the use of genetic engineering to enhance the copper and lead ion pump functions of certain bacteria. This technology has been applied in heavily polluted mining areas, where modified indigenous microorganisms effectively reduced toxic metal concentrations in the environment by enhancing their ion-removal capabilities [95].

Transport proteins mediate the movement of metal ions through various mechanisms, including passive channel proteins and carrier proteins, as well as energy-dependent active transport systems. Channel proteins allow ions to pass through cell membranes following their concentration gradients, while carrier proteins facilitate ion movement by physically binding to ions and undergoing conformational changes to transport them across the membrane [96]. In active transport, symporters and antiporters play crucial roles; symporters co-transport ions along with other molecules, while antiporters simultaneously expel one type of ion while importing another. A study highlighted the use of engineered microbial symport systems to treat chromium (VI)-contaminated wastewater. This system successfully reduced toxic chromium (VI) to the less toxic chromium (III), and the technology has been implemented in industrial wastewater treatment plants [97].

The expression and activity of these ion pumps and transport proteins are tightly regulated by intracellular and extracellular metal ion concentrations, enabling microorganisms to respond to changes in environmental metal levels. Under heavy metal stress, the expression of MRGs is upregulated to reduce the accumulation of toxic metal ions within the cell [98]. This fine-tuned regulation allows microorganisms to survive and thrive in environments with varying metal ion concentrations. Additionally, these mechanisms provide effective biological strategies for addressing and remediating heavy metal pollution in the fields of environmental biotechnology and bioremediation.

5.2. Metal-Binding Proteins

Metalloproteins and sulfur proteins are two key biomolecules that play crucial physiological roles in microorganisms and other organisms by specifically binding metal ions [99].

Metal-binding proteins, through their unique amino acid sequences, contain residues such as cysteine, histidine, aspartic acid, and glutamic acid that can coordinate with metal ions, effectively forming complexes with one or more metal ions [100]. The formation of these complexes helps cells detoxify by reducing the bioavailability and toxicity of metal ions. Additionally, these proteins play a critical role in the storage and transport of essential metal ions within the cell, maintaining the supply of vital metals and participating in signaling processes, such as regulating gene expression and protein activity.

Metallothioneins are a class of low-molecular-weight, cysteine-rich metal-binding proteins characterized by their efficient metal-binding capacity, particularly for copper and zinc ions. These proteins contain 20–30% cysteine which, through their sulfur atoms, form thiol (-SH) coordination bonds with metal ions, creating stable metal-metallothionein complexes [101]. This binding prevents metal ions from interfering with normal cellular functions. Additionally, metallothioneins act as antioxidants, protecting cells from oxidative stress caused by free radicals. Under normal physiological conditions, they are responsible for storing essential trace elements, such as zinc and copper, regulating their bioavailability to ensure proper cellular function.

These proteins play a crucial role in environmental microorganisms, enabling them to adapt to environments with high concentrations of heavy metals, such as polluted water bodies and soils [102]. In environmental biotechnology and bioremediation, enhancing the expression of these proteins in microorganisms through genetic engineering significantly improves their ability to remove specific heavy metals. This provides an effective biological tool for addressing heavy metal pollution, demonstrating substantial potential for applications in ecological restoration and environmental management [103].

5.3. Oxidoreductase and Antioxidant Enzyme Systems

Oxidoreductases and antioxidant enzyme systems play crucial regulatory roles in maintaining redox balance within cells, scavenging free radicals and other reactive oxygen species (ROS), and protecting cells from oxidative stress damage. Oxidoreductases catalyze redox reactions, facilitating electron transfer between molecules, which is vital for cellular energy metabolism, respiration, and photosynthesis [104]. For instance, cytochrome oxidase promotes electron flow in the electron transport chain, aiding ATP synthesis, while peroxidase reduces oxidative stress by converting hydrogen peroxide into water and oxygen. Additionally, NADPH oxidase contributes to immune defense by generating ROS to combat pathogens [105].

The antioxidant enzyme system includes superoxide dismutase (SOD), catalase, and glutathione peroxidase (GPx), which work synergistically to eliminate free radicals and other reactive oxygen species (ROS) within cells. This system protects proteins, lipids, and DNA from oxidative damage. SOD serves as the first line of defense by converting superoxide radicals into hydrogen peroxide, which is then further processed by catalase and GPx. These enzymes ensure the stability of the intracellular environment and maintain the integrity of cellular structures [106].

These systems not only play a role in the normal physiological activities of organisms but also serve as crucial mechanisms for coping with environmental stress. When exposed to environmental stressors such as heavy metal pollution, many microorganisms and plants enhance their defense against oxidative damage by upregulating the expression of antioxidant enzymes [107]. This mechanism enables them to survive and thrive under extreme environmental conditions. For instance, studies have shown that plants subjected to heavy metal stress, such as copper and zinc, increase the activity of superoxide dismutase and catalase to mitigate oxidative stress induced by these metals [108].

Research on redox enzymes and antioxidant enzyme systems provides a deep understanding of how organisms adapt to and cope with various internal and external stresses through these intricate molecular mechanisms. These findings have found wide applications in fields such as medicine, agriculture, and environmental science. For example, genetic engineering can be used to enhance crop resistance to stress, novel antioxidant drugs can be developed to treat related diseases, and the principles of these systems can be applied to design new bioremediation strategies for addressing environmental pollutants.

5.4. Resistance Regulation and Response

In microorganisms and other organisms, resistance regulation and response mechanisms form a complex biomolecular network that enables them to sense and adapt to harmful substances in the environment, such as heavy metals and antibiotics. These mechanisms involve sensor proteins, signal transduction pathways, transcriptional regulators, and gene expression modulation [109]. Sensor proteins, often located on the cell membrane, can directly interact with environmental toxins or detect changes in the intracellular and extracellular environment. Once activated, these sensors trigger a series of signal transduction events, typically involving phosphorylation and dephosphorylation processes, which not only amplify the signal but also effectively transmit it to transcription factors within the cell nucleus [110].

These transcription factors subsequently initiate or enhance the expression of resistance-related genes, which encode proteins such as detoxification enzymes, metal ion pumps, antibiotic efflux pumps, and enzymes that repair damaged DNA, directly engaging in the cell's defense mechanisms [111]. To prevent excessive responses that could lead to excessive energy consumption, cells employ negative feedback mechanisms to regulate the intensity and duration of these responses, maintaining energy balance and physiological stability. For instance, under heavy metal stress, microorganisms utilize specific regulatory systems like the Cus system and the ArsR/SmtB family to control the expression of metal ion pumps and metal-binding proteins, thereby reducing metal toxicity within the cell. Similarly, under antibiotic pressure, they may express specific efflux pumps such as the TetA tetracycline efflux pump or modify antibiotic target sites through β-lactamase enzymes [112].

In addition, organisms respond to temperature and salinity stress by adjusting membrane lipid composition and expressing proteins related to osmotic protection, such as late embryogenesis abundant (LEA) proteins and compatible solutes. These adjustments help maintain membrane fluidity and intracellular osmotic balance. This intricate regulatory network not only ensures survival and reproduction under harsh environmental conditions but also offers valuable biological insights for developing new resistance management strategies in agriculture, medicine, and environmental science [113].

6. Impact of Metal Resistance Genes on Environmental Security

MRGs are widely present in natural environments, especially in areas with severe arsenic pollution, such as mining sites, industrial discharge zones, and high-arsenic groundwater regions, where the abundance of MRGs increases significantly [114]. Microbial communities adapt to arsenic-contaminated environments by carrying these MRGs, enhancing their ability to survive in high-arsenic conditions. Research has shown that the distribution and diversity of MRGs vary significantly among different types of environmental microbiomes [115]. For example, microbial communities in sediments are often rich in arsenic efflux genes (such as arsB), while anaerobic sediments are more likely to harbor genes related to arsenic reduction (such as arsC). In these environments, the synergistic interactions of MRGs among different microbial groups allow the entire microbiome to maintain ecological balance in arsenic-contaminated settings [116].

6.1. Mechanisms of Association Between Metal Resistance Genes and Antibiotic Resistance Genes

MRGs and antibiotic resistance genes (ARGs) are often co-located on the same mobile genetic elements, such as plasmids, transposons, or integrons. This co-location facilitates the simultaneous horizontal transfer of both types of resistance genes, enabling a single genetic event to confer resistance to both metals and antibiotics in a microorganism [117]. This phenomenon is widespread in the environment, particularly in hospital wastewater, agricultural soils, and other environments heavily influenced by human activities.

6.1.1. Co-Localization and Horizontal Gene Transfer

In the fields of environmental microbiology and public health, the co-location and horizontal gene transfer (HGT) of microbial MRGs and ARGs have become increasingly concerning phenomena. MRGs within environmental microbiomes are not confined to individual microbial genomes; they can spread between different microbial populations through HGT [118]. This exacerbates the proliferation of resistance genes in the environment and poses significant potential risks to human health. MRGs and ARGs often coexist on mobile genetic elements such as plasmids and transposons, facilitating their rapid dissemination within microbial communities. HGT mechanisms, including transduction, transformation, and conjugation, enable these resistance genes to spread swiftly, particularly in environments where both antibiotics and heavy metals are present [119]. Such environments, including hospital wastewater and agricultural soils, provide selective pressures that promote the retention and spread of resistance genes. The widespread dissemination of resistance genes alters the structure and function of environmental microbial communities, complicates and increases the cost of antibiotic therapy, and can lead to the transmission of resistance through environmental pathways, such as water sources and food chains, ultimately heightening the challenge of treating medical infections.

6.1.2. Synergistic Effects of Selection Pressure

In environmental microbiology, the association between microbial MRGs and ARGs is particularly noteworthy, especially regarding the synergistic effects of selective pressures. The widespread presence of both antibiotics and heavy metals in the environment imposes not only independent selective pressures but also facilitates the co-location and horizontal gene transfer of these resistance genes within microbial communities [120]. This accelerates the spread of resistance traits, creating a stable resistance gene pool within environmental microbes. Such a gene pool enhances the resilience of environmental microbes to multiple antibiotics. These resistance genes can transfer to human pathogenic microorganisms through various pathways, reducing the effectiveness of antibiotics in clinical settings and increasing the difficulty of treating related infections [121]. This poses a serious threat to public health. Moreover, the rise in antibiotic resistance implies higher healthcare costs and greater treatment challenges, especially as antibiotic resistance has become a global health crisis.

6.2. Ecological and Health Impacts of MRGs

In the intersection of environmental microbiology and public health, the link between microbial MRGs and heavy metal pollution presents a complex challenge to environmental safety. These genes enable microorganisms to survive and thrive in heavy metal-contaminated environments by encoding metal ion pumps, binding proteins, or enzymes that detoxify metal ions [122]. As arsenic heavy metal concentrations in the environment increase, the expression of MRGs within microbial communities rises in response to this survival pressure. This not only alters the composition of native microbial communities by promoting the growth of metal-resistant populations but also facilitates the horizontal transfer of resistance genes to other microorganisms, thereby expanding the size and diversity of the environmental resistome.

The widespread presence and active expression of MRGs in the environment have direct impacts on ecosystem health and functionality. They may disrupt natural nutrient cycling and organic matter decomposition processes, ultimately affecting ecological balance [92]. Additionally, the environmental dissemination of MRGs poses an indirect risk to human health, particularly when these resistance genes enter the human body through the food chain or water resources, potentially rendering conventional treatments for certain infections ineffective. To manage MRGs and heavy metal pollution in the environment, stringent environmental monitoring and effective pollution control technologies are essential. Furthermore, public education and policy support are crucial to reducing pollution sources and limiting the spread of resistance genes [123]. By implementing these comprehensive measures, the risks associated with environmental resistance genes can be mitigated, safeguarding both environmental and public health.

7. Current Research Trends and Future Directions

7.1. Advanced Molecular Techniques

Advancements in molecular techniques, such as metagenomics and metatranscriptomics, have revolutionized the study of microbial arsenic resistance [124]. First, precise manipulation of microorganisms containing MRGs (microbial resistance genes) using gene-editing technologies like CRISPR-Cas aims to enhance their resistance and transformation capabilities for arsenic. Additionally, transcriptomic and proteomic analyses are employed to study how arsenic regulates MRG expression and microbial response mechanisms in depth. Furthermore, the development of biosensors based on MRGs can enable rapid and accurate on-site monitoring of arsenic pollution, improving the efficiency and real-time nature of environmental monitoring. In the field of ecotoxicology, combining laboratory simulations with field data helps investigate the long-term ecological impacts of arsenic on microbial communities containing MRGs and explore the transmission and evolution patterns of these genes in natural environments. Finally, applying computational biology and systems biology approaches allows for the creation of digital models to predict arsenic bioavailability under various environmental conditions and changes in microbial communities, providing scientific decision support for environmental management and pollution remediation. Through the integrated application of these advanced technologies, not only can the understanding of arsenic pollution mechanisms be enhanced, but environmental science can also advance towards more efficient and precise monitoring and remediation solutions.

7.2. Environmental Monitoring

Effective environmental monitoring is essential for assessing the extent and impact of arsenic contamination in sludge and sediment [125]. The development of real-time monitoring technologies, particularly the application of online sensors and automated monitoring stations, has greatly improved the timeliness and accuracy of arsenic pollution monitoring. These systems can provide real-time updates on the spatial distribution of arsenic and, when integrated with a geographic information system (GIS), offer detailed mapping of the dynamic spread of arsenic pollution. In addition, by using PCR and high-throughput sequencing technologies to specifically monitor MRGs (microbial resistance genes) in the environment, it is possible to gain detailed insights into the expression and distribution of resistance genes under varying environmental conditions. Moreover, the use of metagenomics as a bioindicator tool not only allows for the assessment of arsenic bioavailability but also monitors its impact on microbial community structure and function, further evaluating the long-term ecological effects on the environment. By constructing an intelligent monitoring network that transmits and analyzes data from multiple monitoring points via the internet of things (IoT), the dynamic distribution and spread patterns of arsenic and MRGs over large areas can be tracked in real time. Lastly, developing computational models and predictive tools that integrate environmental and biological data can provide deeper insights into the dynamics of arsenic pollution and help predict the long-term effects of different management measures on the spread of arsenic and MRGs. Such predictive modeling is critical for developing effective environmental management strategies and control measures. Through the comprehensive application of these advanced monitoring technologies and strategies, the understanding and management of arsenic pollution and its ecological impacts can be significantly enhanced.

7.3. Remediation Strategies

Developing effective remediation strategies for arsenic-contaminated sites is a critical challenge. Understanding the role of microorganisms and MRGs in arsenic biogeochemistry is essential for designing targeted and sustainable remediation approaches [126]. First, by combining bioremediation and chemical remediation methods, genetically engineered microorganisms with specific MRGs (microbial resistance genes) can be used to enhance arsenic bioavailability, thereby improving the efficiency of traditional chemical treatment methods such as precipitation and adsorption. At the same time, ecological engineering applications based on MRGs can be developed to optimize microbial communities containing these genes, thereby enhancing the self-purification ability of polluted environments through ecological engineering techniques. Additionally, high-throughput screening technologies can be employed to select microorganisms from natural environments or construct genetically engineered strains that can efficiently treat arsenic, providing targeted bioremediation solutions for complex polluted environments. Lastly, advanced monitoring tools, such as metagenomics and bioinformatics technologies, can be developed and applied to track the spread and changes of MRGs in the environment, assess their impact on the remediation process, and adjust remediation strategies in real time to prevent the adverse spread of resistance genes. By implementing these strategies, not only can the efficiency of arsenic pollution remediation be significantly improved, but the long-term stability and health of ecosystems can also be ensured.

7.4. Risk Assessment and Policy

Assessing the risks associated with microbial arsenic resistance and MRGs in contaminated environments is essential for informed decision-making and policy development. Integrating scientific research into risk assessment frameworks can help prioritize management strategies and regulatory actions. Policymakers must consider the potential long-term consequences of microbial arsenic resistance for environmental and human health when formulating regulations and guidelines. First, developing integrated risk assessment models for arsenic and MRGs (microbial resistance genes) is crucial. These models should predict the behavior of arsenic in the environment as well as the spread and impact of MRGs, guiding the formulation of risk management and monitoring strategies. Using geographic information systems (GISs) and ecological models for risk mapping can effectively identify high-risk polluted areas and implement appropriate prevention and control measures. Additionally, establishing scientific monitoring policies is essential to ensure systematic and regular monitoring of arsenic and MRGs, while advanced biotechnologies such as high-throughput sequencing and real-time PCR can improve monitoring accuracy and efficiency. Risk communication and public participation are equally indispensable; through education and information transparency, the public's awareness and ability to respond to the risks of arsenic and MRGs can be enhanced. Lastly, updating and improving relevant environmental regulations and policies—particularly emission and treatment standards—should reflect the latest scientific research and technological advancements, while promoting international cooperation to develop unified cross-border policies and enforcement standards. By implementing these comprehensive strategies, not only can the environmental and health risks posed by arsenic and MRGs be effectively managed, but sustainable environmental management practices can also be promoted.

8. Conclusions

Arsenic contamination in sludge and sediment is a complex and multifaceted environmental issue with global implications. Microorganisms and their associated resistance genes play a central role in the biogeochemical cycling of arsenic, influencing its mobility, toxicity, and environmental fate. Understanding the genetic basis of microbial arsenic resistance, its evolution, and its ecological and health implications is essential for effective mitigation and remediation efforts. This review has provided a comprehensive overview of the sources and distribution of arsenic contamination, the mechanisms of microbial arsenic resistance, and the environmental and health consequences of microbial arsenic resistance. It has also highlighted current research trends and areas requiring further investigation. By advancing our understanding of the interplay between arsenic contamination, microorganisms, and MRGs, we can develop more informed strategies for protecting our environment and safeguarding human health in the face of this pressing global challenge.

Author Contributions: Conceptualization, M.X. and D.Y.; writing—original draft preparation, M.X. and D.Y.; writing—review and editing, M.H., Y.Z., Z.Z. and F.L. All authors have read and agreed to the published version of the manuscript.

Funding: The authors are grateful for the support from Key Research and Development Program of Shandong Province (2023TSGC0430), Key Research and Development Program of Shandong Province (2022SFGC0302), Hainan Provincial Natural Science Foundation of China (423CXTD384), Natural Science Foundation of Shandong Province (ZR2020ME222), National Natural Science Foundation of China (52070122), Key Research and Development Project of Shandong (No. 2020CXGCO11404) and Innovation Fund for Jinan High Education's 20 policies (202228056), Taishan Scholars Project.

Data Availability Statement: No new data were created or analyzed in this study. Data sharing is not applicable to this article.

Acknowledgments: The authors would like to thank all the anonymous referees for their constructive comments and suggestions.

Conflicts of Interest: The authors have no competing interests to declare that are relevant to the content of this article.

References

1. Kumar, M.; Bolan, N.S.; Hoang, S.A.; Sawarkar, A.D.; Jasemizad, T.; Gao, B.; Keerthanan, S.; Padhye, L.P.; Singh, L.; Kumar, S.; et al. Remediation of soils and sediments polluted with polycyclic aromatic hydrocarbons: To immobilize, mobilize, or degrade? *J. Hazard. Mater.* **2021**, *420*, 126534. [CrossRef] [PubMed]
2. Chen, R.; Zhang, Q.; Chen, H.; Yue, W.; Teng, Y. Source apportionment of heavy metals in sediments and soils in an interconnected river-soil system based on a composite fingerprint screening approach. *J. Hazard. Mater.* **2021**, *411*, 125125. [CrossRef] [PubMed]
3. ur Rehman, I.; Ishaq, M.; Muhammad, S.; Din, I.U.; Khan, S.; Yaseen, M. Evaluation of arsenic contamination and potential risks assessment through water, soil and rice consumption. *Environ. Technol. Innov.* **2020**, *20*, 101155. [CrossRef]
4. Amarasekara, N.R.; Mafiz, A.I.; Qian, X.; Tiedje, J.M.; Hao, W.; Zhang, Y. Exploring the co-occurrence of antibiotic, metal, and biocide resistance genes in the urban agricultural environment. *J. Agric. Food Res.* **2023**, *11*, 100474. [CrossRef]
5. Raju, N.J. Arsenic in the geo-environment: A review of sources, geochemical processes, toxicity and removal technologies. *Environ. Res.* **2022**, *203*, 111782. [CrossRef] [PubMed]
6. Khan, S.S.; Flora, S.J.S. Arsenic: Chemistry, occurrence, and exposure. In *Handbook of Arsenic Toxicology*; Elsevier: Amsterdam, The Netherlands, 2023; pp. 1–49.
7. Han, L. Arsenic pollution of sediments in China: An assessment by geochemical baseline. *Sci. Total Environ.* **2019**, *651*, 1983–1991. [CrossRef] [PubMed]
8. Cai, Y.; He, Y.; Wang, G.; Li, Z.; Pei, Y.; Yang, W.; Rao, W.; Li, J. Machine learning models in the process of metal organic chemical vapor deposition epitaxial manufacturing of Gallium Arsenide. *Mater. Today Commun.* **2024**, *40*, 109808. [CrossRef]
9. Sun, J.; Wu, L.; Wu, M.; Liu, Q.; Cao, H. Non-coding RNA therapeutics: Towards a new candidate for arsenic-induced liver disease. *Chem. Interact.* **2023**, *382*, 110626. [CrossRef]
10. Cann, J.L.; De Luca, A.; Dunand, D.C.; Dye, D.; Miracle, D.B.; Oh, H.S.; Olivetti, E.A.; Pollock, T.M.; Poole, W.J.; Yang, R.; et al. Sustainability through alloy design: Challenges and opportunities. *Prog. Mater. Sci.* **2021**, *117*, 100722. [CrossRef]
11. Amen, R.; Bashir, H.; Bibi, I.; Shaheen, S.M.; Niazi, N.K.; Shahid, M.; Hussain, M.M.; Antoniadis, V.; Shakoor, M.B.; Al-Solaimani, S.G.; et al. A critical review on arsenic removal from water using biochar-based sorbents: The significance of modification and redox reactions. *Chem. Eng. J.* **2020**, *396*, 125195. [CrossRef]

12. Huq, M.E.; Fahad, S.; Shao, Z.; Sarven, M.S.; Khan, I.A.; Alam, M.; Saeed, M.; Ullah, H.; Adnan, M.; Saud, S.; et al. Arsenic in a groundwater environment in Bangladesh: Occurrence and mobilization. *J. Environ. Manag.* **2020**, *262*, 110318. [CrossRef]
13. Tabelin, C.B.; Corpuz, R.D.; Igarashi, T.; Villacorte-Tabelin, M.; Alorro, R.D.; Yoo, K.; Raval, S.; Ito, M.; Hiroyoshi, N. Acid mine drainage formation and arsenic mobility under strongly acidic conditions: Importance of soluble phases, iron oxyhydroxides/oxides and nature of oxidation layer on pyrite. *J. Hazard. Mater.* **2020**, *399*, 122844. [CrossRef]
14. Aftabtalab, A.; Rinklebe, J.; Shaheen, S.M.; Niazi, N.K.; Moreno-Jiménez, E.; Schaller, J.; Knorr, K.-H. Review on the interactions of arsenic, iron (oxy)(hydr)oxides, and dissolved organic matter in soils, sediments, and groundwater in a ternary system. *Chemosphere* **2022**, *286*, 131790. [CrossRef] [PubMed]
15. Byeon, E.; Kang, H.-M.; Yoon, C.; Lee, J.-S. Toxicity mechanisms of arsenic compounds in aquatic organisms. *Aquat. Toxicol.* **2021**, *237*, 105901. [CrossRef] [PubMed]
16. Hare, V.; Chowdhary, P.; Boudh, S.; Singh, A.K.; Mani, S.; Kumar, A. Effect of arsenic on human health and its removal through physiochemical techniques. In *Contaminants and Clean Technologies*; CRC Press: Boca Raton, FL, USA, 2020; pp. 93–110.
17. Jain, N.; Maiti, A. Fe-Mn-Al metal oxides/oxyhydroxides as As(III) oxidant under visible light and adsorption of total arsenic in the groundwater environment. *Sep. Purif. Technol.* **2022**, *302*, 122170. [CrossRef]
18. Schreiber, M.E.; Cozzarelli, I.M. Arsenic release to the environment from hydrocarbon production, storage, transportation, use and waste management. *J. Hazard. Mater.* **2021**, *411*, 125013. [CrossRef]
19. Wang, J.; Li, Z.; Zhu, Q.; Wang, C.; Tang, X. Review on arsenic environment behaviors in aqueous solution and soil. *Chemosphere* **2023**, *333*, 138869. [CrossRef] [PubMed]
20. Huang, J.; Jones, A.; Waite, T.D.; Chen, Y.; Huang, X.; Rosso, K.M.; Kappler, A.; Mansor, M.; Tratnyek, P.G.; Zhang, H. Fe(II) Redox Chemistry in the Environment. *Chem. Rev.* **2021**, *121*, 8161–8233. [CrossRef] [PubMed]
21. Moreira, V.; Lebron, Y.; Santos, L.; De Paula, E.C.; Amaral, M. Arsenic contamination, effects and remediation techniques: A special look onto membrane separation processes. *Process Saf. Environ. Prot.* **2021**, *148*, 604–623. [CrossRef]
22. Borah, P.; Kumar, M.; Devi, P. Types of inorganic pollutants: Metals/metalloids, acids, and organic forms. In *Inorganic Pollutants in Water*; Elsevier: Amsterdam, The Netherlands, 2020; pp. 17–31.
23. Su, Q.; He, Y.; Pan, H.; Liu, H.; Mehmood, K.; Tang, Z.; Hu, L. Toxicity of inorganic arsenic to animals and its treatment strategies. *Comp. Biochem. Physiol. Part C Toxicol. Pharmacol.* **2023**, *271*, 109654. [CrossRef] [PubMed]
24. Garkal, A.; Sarode, L.; Bangar, P.; Mehta, T.; Singh, D.P.; Rawal, R. Understanding arsenic toxicity: Implications for environmental exposure and human health. *J. Hazard. Mater. Lett.* **2023**, *5*, 100090.
25. Tsai, T.-L.; Kuo, C.-C.; Hsu, L.-I.; Tsai, S.-F.; Chiou, H.-Y.; Chen, C.-J.; Hsu, K.-H.; Wang, S.-L. Association between arsenic exposure, DNA damage, and urological cancers incidence: A long-term follow-up study of residents in an arseniasis endemic area of northeastern Taiwan. *Chemosphere* **2021**, *266*, 129094. [CrossRef] [PubMed]
26. Tagliabue, M.; Bagatin, R.; Conte, A.; Congiu, A.; Perucchini, S.; Zanardi, S.; Bellettato, M.; Carati, A. Metal-rich sludge from mine water treatment: From waste to effective arsenate adsorbent. *Desalin. Water Treat.* **2017**, *69*, 294–301. [CrossRef]
27. Osuna-Martínez, C.C.; Armienta, M.A.; Bergés-Tiznado, M.E.; Páez-Osuna, F. Arsenic in waters, soils, sediments, and biota from Mexico: An environmental review. *Sci. Total Environ.* **2021**, *752*, 142062. [CrossRef] [PubMed]
28. Darma, A.; Yang, J.; Bloem, E.; Możdżen, K.; Zandi, P. Arsenic biotransformation and mobilization: The role of bacterial strains and other environmental variables. *Environ. Sci. Pollut. Res.* **2021**, *29*, 1763–1787. [CrossRef]
29. Rokonuzzaman, M.; Li, W.C.; Wu, C.; Ye, Z. Human health impact due to arsenic contaminated rice and vegetables consumption in naturally arsenic endemic regions. *Environ. Pollut.* **2022**, *308*, 119712. [CrossRef]
30. Zhang, J.; Hamza, A.; Xie, Z.; Hussain, S.; Brestic, M.; Tahir, M.A.; Ulhassan, Z.; Yu, M.; Allakhverdiev, S.I.; Shabala, S. Arsenic transport and interaction with plant metabolism: Clues for improving agricultural productivity and food safety. *Environ. Pollut.* **2021**, *290*, 117987. [CrossRef]
31. Bali, A.S.; Sidhu, G.P.S. Arsenic acquisition, toxicity and tolerance in plants-From physiology to remediation: A review. *Chemosphere* **2021**, *283*, 131050. [CrossRef]
32. Zulfiqar, F.; Ashraf, M. Antioxidants as modulators of arsenic-induced oxidative stress tolerance in plants: An overview. *J. Hazard. Mater.* **2022**, *427*, 127891. [CrossRef]
33. Hussain, M.M.; Wang, J.; Bibi, I.; Shahid, M.; Niazi, N.K.; Iqbal, J.; Mian, I.A.; Shaheen, S.M.; Bashir, S.; Shah, N.S.; et al. Arsenic speciation and biotransformation pathways in the aquatic ecosystem: The significance of algae. *J. Hazard. Mater.* **2021**, *403*, 124027. [CrossRef] [PubMed]
34. Rahaman, M.S.; Rahman, M.M.; Mise, N.; Sikder, M.T.; Ichihara, G.; Uddin, M.K.; Kurasaki, M.; Ichihara, S. Environmental arsenic exposure and its contribution to human diseases, toxicity mechanism and management. *Environ. Pollut.* **2021**, *289*, 117940. [CrossRef]
35. Zhang, X.; Fan, H.; Yuan, J.; Tian, J.; Wang, Y.; Lu, C.; Han, H.; Sun, W. The application and mechanism of iron sulfides in arsenic removal from water and wastewater: A critical review. *J. Environ. Chem. Eng.* **2022**, *10*, 108856. [CrossRef]
36. Morales-Simfors, N.; Bundschuh, J.; Herath, I.; Inguaggiato, C.; Caselli, A.T.; Tapia, J.; Choquehuayta, F.E.A.; Armienta, M.A.; Ormachea, M.; Joseph, E.; et al. Arsenic in Latin America: A critical overview on the geochemistry of arsenic originating from geothermal features and volcanic emissions for solving its environmental consequences. *Sci. Total Environ.* **2020**, *716*, 135564. [CrossRef] [PubMed]

37. Baba, A.; Uzelli, T.; Sozbilir, H. Distribution of geothermal arsenic in relation to geothermal play types: A global review and case study from the Anatolian plate (Turkey). *J. Hazard. Mater.* **2021**, *414*, 125510. [CrossRef] [PubMed]
38. Shaji, E.; Santosh, M.; Sarath, K.; Prakash, P.; Deepchand, V.; Divya, B. Arsenic contamination of groundwater: A global synopsis with focus on the Indian Peninsula. *Geosci. Front.* **2021**, *12*, 101079. [CrossRef]
39. Li, X.; Xiao, J.; Chai, N.; Jin, Z.J. Controlling mechanism and water quality assessment of arsenic in China's Yellow River Basin. *J. Clean. Prod.* **2023**, *418*, 137953. [CrossRef]
40. Morales-Simfors, N.; Bundschuh, J. Arsenic-rich geothermal fluids as environmentally hazardous materials—A global assessment. *Sci. Total Environ.* **2022**, *817*, 152669. [CrossRef]
41. Wang, Z.; Guo, H.-M.; Liu, H.-Y.; Zhang, W.-M. Source, migration, distribution, toxicological effects and remediation technologies of arsenic in groundwater in China. *China Geol.* **2023**, *6*, 476–493.
42. Alam, M.A.; Mukherjee, A.; Bhattacharya, P.; Bundschuh, J. An appraisal of the principal concerns and controlling factors for Arsenic contamination in Chile. *Sci. Rep.* **2023**, *13*, 11168. [CrossRef]
43. Benis, K.Z.; Damuchali, A.M.; Soltan, J.; McPhedran, K.N. Treatment of aqueous arsenic—A review of biochar modification methods. *Sci. Total Environ.* **2020**, *739*, 139750. [CrossRef] [PubMed]
44. Moulick, D.; Samanta, S.; Sarkar, S.; Mukherjee, A.; Pattnaik, B.K.; Saha, S.; Awasthi, J.P.; Bhowmick, S.; Ghosh, D.; Samal, A.C.; et al. Arsenic contamination, impact and mitigation strategies in rice agro-environment: An inclusive insight. *Sci. Total Environ.* **2021**, *800*, 149477. [CrossRef]
45. Bundschuh, J.; Schneider, J.; Alam, M.A.; Niazi, N.K.; Herath, I.; Parvez, F.; Tomaszewska, B.; Guilherme, L.R.G.; Maity, J.P.; López, D.L.; et al. Seven potential sources of arsenic pollution in Latin America and their environmental and health impacts. *Sci. Total Environ.* **2021**, *780*, 146274. [CrossRef]
46. Rezende-Teixeira, P.; Dusi, R.G.; Jimenez, P.C.; Espindola, L.S.; Costa-Lotufo, L.V. What can we learn from commercial insecticides? Efficacy, toxicity, environmental impacts, and future developments. *Environ. Pollut.* **2022**, *300*, 118983. [CrossRef] [PubMed]
47. Rokonuzzaman, M.; Ye, Z.; Wu, C.; Li, W. Arsenic accumulation in rice: Alternative irrigation regimes produce rice safe from arsenic contamination. *Environ. Pollut.* **2022**, *310*, 119829. [CrossRef] [PubMed]
48. Genchi, G.; Lauria, G.; Catalano, A.; Carocci, A.; Sinicropi, M.S. Arsenic: A Review on a Great Health Issue Worldwide. *Appl. Sci.* **2022**, *12*, 6184. [CrossRef]
49. Chen, Y.; Zhu, S.; Taskinen, P.; Peng, N.; Peng, B.; Jokilaakso, A.; Liu, H.; Liang, Y.; Zhao, Z.; Wang, Z. Treatment of high-arsenic copper smelting flue dust with high copper sulfate: Arsenic separation by low temperature roasting. *Miner. Eng.* **2021**, *164*, 106796. [CrossRef]
50. Bose-O'Reilly, S.; Landrigan, P.J. Metal toxicology in low-income and lower-middle-income countries. In *Handbook on the Toxicology of Metals*; Elsevier: Amsterdam, The Netherlands, 2022; pp. 705–729.
51. Yadav, M.K.; Saidulu, D.; Gupta, A.K.; Ghosal, P.S.; Mukherjee, A. Status and management of arsenic pollution in groundwater: A comprehensive appraisal of recent global scenario, human health impacts, sustainable field-scale treatment technologies. *J. Environ. Chem. Eng.* **2021**, *9*, 105203. [CrossRef]
52. Alcaine, A.A.; Schulz, C.; Bundschuh, J.; Jacks, G.; Thunvik, R.; Gustafsson, J.-P.; Mörth, C.-M.; Sracek, O.; Ahmad, A.; Bhattacharya, P. Hydrogeochemical controls on the mobility of arsenic, fluoride and other geogenic co-contaminants in the shallow aquifers of northeastern La Pampa Province in Argentina. *Sci. Total Environ.* **2020**, *715*, 136671. [CrossRef]
53. Spaur, M.; Lombard, M.A.; Ayotte, J.D.; Harvey, D.E.; Bostick, B.C.; Chillrud, S.N.; Navas-Acien, A.; Nigra, A.E. Associations between private well water and community water supply arsenic concentrations in the conterminous United States. *Sci. Total Environ.* **2021**, *787*, 147555. [CrossRef] [PubMed]
54. Alarcón-Herrera, M.T.; Martin-Alarcon, D.A.; Gutiérrez, M.; Reynoso-Cuevas, L.; Martín-Domínguez, A.; Olmos-Márquez, M.A.; Bundschuh, J. Co-occurrence, possible origin, and health-risk assessment of arsenic and fluoride in drinking water sources in Mexico: Geographical data visualization. *Sci. Total Environ.* **2020**, *698*, 134168. [CrossRef]
55. Sevak, P.; Pushkar, B. Arsenic pollution cycle, toxicity and sustainable remediation technologies: A comprehensive review and bibliometric analysis. *J. Environ. Manag.* **2024**, *349*, 119504. [CrossRef] [PubMed]
56. Bundschuh, J.; Niazi, N.K.; Alam, M.A.; Berg, M.; Herath, I.; Tomaszewska, B.; Maity, J.P.; Ok, Y.S. Global arsenic dilemma and sustainability. *J. Hazard. Mater.* **2022**, *436*, 129197. [CrossRef]
57. Alarcón-Herrera, M.T.; Gutiérrez, M. Geogenic arsenic in groundwater: Challenges, gaps, and future directions. *Curr. Opin. Environ. Sci. Health* **2022**, *27*, 100349. [CrossRef]
58. Bianchini, G.; Brombin, V.; Marchina, C.; Natali, C.; Godebo, T.R.; Rasini, A.; Salani, G.M. Origin of Fluoride and Arsenic in the Main Ethiopian Rift Waters. *Minerals* **2020**, *10*, 453. [CrossRef]
59. Owusu, C.; Ofori, A.; Adusei-Mensah, F.; Dodoo, D.K.; Essumang, D.K. Water quality, physicochemical, heavy metal content, and health risk assessment of borehole-water from selected mining communities in Ghana. *Groundw. Sustain. Dev.* **2024**, *26*, 101230. [CrossRef]
60. Balasooriya, B.K.; Rajapakse, J.; Gallage, C. A review of drinking water quality issues in remote and indigenous communities in rich nations with special emphasis on Australia. *Sci. Total Environ.* **2023**, *903*, 166559. [CrossRef] [PubMed]
61. Medunić, G.; Fiket, Ž.; Ivanić, M. Arsenic contamination status in Europe, Australia, and other parts of the world. In *Arsenic in Drinking Water and Food*; Springer: Berlin/Heidelberg, Germany, 2020; pp. 183–233.

62. Uppal, J.S.; Zheng, Q.; Le, X.C. Arsenic in drinking water—Recent examples and updates from Southeast Asia. *Curr. Opin. Environ. Sci. Health* **2019**, *7*, 126–135. [CrossRef]
63. Hussain, M.M.; Bibi, I.; Niazi, N.K.; Shahid, M.; Iqbal, J.; Shakoor, M.B.; Ahmad, A.; Shah, N.S.; Bhattacharya, P.; Mao, K.; et al. Arsenic biogeochemical cycling in paddy soil-rice system: Interaction with various factors, amendments and mineral nutrients. *Sci. Total Environ.* **2021**, *773*, 145040. [CrossRef]
64. Shi, W.; Song, W.; Zheng, J.; Luo, Y.; Qile, G.; Lü, S.; Lü, X.; Zhou, B.; Lü, C.; He, J. Factors and pathways regulating the release and transformation of arsenic mediated by reduction processes of dissimilated iron and sulfate. *Sci. Total Environ.* **2021**, *768*, 144697. [CrossRef]
65. Anand, V.; Kaur, J.; Srivastava, S.; Bist, V.; Singh, P.; Srivastava, S. Arsenotrophy: A pragmatic approach for arsenic bioremediation. *J. Environ. Chem. Eng.* **2022**, *10*, 107528. [CrossRef]
66. Hare, V.; Chowdhary, P.; Singh, A.K. Arsenic toxicity: Adverse effect and recent advance in microbes mediated bioremediation. In *Microorganisms for Sustainable Environment and Health*; Elsevier: Amsterdam, The Netherlands, 2020; pp. 53–80.
67. Zhao, Y.; Liu, Y.; Cao, S.; Hao, Q.; Liu, C.; Li, Y. Anaerobic oxidation of methane driven by different electron acceptors: A review. *Sci. Total Environ.* **2024**, *946*, 174287. [CrossRef]
68. Liang, D.; Liu, X.; Woodard, T.L.; Holmes, D.E.; Smith, J.A.; Nevin, K.P.; Feng, Y.; Lovley, D.R. Extracellular Electron Exchange Capabilities of *Desulfovibrio ferrophilus* and *Desulfopila corrodens*. *Environ. Sci. Technol.* **2021**, *55*, 16195–16203. [CrossRef]
69. Raturi, G.; Chaudhary, A.; Rana, V.; Mandlik, R.; Sharma, Y.; Barvkar, V.; Salvi, P.; Tripathi, D.K.; Kaur, J.; Deshmukh, R.; et al. Microbial remediation and plant-microbe interaction under arsenic pollution. *Sci. Total Environ.* **2023**, *864*, 160972. [CrossRef] [PubMed]
70. Altowayti, W.A.H.; Almoalemi, H.; Shahir, S.; Othman, N. Comparison of culture-independent and dependent approaches for identification of native arsenic-resistant bacteria and their potential use for arsenic bioremediation. *Ecotoxicol. Environ. Saf.* **2020**, *205*, 111267. [CrossRef] [PubMed]
71. Hao, X.; Zhu, J.; Rensing, C.; Liu, Y.; Gao, S.; Chen, W.; Huang, Q.; Liu, Y.-R. Recent advances in exploring the heavy metal(loid) resistant microbiome. *Comput. Struct. Biotechnol. J.* **2021**, *19*, 94–109. [CrossRef]
72. Mondal, S.; Pramanik, K.; Ghosh, S.K.; Pal, P.; Mondal, T.; Soren, T.; Maiti, T.K. Unraveling the role of plant growth-promoting rhizobacteria in the alleviation of arsenic phytotoxicity: A review. *Microbiol. Res.* **2021**, *250*, 126809. [CrossRef] [PubMed]
73. Ulhassan, Z.; Bhat, J.A.; Zhou, W.; Senan, A.M.; Alam, P.; Ahmad, P. Attenuation mechanisms of arsenic induced toxicity and its accumulation in plants by engineered nanoparticles: A review. *Environ. Pollut.* **2022**, *302*, 119038. [CrossRef] [PubMed]
74. Sevak, P.; Pushkar, B. Bacterial responses towards arsenic toxicity and in-depth analysis of molecular mechanism along with possible on-field application. *J. Environ. Chem. Eng.* **2023**, *11*, 110187. [CrossRef]
75. Feng, F.; Jiang, Y.; Jia, Y.; Lian, X.; Shang, C.; Zhao, M. Exogenous-organic-matter-driven mobilization of groundwater arsenic. *Environ. Sci. Ecotechnol.* **2023**, *15*, 100243. [CrossRef]
76. Nabi, A.; Naeem, M.; Aftab, T.; Khan, M.M.A.; Ahmad, P. A comprehensive review of adaptations in plants under arsenic toxicity: Physiological, metabolic and molecular interventions. *Environ. Pollut.* **2021**, *290*, 118029. [CrossRef] [PubMed]
77. Chaolei, Y.; Qi, L.; Zhaoyang, S.; Zhang, W.; Jiangrong, C.; Zheng, C.; Shuo, N.; Hongwen, S. Chemical oxidation of arsenic in the environment and its application in remediation: A mini review. *Pedosphere* **2023**, *33*, 185–193.
78. Wang, D.; Rensing, C.; Zheng, S. Microbial reduction and resistance to selenium: Mechanisms, applications and prospects. *J. Hazard. Mater.* **2022**, *421*, 126684. [CrossRef] [PubMed]
79. Glodowska, M.; Welte, C.U.; Kurth, J.M. Metabolic potential of anaerobic methane oxidizing archaea for a broad spectrum of electron acceptors. *Adv. Microb. Physiol.* **2022**, *80*, 157–201.
80. Shakya, A.K.; Rajput, P.; Ghosh, P.K. Investigation on stability and leaching characteristics of mixtures of biogenic arsenosulphides and iron sulphides formed under reduced conditions. *J. Hazard. Mater.* **2018**, *353*, 320–328. [CrossRef] [PubMed]
81. Behera, S.; Das, S. Potential and prospects of Actinobacteria in the bioremediation of environmental pollutants: Cellular mechanisms and genetic regulations. *Microbiol. Res.* **2023**, *273*, 127399. [CrossRef]
82. Gong, Y.-Z.; Niu, Q.-Y.; Liu, Y.-G.; Dong, J.; Xia, M.-M. Development of multifarious carrier materials and impact conditions of immobilised microbial technology for environmental remediation: A review. *Environ. Pollut.* **2022**, *314*, 120232. [CrossRef]
83. Yin, S.; Zhang, X.; Yin, H.; Zhang, X. Current knowledge on molecular mechanisms of microorganism-mediated bioremediation for arsenic contamination: A review. *Microbiol. Res.* **2022**, *258*, 126990. [CrossRef]
84. Zhao, X.; Xie, Z.; Qiu, R.; Peng, W.; Ju, F.; Zhong, F. Collaborating iron and manganese for enhancing stability of natural arsenic sinks in groundwater: Current knowledge and future perspectives. *Chem. Geol.* **2024**, *669*, 122369. [CrossRef]
85. Liu, X.; Ju, Y.; Mandzhieva, S.; Pinskii, D.; Minkina, T.; Rajput, V.D.; Roane, T.; Huang, S.; Li, Y.; Ma, L.Q.; et al. Sporadic Pb accumulation by plants: Influence of soil biogeochemistry, microbial community and physiological mechanisms. *J. Hazard. Mater.* **2023**, *444*, 130391. [CrossRef]
86. Kaya, C.; Uğurlar, F.; Ashraf, M.; Hou, D.; Kirkham, M.B.; Bolan, N. Microbial consortia-mediated arsenic bioremediation in agricultural soils: Current status, challenges, and solutions. *Sci. Total Environ.* **2024**, *917*, 170297. [CrossRef] [PubMed]
87. Wang, L.; Li, Z.; Wang, Y.; Brookes, P.C.; Wang, F.; Zhang, Q.; Xu, J.; Liu, X. Performance and mechanisms for remediation of Cd(II) and As(III) co-contamination by magnetic biochar-microbe biochemical composite: Competition and synergy effects. *Sci. Total Environ.* **2021**, *750*, 141672. [CrossRef] [PubMed]

88. Qu, J.; Shi, J.; Wang, Y.; Tong, H.; Zhu, Y.; Xu, L.; Wang, Y.; Zhang, B.; Tao, Y.; Dai, X.; et al. Applications of functionalized magnetic biochar in environmental remediation: A review. *J. Hazard. Mater.* **2022**, *434*, 128841. [CrossRef] [PubMed]
89. Baskar, A.V.; Bolan, N.; Hoang, S.A.; Sooriyakumar, P.; Kumar, M.; Singh, L.; Jasemizad, T.; Padhye, L.P.; Singh, G.; Vinu, A.; et al. Recovery, regeneration and sustainable management of spent adsorbents from wastewater treatment streams: A review. *Sci. Total Environ.* **2022**, *822*, 153555. [CrossRef] [PubMed]
90. Wei, X.; Zhou, Y.; Tsang, D.C.; Song, L.; Zhang, C.; Yin, M.; Liu, J.; Xiao, T.; Zhang, G.; Wang, J. Hyperaccumulation and transport mechanism of thallium and arsenic in brake ferns (*Pteris vittata* L.): A case study from mining area. *J. Hazard. Mater.* **2020**, *388*, 121756. [CrossRef] [PubMed]
91. Yang, C.; Han, N.; Inoue, C.; Yang, Y.-L.; Nojiri, H.; Ho, Y.-N.; Chien, M.-F. Rhizospheric plant-microbe synergistic interactions achieve efficient arsenic phytoextraction by Pteris vittata. *J. Hazard. Mater.* **2022**, *434*, 128870. [CrossRef] [PubMed]
92. He, Z.; Shen, J.; Li, Q.; Yang, Y.; Zhang, D.; Pan, X. Bacterial metal(loid) resistance genes (MRGs) and their variation and application in environment: A review. *Sci. Total Environ.* **2023**, *871*, 162148. [CrossRef]
93. Ayala-Muñoz, D.; Burgos, W.D.; Sánchez-España, J.; Couradeau, E.; Falagán, C.; Macalady, J.L. Metagenomic and Metatranscriptomic Study of Microbial Metal Resistance in an Acidic Pit Lake. *Microorganisms* **2020**, *8*, 1350. [CrossRef] [PubMed]
94. Rilstone, V.; Vignale, L.; Craddock, J.; Cushing, A.; Filion, Y.; Champagne, P. The role of antibiotics and heavy metals on the development, promotion, and dissemination of antimicrobial resistance in drinking water biofilms. *Chemosphere* **2021**, *282*, 131048. [CrossRef] [PubMed]
95. Yaashikaa, P.; Palanivelu, J.; Hemavathy, R. Sustainable approaches for removing toxic heavy metal from contaminated water: A comprehensive review of bioremediation and biosorption techniques. *Chemosphere* **2024**, *357*, 141933. [CrossRef]
96. Wu, J.; Zhao, R.; Zhao, L.; Xu, Q.; Lv, J.; Ma, F. Sorption of petroleum hydrocarbons before transmembrane transport and the structure, mechanisms and functional regulation of microbial membrane transport systems. *J. Hazard. Mater.* **2023**, *441*, 129963. [CrossRef]
97. Yan, Z.; Liu, C.; Liu, Y.; Tan, X.; Li, X.; Shi, Y.; Ding, C. The interaction of ZnO nanoparticles, Cr (VI), and microorganisms triggers a novel ROS scavenging strategy to inhibit microbial Cr (VI) reduction. *J. Hazard. Mater.* **2023**, *443*, 130375. [CrossRef]
98. Li, X.; Sun, M.; Zhang, L.; Finlay, R.D.; Liu, R.; Lian, B. Widespread bacterial responses and their mechanism of bacterial metallogenic detoxification under high concentrations of heavy metals. *Ecotoxicol. Environ. Saf.* **2022**, *246*, 114193. [CrossRef] [PubMed]
99. Domán, A.; Dóka, É.; Garai, D.; Bogdándi, V.; Balla, G.; Balla, J.; Nagy, P. Interactions of reactive sulfur species with metalloproteins. *Redox Biol.* **2023**, *60*, 102617. [CrossRef] [PubMed]
100. Ashaolu, T.J.; Lee, C.C.; Ashaolu, J.O.; Pourjafar, H.; Jafari, S.M. Metal-binding peptides and their potential to enhance the absorption and bioavailability of minerals. *Food Chem.* **2023**, *428*, 136678. [CrossRef] [PubMed]
101. Samuel, M.S.; Datta, S.; Khandge, R.S.; Selvarajan, E. A state of the art review on characterization of heavy metal binding metallothioneins proteins and their widespread applications. *Sci. Total Environ.* **2021**, *775*, 145829. [CrossRef]
102. Priya, A.; Gnanasekaran, L.; Dutta, K.; Rajendran, S.; Balakrishnan, D.; Soto-Moscoso, M. Biosorption of heavy metals by microorganisms: Evaluation of different underlying mechanisms. *Chemosphere* **2022**, *307*, 135957. [CrossRef]
103. Sarker, A.; Al Masud, M.A.; Deepo, D.M.; Das, K.; Nandi, R.; Ansary, M.W.R.; Islam, A.R.M.T.; Islam, T. Biological and green remediation of heavy metal contaminated water and soils: A state-of-the-art review. *Chemosphere* **2023**, *332*, 138861. [CrossRef]
104. Zandi, P.; Schnug, E. Reactive Oxygen Species, Antioxidant Responses and Implications from a Microbial Modulation Perspective. *Biology* **2022**, *11*, 155. [CrossRef] [PubMed]
105. George, S.; Abrahamse, H. Redox Potential of Antioxidants in Cancer Progression and Prevention. *Antioxidants* **2020**, *9*, 1156. [CrossRef]
106. Pisoschi, A.M.; Pop, A.; Iordache, F.; Stanca, L.; Predoi, G.; Serban, A.I. Oxidative stress mitigation by antioxidants—An overview on their chemistry and influences on health status. *Eur. J. Med. Chem.* **2021**, *209*, 112891. [CrossRef] [PubMed]
107. Nakamura, K.; Nakamura, Y.; Kataoka, N. A hypothalamomedullary network for physiological responses to environmental stresses. *Nat. Rev. Neurosci.* **2022**, *23*, 35–52. [CrossRef] [PubMed]
108. Singh, A.; Mehta, S.; Yadav, S.; Nagar, G.; Ghosh, R.; Roy, A.; Chakraborty, A.; Singh, I.K. How to Cope with the Challenges of Environmental Stresses in the Era of Global Climate Change: An Update on ROS Stave off in Plants. *Int. J. Mol. Sci.* **2022**, *23*, 1995. [CrossRef] [PubMed]
109. Pal, A.; Bhattacharjee, S.; Saha, J.; Sarkar, M.; Mandal, P. Bacterial survival strategies and responses under heavy metal stress: A comprehensive overview. *Crit. Rev. Microbiol.* **2022**, *48*, 327–355. [CrossRef]
110. Bahar, M.E.; Kim, H.J.; Kim, D.R. Targeting the RAS/RAF/MAPK pathway for cancer therapy: From mechanism to clinical studies. *Signal Transduct. Target. Ther.* **2023**, *8*, 455. [CrossRef] [PubMed]
111. Zhang, C.; Kong, Y.; Xiang, Q.; Ma, Y.; Guo, Q. Bacterial memory in antibiotic resistance evolution and nanotechnology in evolutionary biology. *iScience* **2023**, *26*, 107433. [CrossRef] [PubMed]
112. Dhanda, G.; Acharya, Y.; Haldar, J. Antibiotic Adjuvants: A Versatile Approach to Combat Antibiotic Resistance. *ACS Omega* **2023**, *8*, 10757–10783. [CrossRef] [PubMed]
113. Nayak, L.; Lal, M.K.; Tiwari, R.K.; Kumar, R.; Lal, P.; Das, R.; Behera, B.; Thakur, P.; Altaf, M.A.; Kumar, A. A balancing act: Exploring the interplay between HSPs and osmoprotectants in temperature stress responses. *S. Afr. J. Bot.* **2023**, *162*, 64–71. [CrossRef]

114. Qi, R.; Xue, N.; Wang, S.; Zhou, X.; Zhao, L.; Song, W.; Yang, Y. Heavy metal(loid)s shape the soil bacterial community and functional genes of desert grassland in a gold mining area in the semi-arid region. *Environ. Res.* **2022**, *214*, 113749. [CrossRef]
115. Zeng, J.; Pan, Y.; Hu, R.; Liu, F.; Gu, H.; Ding, J.; Liu, S.; Liu, S.; Yang, X.; Peng, Y.; et al. The vertically-stratified resistomes in mangrove sediments was driven by the bacterial diversity. *J. Hazard. Mater.* **2023**, *458*, 131974. [CrossRef] [PubMed]
116. Singh, C.K.; Sodhi, K.K.; Shree, P.; Nitin, V. Heavy Metals as Catalysts in the Evolution of Antimicrobial Resistance and the Mechanisms Underpinning Co-selection. *Curr. Microbiol.* **2024**, *81*, 148. [CrossRef] [PubMed]
117. Kang, Y.; Wang, J.; Wang, Y.; Li, Z. Profiles of phage in global hospital wastewater: Association with microbial hosts, antibiotic resistance genes, metal resistance genes, and mobile genetic elements. *Sci. Total Environ.* **2024**, *926*, 171766. [CrossRef] [PubMed]
118. James, C.; James, S.J.; Onarinde, B.A.; Dixon, R.A.; Williams, N. A Critical Review of AMR Risks Arising as a Consequence of Using Biocides and Certain Metals in Food Animal Production. *Antibiotics* **2023**, *12*, 1569. [CrossRef] [PubMed]
119. Feng, G.; Huang, H.; Chen, Y. Effects of emerging pollutants on the occurrence and transfer of antibiotic resistance genes: A review. *J. Hazard. Mater.* **2021**, *420*, 126602. [CrossRef] [PubMed]
120. Liu, C.; Chen, J.; Shan, X.; Yang, Y.; Song, L.; Teng, Y.; Chen, H. Meta-analysis addressing the characterization and risk identification of antibiotics and antibiotic resistance genes in global groundwater. *Sci. Total Environ.* **2023**, *860*, 160513. [CrossRef] [PubMed]
121. Uddin, T.M.; Chakraborty, A.J.; Khusro, A.; Zidan, B.R.M.; Mitra, S.; Emran, T.B.; Dhama, K.; Ripon, M.K.H.; Gajdács, M.; Sahibzada, M.U.K.; et al. Antibiotic resistance in microbes: History, mechanisms, therapeutic strategies and future prospects. *J. Infect. Public Health* **2021**, *14*, 1750–1766. [CrossRef] [PubMed]
122. Di Cesare, A.; Sabatino, R.; Sbaffi, T.; Fontaneto, D.; Brambilla, D.; Beghi, A.; Pandolfi, F.; Borlandelli, C.; Fortino, D.; Biccai, G.; et al. Anthropogenic pollution drives the bacterial resistome in a complex freshwater ecosystem. *Chemosphere* **2023**, *331*, 138800. [CrossRef]
123. Samreen; Ahmad, I.; Malak, H.A.; Abulreesh, H.H. Environmental antimicrobial resistance and its drivers: A potential threat to public health. *J. Glob. Antimicrob. Resist.* **2021**, *27*, 101–111. [CrossRef]
124. Sharma, P.; Kumar, S.; Pandey, A. Bioremediated techniques for remediation of metal pollutants using metagenomics approaches: A review. *J. Environ. Chem. Eng.* **2021**, *9*, 105684. [CrossRef]
125. Wang, Z.; Luo, P.; Zha, X.; Xu, C.; Kang, S.; Zhou, M.; Nover, D.; Wang, Y. Overview assessment of risk evaluation and treatment technologies for heavy metal pollution of water and soil. *J. Clean. Prod.* **2022**, *379*, 134043. [CrossRef]
126. Du, C.; Yang, F.; Li, X.; Liao, H.; Li, Z.; Gao, J.; Zhang, L. Metagenomic analysis of microbial community structure and distribution of resistance genes in Daihai Lake, China. *Environ. Pollut.* **2022**, *302*, 119065. [CrossRef]

water

MDPI

Review

Green Roof Systems for Rainwater and Sewage Treatment

Jing Yan [1,†], Pu Yang [2,†], Binjie Wang [2], Suqing Wu [2,3,4], Min Zhao [2,3,4], Xiangyong Zheng [2,3,4], Zhiquan Wang [2,3,4], Yejian Zhang [2,3,*] and Chunzhen Fan [2,3,4,*]

1 North China Sea Survey Center, Ministry of Natural Resources, Qingdao 266100, China; yanjing150@163.com
2 College of Life and Environmental Science, Wenzhou University, Wenzhou 325000, China; yangpu1130@163.com (P.Y.); w21211270127@163.com (B.W.); wusuqing@wzu.edu.cn (S.W.); zmzjcn@sina.com (M.Z.); x.zheng@wzu.edu.cn (X.Z.); wangzhiquan@wzu.edu.cn (Z.W.)
3 National and Local Joint Engineering Research Center of Ecological Treatment Technology for Urban Water Pollution, Wenzhou University, Wenzhou 325000, China
4 Institute for Eco-Environmental Research of Sanyang Wetland, Wenzhou University, Wenzhou 325014, China
* Correspondence: yjzhang@wzu.edu.cn (Y.Z.); fanchunzhen@wzu.edu.cn (C.F.)
† These authors contributed equally to this work.

Abstract: Green roof systems are regarded as a viable solution for mitigating urban environmental challenges and offering a multitude of environmental benefits. Currently, green roofs are increasingly being utilized for the management of rainwater runoff and wastewater. The integration of decentralized rainwater and sewage on-site treatment technology with urban green buildings is being gradually promoted. Green roofs can also be considered as a form of decentralized rainwater and sewage on-site technology, which holds great potential for widespread adoption in the future. Several studies have suggested that green roofs may serve as a potential source of pollutants; however, there are also studies that clearly demonstrate the efficient removal of nutrients and organic pollutants by green roofs. This article critically examines the existing literature on water treatment aspects associated with green roofs and elucidates their classification and operational mechanisms. Through an analysis of previous research cases, it becomes evident that both substrate and vegetation play a significant role in influencing the treatment performance of green roofs. By designing and configuring appropriate substrate and vegetation, green roofs can play a pivotal role in the purification of water quality. Finally, a brief outlook is presented for the future research directions of green roofs, with the anticipation that green roofs will feature more innovative and environmentally friendly designs, as well as expanded prospects for application.

Keywords: green roof; wetland roof; rainwater and wastewater treatment

1. Introduction

The development of green roofs has a rich historical background, tracing its roots back to ancient rooftop gardens. Over 2000 years ago, the initial concept primarily involved the placement of conventional green plants on rooftops. The earliest documented evidence pertaining to green roofs can be found in the Hanging Gardens of Semiramis, located in present-day Syria [1]. In ancient Rome over 2000 years ago, trees were imported for use on local green roofs. Several centuries ago in Scandinavian countries, locals laid grass on rooftops for wind and rain protection, using seaweed as a separator. These early forms of green roofs primarily emphasized aesthetics and practicality in architecture. However, with the economic and social development, developed regions have witnessed a rapid decline in green spaces, resulting in deteriorating environmental conditions. In response to this issue, some countries led by Germany have initiated research on the ecological and environmental aspects of green roofs. In recent years, green roofs have been widely recognized as an effective approach to mitigate urban environmental issues and offer a multitude of ecological benefits [2]. These ecological advantages encompass temperature regulation [3], enhancement of air quality [4], alleviation of the urban heat island effect [5], and promotion of biodiversity [6]. Currently, researchers have conducted several studies on the utilization of green roofs for rainwater runoff management and wastewater treatment. It is anticipated that the future implementation of green roofs for water purification purposes will yield both environmental and economic benefits.

In urban areas, the predominant mode of sewage treatment is centralized water treatment, whereby greywater (wastewater from bathrooms, kitchens, and laundry) and blackwater (wastewater from toilets) are conveyed through sewage pipelines to sewage treatment plants [7]. While this approach is efficient and straightforward, it suffers from low water resource utilization rates as well as high infrastructure construction costs. Furthermore, long-distance transportation of sewage can result in blockages or even leakage that may lead to malodorous water requiring extensive maintenance requirements [8]. Distributed wastewater treatment presents a viable alternative to centralized water treatment, offering enhanced device flexibility and eliminating the need for long-distance water transportation [9]. Moreover, distributed water treatment systems can alleviate the burden on conventional sewage treatment plants [10] as they capitalize on the proximity of wastewater to the treatment system, resulting in reduced pumping requirements and significantly lowered construction and operational costs [11]. Common decentralized water treatment systems encompass constructed wetlands (CWs), aerobic treatment systems, and anaerobic treatment systems. Constructed wetlands are a natural treatment system that employs vegetation, substrate, and biological processes for wastewater treatment. However, the operational feasibility of this system is limited in urban areas with spatial constraints. In anaerobic treatment systems, the lower efficiency of anaerobic bacteria compared to aerobic bacteria results in suboptimal effluent quality attainment, leading to comparatively lower levels of wastewater treatment than other systems [12]. Membrane bioreactor (MBR) technology is widely employed in decentralized water treatment systems. MBR integrates membrane processes with biological wastewater treatment processes to effectively eliminate pollutants from wastewater. Nevertheless, practical implementation of this technology often encounters issues related to membrane fouling, which significantly impairs the efficiency of wastewater treatment processes [13].

In addition to the aforementioned environmental benefits, the implementation of green roof systems can also serve as a decentralized water treatment solution. In comparison to traditional centralized sewage treatment systems, green roof systems offer a more cost-effective and efficient on-site collection and treatment system for rainwater and wastewater. The green roof system can effectively purify effluent water by utilizing vegetation, substrate, and other mechanisms to absorb and filter pollutants. However, several studies have reported that green roofs can potentially act as sources of pollutants, such as nitrogen, phosphorus, and heavy metals [14,15]. Conversely, other research has explicitly highlighted the capacity of green roofs to function as sinks for nitrogen, phosphorus, and certain heavy metals [16,17], thereby mitigating the concentration of pollutants in runoff. According to Chen [18], various studies have indicated that common roofs can contribute to higher levels of runoff pollution compared to green roofs, whereas green roofs can mitigate this pollution to some extent. While some studies have indicated the potential for water pollution caused by green roofs, there is a lack of comprehensive analysis and conclusive findings on this matter. Facing the different results of rainwater and wastewater treatment in green roof systems, this article reviews existing research cases on rainwater runoff management and wastewater purification for reuse. It summarizes the types, operational mechanisms, and designs of current green roof systems. By conducting a thorough analysis and summarization, our aim is to design efficient green roof systems that effectively treat rainwater and sewage for high-quality water purification while minimizing pollution risks. Numerous studies have been conducted on green roof systems for rainwater and sewage treatment, demonstrating that green roof systems have been one of the important techniques for decentralized wastewater treatment. However, the influencing factors of green roof systems are not comprehensively summarized, and discussions on their influencing mechanisms for pollutant removal are still insufficient, which is not beneficial for the development and application of green roof systems. This situation stimulates us to write this comprehensive review article to provide insights and directions for future studies and the design of green roof systems.

2. Methods

To select the most relevant papers on the topic of this review, a comprehensive search was conducted on 15 May 2024, using the keywords 'green roof' (or 'wetland roof') combined with 'rainwater treatment' and 'sewage treatment' in the Web of Science (WoS) database. A total of 69,135,14, and 50 papers related to the combined keywords of 'green roof and rainwater treatment', 'green roof and sewage treatment', 'wetland roof and rainwater treatment', and 'wetland roof and sewage treatment' were obtained, respectively. By limiting article types to "Article" and "Review", the number of relevant papers decreased to 68,130,10 and 48, respectively. The emphasis of this review was to summarize the influencing factors of green roof systems on water treatment and discuss the influencing mechanisms. Thus, a detailed analysis was conducted on both the influent and effluent water quality of green roof systems, such as nitrogen, phosphorus, and heavy metals. Based on the literature obtained from the search results, the impact of each factor on both influent and effluent water quality was individually discussed through appropriate comparisons and analyses.

3. Green Roof Systems

3.1. Definition and Classification

The concept of green roofs involves the installation of vegetation on building structures, such as roofs, balconies, or terraces [19]. Similar to conventional green spaces and gardens, these areas are directly exposed to natural elements, including sunlight and rainwater. However, the key distinction lies in the independent soil substrate required for green roofs, which is situated above a designated space rather than being connected to natural soil. In scientific literature, alternative terms used for green roofs include living roofs, eco-roofs, or vegetated roofs [20].

Green roofs often exhibit regional variations in their construction but are typically comprised of five main components: a vegetation layer, a substrate layer, a filter layer, a drainage layer, and a protection layer [21]. The vegetation layer serves as the fundamental element of the green roof and plays a pivotal role in delivering ecological benefits. The substrate layer provides essential nutrients for plant growth while also facilitating rainwater absorption and retention. The filter layer acts as a barrier between the substrate and drainage layers to prevent soil particles from obstructing the drainage system. Serving as both a water storage and drainage system, the drainage layer effectively removes excess water during heavy rainfall while supplying water to plants during dry periods. Lastly, the protection layer safeguards against damage caused by excessive root growth while simultaneously acting as a protective barrier between water and the building's roof surface. These key components are visually depicted in Figure 1.

Figure 1. Five main components of the typical green roof. Reproduced with permission from [22]. Copyright 2015, Elsevier.

Currently, the classification of green roofs primarily relies on substrate thickness, categorizing them into extensive and intensive green roofs. A concise overview of these two roof types is provided based on pertinent studies [23,24]. The two types of green roof patterns are shown in Figure 2.

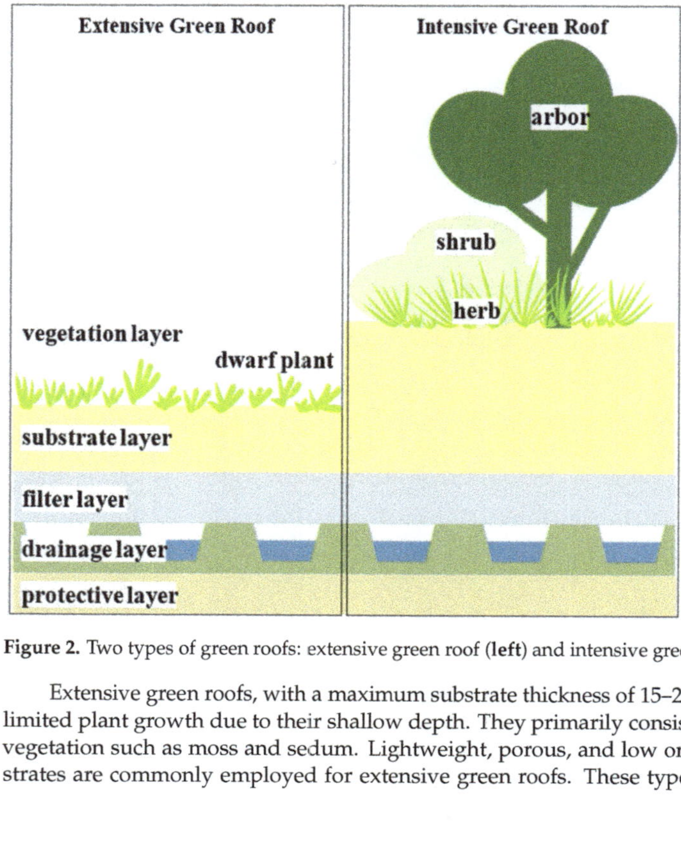

Figure 2. Two types of green roofs: extensive green roof (**left**) and intensive green roof (**right**).

Extensive green roofs, with a maximum substrate thickness of 15–20 cm, can support limited plant growth due to their shallow depth. They primarily consist of low-growing vegetation such as moss and sedum. Lightweight, porous, and low organic matter substrates are commonly employed for extensive green roofs. These types of roofs can be installed on steep slopes with angles up to 45° and are frequently utilized for fire prevention, insulation, and rainwater management purposes. The thinness, drought resistance, and low load-bearing capacity of the vegetation require no special maintenance at a later stage, with relatively low installation and operation costs.

Intensive green roofs have a substrate thickness of more than 15–20 cm, allowing for a wider range of plant selection. Perennial herbaceous plants, shrubs, or arbors can be chosen for cultivation. The materials utilized in this roofing type are lightweight and possess a low organic concentration. However, due to the increased load resulting from thicker substrates, additional support structures are necessary. Moreover, the installation slope of intensive green roofs is limited and generally remains below 10°. This type of roof is commonly utilized for aesthetic, entertainment, social, and leisure purposes, akin to a rooftop garden. Additionally, it can support a greater variety of organisms and exhibit enhanced biodiversity. However, this particular form necessitates substantial irrigation and maintenance requirements, as well as stringent structural support prerequisites, resulting in elevated construction and operational expenses.

In recent years, researchers have developed innovative green roof systems that differ from typical ones. By integrating horizontal subsurface flow constructed wetlands with green roofs, a shallow wetland roof system has been devised to fully exploit the benefits of constructed wetlands [25,26]. Xu et al. have introduced the hydroponic green roof system, replacing the conventional soil cultivation method with hydroponics [27] and incorporating lightweight fillers to address the issue of heavy substrates in typical green roofs. These shallow bed wetland roof systems and hydroponic green roof systems can collectively be referred to as "wetland roofs". According to the definition of green roofs, wetland roofs still fall within the conceptual framework of green roofs. Based on their actual construction characteristics and system design, green roofs can be categorized accordingly (Figure 3).

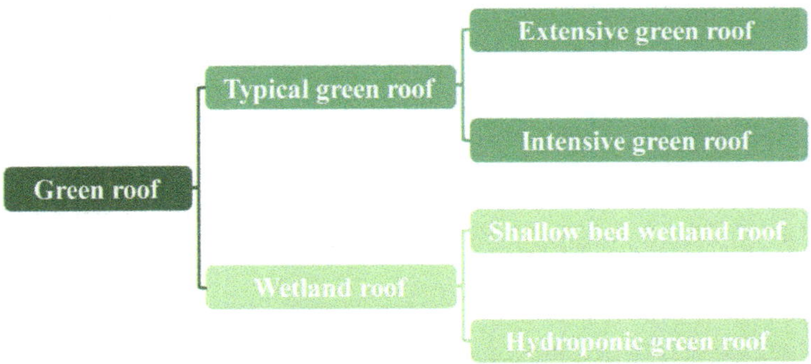

Figure 3. Green roofs classification based on actual construction characteristics and system design.

Compared to conventional green roofs, wetland roofs not only facilitate rainwater collection but also offer the potential for wastewater treatment (greywater or blackwater). Wetland roofs share similar advantages with traditional green roofs, as depicted in Figure 4. Therefore, when designing and installing various types of green roofs, it is essential to consider the specific purpose and application in accordance with local climate conditions and the characteristics of the building.

Figure 4. Potential advantages of implementing a wetland roof. Reproduced with permission from [28]. Copyright 2019, Elsevier.

3.2. Operating Mechanism

Green roof systems are recognized as efficient and cost-effective environmental management solutions for rainwater and wastewater treatment, relying on the synergistic effects of plants, substrates, microorganisms, and other factors [29]. The principles underlying water treatment in green roof systems bear resemblance to those of artificial wetlands [30], primarily achieved through physical processes such as sedimentation and filtration, chemical processes including absorption, reaction, and precipitation, as well as biological processes involving microbial activities to accomplish water purification objectives [31].

Plants necessitate a substantial quantity of nutrients throughout their developmental, growth, and reproductive processes. They have the capacity to assimilate nitrogen and phosphorus nutrients via their root system [25]. Moreover, specific plant species possess the capability to remediate water pollutants, particularly metal contaminants, by converting them into non-toxic compounds through intrinsic mechanisms [32]. Additionally, plants can augment pollutant removal by modulating microbial activity in the substrate and modifying its physicochemical composition. The exudation of specific plant root exudates can activate rhizobacteria and enhance pollutant degradation. Well-developed plant roots, along with the biofilm adhering to their surface, often secrete copious amounts of enzymes that expedite the decomposition of water pollutants, thereby accomplishing water purification objectives [33].

The substrate plays a pivotal role in pollutant removal mechanisms, encompassing physical–chemical and biochemical processes. As an integral component supporting plant growth and microbial existence, the substrate exhibits the capacity to absorb and degrade pollutants, thereby exerting a significant influence on the water quality of green roofs [34]. The physical–chemical mechanisms of the substrate involve the initial interception and adsorption of pollutants, followed by potential chemical reactions in certain substrate materials to eliminate corresponding pollutants. With regard to biochemical mechanisms, the substrate provides attachment surfaces for microbial colonization. Microbes form biofilms on these surfaces and employ their metabolic activities to degrade pollutants effectively, consequently enhancing wastewater treatment performance [35].

Microorganisms also play a pivotal role in the remediation of pollutants. Bacteria are essential for nitrogen removal through processes such as assimilation and nitrification–denitrification. The introduction of mycorrhizal fungi into green roof systems not only enhances plant water use efficiency and mitigates drought damage but also reduces nitrogen and phosphorus runoff concentrations. Moreover, they possess the capability to absorb and accumulate heavy metals, thereby enhancing water quality [36]. The specific treatment process of plants, substrates, and microorganisms is summarized in Figure 5.

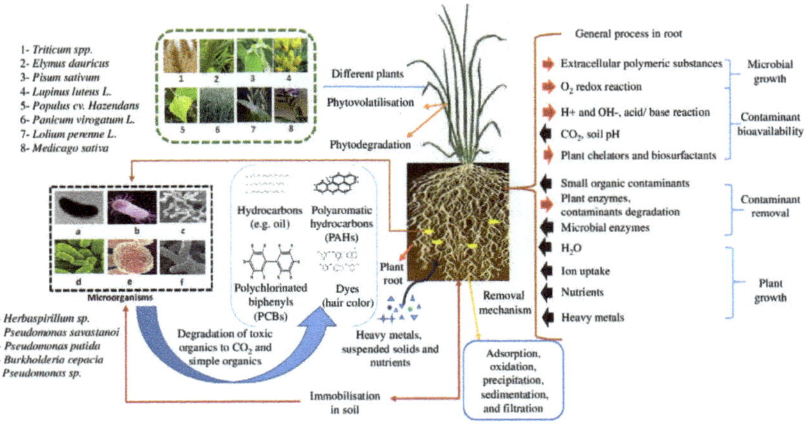

Figure 5. Removal process of plants, substrates, and microorganisms. Reproduced with permission from [37]. Copyright 2019, Elsevier.

3.3. Design and Construction

In the construction of green roof systems, careful consideration must be given to the structural load-bearing capacity of the building. Due to limited load standards for roofs, it is necessary to restrict the substrate depth and total weight of green roofs in order to prevent damage caused by excessive weight. In the design and construction process, there is a need for new requirements regarding lightweight materials with low density for green roof substrates in order to minimize weight [38]. During the architectural design process, meticulous planning of drainage systems should be conducted for green roofs to ensure the timely removal of excess water and avoid waterlogging and root-level plant damage. Green roof systems typically consist of drainage layers and protective layers (usually waterproof), which offer some protection but still pose a risk of moisture penetration into the building's interior. Therefore, when constructing green roofs, emphasis should be placed on ensuring the integrity and long-term feasibility of drainage layers and protective layers. Additionally, sufficient space should be provided for plant root growth in order to prevent structural damage caused by roots [39]. The design and construction of green roofs are influenced by various factors, such as different building structures, original roof slopes, and shapes. Higher slopes often result in increased drainage difficulties, while irregularly shaped roofs add complexity during construction; all these factors must be taken into account during the construction process [40]. However, initial construction costs remain a limiting factor in implementing green roofs despite their potential offset through energy-saving measures over time as well as reduced rainwater management expenses [41].

4. Research Status

4.1. Vegetation

Vegetation constitutes a pivotal element of green roof systems and exerts a significant influence on the water quality of runoff. Owing to their distinct nutrient requirements, utilization efficiency, and impact on nutrient mineralization within the system, different plant types can yield diverse effects on the discharge's water quality. Due to the unique rooftop environment, characterized by strong winds, intense sunlight, and temperature fluctuations, certain limitations exist for cultivating plants on rooftops. Not all plant species can effectively adapt to such conditions. Moreover, not all plants can endure irrigation with rainwater or wastewater, further narrowing down the available choices for rooftop cultivation. In recent years, the types of plants planted on green roofs are shown in Table 1.

4.1.1. The Influence of Vegetation

In green roof systems, the presence of vegetation exerts a discernible influence on the water quality of the system. Under identical substrate type and thickness conditions, green roofs planted with *Ophiopogon japonicus (L. f.) Ker Gawl.* exhibit significantly higher effluent TN concentrations compared to unplanted green roofs ($p < 0.05$) [34]. Furthermore, the TSS concentration of non-planted green roofs (149.11 mg L^{-1}) was found to be three times higher than that of planted green roofs (50.83 mg L^{-1}), and a statistically significant difference between the two was observed ($p < 0.01$). However, no statistically significant difference was observed in the TP concentration between planted green roofs (0.031 mg L^{-1}) and unplanted counterparts (0.023 mg L^{-1}).

The study conducted by Liao et al. focused on investigating the correlation between plant biomass and the presence of TN and TSS in effluent samples [42]. The findings revealed a positive correlation between plant biomass and TN concentration in the effluent, implying that this association may be attributed to an increased quantity of fallen leaves during the later stages of plant growth. Rapid decomposition of leaf litter and nitrogen mineralization leads to an increase in the TN concentration in effluent. Additionally, it was observed that an increase in plant biomass led to a decrease in the TSS concentration, indicating that larger plants can intercept more runoff, thereby reducing the concentration of suspended solids. In the final measurement, the green roof planted with *Agastache foeniculum (Pursh) Kuntze* exhibited a significant reduction in TN, dissolved P, dissolved K, dissolved Ca, and dissolved Mg loads by 50%, 28%, 27%, 18%, and 19%, respectively, compared to the control group without plants ($p < 0.05$).

The study conducted by Park et al. also examined the impact of vegetation on heavy metal concentrations and revealed that green roofs with plant cover exhibited significantly reduced levels of copper, zinc, magnesium, and cadmium in their runoff compared to unplanted green roofs ($p < 0.05$) [24]. The reduction in heavy metals can be attributed to the plants' capacity for absorption, transformation, and volatilization of these pollutants, thereby effectively eliminating or immobilizing them within the system. The process of plants removing metal elements in green roofs is shown in Figure 6.

Liu et al. conducted a comprehensive comparison between green roofs with vegetative cover and exposed substrates, revealing that the presence of plants reduced substrate loss by 5.14% ($p < 0.05$). Furthermore, it preserves the physical and chemical properties of the substrate, as well as the microbial conditions [44]. Maintaining adequate plant coverage and preventing substrate exposure is crucial for retaining nutrients in the substrate and enhancing microbial biomass in green roofs, thereby indirectly influencing water quality emissions from the substrate. Therefore, ensuring optimal plant coverage is imperative during green roof operation.

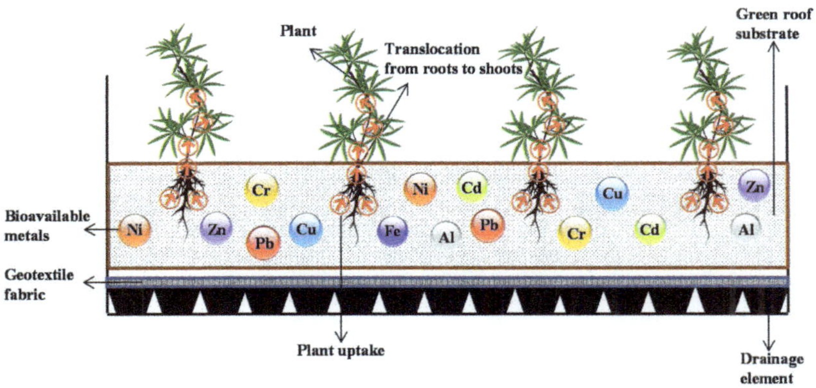

Figure 6. Schematic diagram of the metal removal process by plants on green roofs. Reproduced with permission from [43]. Copyright 2019, Elsevier.

4.1.2. The Influence of a Singular Vegetation Type

The effluent water quality of systems is influenced by different plant types, with the effectiveness of pollutant removal in green roof runoff varying depending on the specific plant species [45]. Current research primarily focuses on comparing the respective contributions of various individual species in green roof ecosystems.

The study conducted by Thi-Dieu-Hien et al. involved a 30-day experiment on wetland roofs using eight different plant species [26]. Two distinct hydraulic loading rate (HLR) conditions were established, and *Kyllinga brevifolia Rottb* consistently exhibited the highest growth rates in terms of both fresh biomass and dry biomass. *Kyllinga brevifolia Rottb, Cyperusjavanicus Houtt*, and *Imperata cylindrical* ranked higher in terms of leaf area, indicating their potential to enhance urban green coverage. The results demonstrated a positive correlation between plant biomass and TN and TP removal rates, with the aforementioned three species exhibiting TN levels below 10 mg L^{-1} in the effluent water, which is lower than other wetland roof systems. Regarding TP removal, all plants displayed similar phosphorus removal rates within the wetland roof system. Overall analysis suggests that *Kyllinga brevifolia Rottb, Cyperusjavanicus Houtt*, and *Imperata cylindrical* are considered to have significant greening effects and wastewater treatment capabilities.

Chai et al. conducted a three-year green roof experiment from July to September, focusing on the summer months and utilizing two distinct plant species [46]. The experimental findings revealed that the choice of plants exerted a significant influence on chemical oxygen demand (COD) levels ($p < 0.01$). *Ophiopogon japonicus (L. f.) Ker Gawl.* exhibited superior COD control compared to *Ophiopogon japonicus 'Nanus'*. Notably, *Ophiopogon japonicus (L. f.) Ker Gawl* demonstrated remarkable environmental adaptability, while *Ophiopogon japonicus 'Nanus'* struggled with adaptation issues and displayed higher mortality rates. The absence of proper root fixation on green roofs resulted in reduced resistance against rainwater erosion, further compounded by soil subsidence and substrate blockage in lower layers, ultimately leading to a substantial increase in COD concentration.

Liu et al. conducted separate studies on the treatment efficiency of C4, C3, and CAM plants in green roofs for wastewater [47]. The research findings demonstrated that TN removal rates ranged from 65.26% to 90.52%, NO_3^--N removal rates ranged from 77.83% to 93.97%, NH_4^+-N removal rates ranged from 83.32% to 96.31%, TP removal rates ranged from 93.77% to 98.94%, PO_4^{3-}-P removal rates ranged from 96.36% to 99.43%, TSS removal rates ranged from 79.27% to 97.38%, and COD removal rates ranged from 79.94% to 98.92%. Moreover, a comparison revealed significantly higher TN, NO_3^--N, NH_4^+-N, TP, PO_4^{3-}-P, TSS, and COD removal efficiencies in C4 plants (*Eremochloa ophiuroides* and *Cynodon dactylo*) and C3 plants (*Poa pratensis* and *Festuca arundinacea*), compared with CAM plants (*Sedum lineare* and *Callisia repens*).

Gong et al. conducted a study on the purification effects of four plant species belonging to the Sedum genus (*Sedum aizoon L., Sedum lineare Thunb., Sedum spurium cv. Coccineum*, and *Sedum spectabile*) on green roofs [48]. The results revealed no significant variations in TN and NH_4^+-N concentrations among different plant species. However, Liu et al. observed notable differences in the TN concentration between *Sedum spectabile* and *Ophiopogon japonicas (Linn. f.) Ker-Gawl* in their investigation, suggesting higher nitrogen utilization efficiency by plants within the Sedum genus [34]. Considering that all four plants utilized in Gong's experiment belong to the same genus, it is more plausible that distinctions exist between plants from diverse families. Furthermore, there were no significant differences in TP concentrations among individual plant groups, providing support for the notion that vegetation types do not significantly impact TP absorption.

In a subsequent study, Thomaidi et al. investigated the effects of *Atriplex halimus, Polygala myrtifolia*, and *Geranium zonale* on greywater treatment in green roofs [49]. In a system filled with 10 cm of gravel without any plants, the average removal rate of COD was 70%. However, when *Polygala myrtifolia* and *Atriplex halimus* were introduced into the system as vegetation cover, the average COD removal rates significantly increased to 78% and 82%, respectively. Nevertheless, as the substrate thickness increased to 20 cm, the contribution of plants to organic matter removal became limited. Plant roots can enhance filtration by obstructing particulate matter and providing attachment surfaces for microorganisms to facilitate degradation [50]. However, the substrate in the vertical flow system also possesses the above-mentioned mechanisms, which can enhance treatment efficiency. It is evident that with increasing substrate thicknesses, the role of plants in removing organic matter and TSS diminishes. Therefore, for the effective removal of organic matter and TSS from water bodies on green roofs, plant presence is more crucial in extensive green roofs with smaller thicknesses compared to intensive green roofs with larger thicknesses. Additionally, in this study, TP removal was not influenced by plant presence since adsorption and chemical reactions between phosphorus and substrates are considered primary mechanisms for TP removal rather than plant uptake.

4.1.3. The Influence of Vegetation Combinations

In recent years, researchers have shifted their focus from studying the impact of individual plant species on the water quality of green roofs to investigating synergistic combinations of diverse plant species. The research findings demonstrate that mixed planting in green roof systems exhibits superior performance in terms of plant survival, canopy density, plant height, and growth when compared to single-species planting. Incorporating mixed planting on green roofs effectively enhances plants' adaptability and resilience towards diverse climatic conditions [51].

Caceres et al. conducted a study to evaluate the green coverage and survival rate under mixed conditions on green roofs by selecting four species (*Phyla nodiflora, Grindelia cabrerae, Eustachys retusa*, and *Sedum mexicanum*) with different growth forms and stress resistance for mixed planting [52]. A total of 11 experimental groups were established using combinations of two, three, or four species. After one year, significant differences in both green coverage and survival rate were observed among different plant combinations, with nine combinations having a total green coverage of >80% and six combinations having a total survival rate of >80%. By the end of the second year, only five combinations (*P. nodiflora* and *E. retusa*; *G. cabrerae* and *E. retusa*; *G. cabrerae, E. retusa* and *S. mexicanum*; *P. nodiflora, E. retusa* and *S. mexicanum*; *P. nodiflora, G. cabrerae, E. retusa* and *S. mexicanum*) maintained a total green coverage and survival rate between 60 and 80%. The findings suggest that there are variations in green coverage and survival rates among different mixed planting combinations on green roofs. Therefore, when selecting plant combinations for green roofs, it is crucial to consider species persistence and colonization mechanisms as well as understand the spatial heterogeneity of plants to establish long-term stable plant diversity. Furthermore, increasing species richness and enhancing plant diversity can improve nitrogen retention in water quality purification systems, which is essential for maintaining water quality and sustaining ecosystem health [53].

Liao et al. established two types of green roofs: one with a predominant coverage of *Phedimus kamtschaticus (Fisch.) t Hart*, encompassing over 95% of the area, along with a small proportion of *sedum sexangulare L.* and *sedum album L.* (referred to as sedum mats), while the other group consisted of 14 species of non-grass herbaceous plants native to the eastern and central regions of North America (referred to as native plants) [54]. The study assessed the levels of TN, dissolved P, dissolved K, dissolved Ca, dissolved Na, and dissolved Mg loads. It was observed that compared to the group planted with native plants, the sedum mats group exhibited significantly lower nutrient load values ($p < 0.05$), resulting in reductions ranging from 21 to 64%. Furthermore, in comparison to the native plants group, the sedum mats group demonstrated decreased turbidity, EC, TSS loads, TDS concentrations, and load values, which were reduced by 41%, 19%, 32%, 19%, and 41%, respectively ($p < 0.05$). The pH emissions from the sedum mats group were higher than those emitted by the native plants group (7.39) ($p < 0.05$). These findings suggest that when it comes to reducing nutrient leaching from green roofs, the sedum mats group outperforms its counterpart planted with native plants. Possible reasons include, firstly, that sedum mats exhibit higher vegetation coverage, thereby enhancing nutrient absorption. They possess a greater capacity to intercept runoff, consequently reducing discharge levels to a certain extent. Secondly, the composition of vegetation within each group plays a significant role. Sedum mats comprise perennial succulents that thrive throughout the year; in contrast, native plants predominantly consist of annuals and short-lived perennials that perish during winter months. Plants with shorter lifespans tend to generate plant parts with elevated nitrogen and phosphorus concentration [55], which can result in increased nutrient concentration upon decomposition. Thirdly, the dense root network of sedum mat plants effectively impedes particulate matter and minimizes particle loss on green roofs.

Hu et al. established four types of mixed turf grasses, consisting of the following specific types and ratios: Group 1—*Cynodon dactylon*: *Zoysia japonica*: *Lolium perenne* = 7:2:1; Group 2—*Poa pratensis*: *Agrostis matsumurae*: *Lolium perenne* = 5:4:1; Group 3—*Poa pratensis*: *Festuca elata*: *Lolium perenne* = 5:3:2; Group 4—*Zoysia japonica*: *Cynodon dactylon* = 2:1 [56]. Simultaneously, the experiment employed four distinct substrate proportions denoted as groups A, B, C, and D. By utilizing an orthogonal design approach, a total of 16 experimental groups were established through the combination of these four plant combinations and substrates. The ASFV method was employed to comprehensively assess the removal efficiency of NH_4^+-N, SS, COD, TP, and TN for each combination and prioritize them accordingly. The results revealed significant variations in purification effects among different plant combinations at the same substrate level. Combination A2 exhibited a superior stormwater runoff purification effect, while combination A3 ranked fifteenth out of all sixteen tested combinations. In substrate A and D groups, plants in Group 2 showed the best purification effect. In the substrate C group, only when combined with plants in Group 3, can a better water purification effect be achieved. Therefore, the water purification process is significantly influenced by different combinations of plants. However, when considering various plant combinations, it is imperative to also account for the interaction between mixed planting and diverse substrate types in order to attain optimal effects on water purification.

Table 1. Plant species used for green roof planting in recent years.

Plants	References
Ophiopogon japonicus (L. f.) Ker Gawl., Ophiopogon japonicus 'Nanus'	[46]
Cynodon dactylon, Cyperus javanicus Houtt Cyperus rotundus L., Eleusine indica (L.) Gaertn Imperata cylindrical, Kyllinga brevifolia Rottb Struchium sparganophorum (L.) Kuntze, Zenith zoysia grass	[26]
Hylotelephium erythrostictum (Miq.) H. Ohba, Iris tectorum Maxim. Ophiopogon japonicus (L. f.) Ker Gawl.	[34,45]
Callisia repens L., Cynodon dactylon (L.) Persoon Eremochloa ophiuroides (Munro) Hack. Festuca arundinacea Schreb., Poa pratensis L., Sedum lineare Thunb.	[47]
Sedum aizoon L., Sedum lineare Thunb. Sedum spurium cv.Coccineum, Sedum spectabile	[48]
Briza maxima, Conyza sp.Digitaria sanguinalis, Dittrichia viscosa Filago pyramidata, Gomphocarpus fruticosus Illecebrum verticillatum, Lavandula stoechas subsp. luisieri Pleurochaete squarrosa, Sedum sediforme Teucrium scorodonia, Trifolium angustifolium, Vulpia geniculata	[57]
Axonopus Compressus, Wedelia Trilobata	[58]
Atriplex halimus, Geranium zonale, Polygala myrtifolia	[49]
Agrostis matsumurae, Cynodondactylon, Festuca elata Lolium perenne, Poa pratensis, Zoysia japonica	[56]
Eustachys retusa, Grindelia cabrerae Phyla nodiflora, Sedum mexicanum	[52]

4.2. Substrate

As a crucial component of green roofs, substrates have a significant impact on the water quality of green roof runoff. It is widely recognized that substrates possess adsorption and filtration capabilities, enabling the direct absorption and filtration of nitrogen, phosphorus, chemical oxygen demand, suspended solids, etc., thereby reducing water pollutants. Additionally, microorganisms attached to the substrate contribute to biodegradation and pollutant removal from the water. However, certain studies have indicated that green roof substrates may also act as sources of pollutants due to nutrient leaching and eutrophication processes, leading to adverse effects on runoff water quality. Moreover, substrates indirectly influence plant growth and function while affecting water quality as well. Depending on the type of growing substrate used [59], green roofs can serve both as sources and sinks for pollutants. Therefore, when constructing green roofs, careful consideration should be given to factors such as the choice of materials, appropriate proportions, substrate thickness, and the presence of amendment. The composition of green roof substrate components and their substrate depth settings in rainwater and sewage treatment applications from 2014 to 2024 are presented in Table 2.

4.2.1. Substrate Composition and Proportion

The study suggests that the substrate components can be categorized into inorganic and organic constituents [60]. The inorganic components serve as the source of mineral elements, which are beneficial for enhancing cation exchange and providing trace nutrients to plants. The organic components also provide nutrition for plant growth [61]. In the design of actual substrates, it is common to have the simultaneous presence of both inorganic and organic components. Given the variations in physical properties and chemical characteristics among different components, it is typically imperative to experimentally determine the proportions and combinations for achieving optimal performance in green roof treatments.

In recent years, Peczkowski et al. have developed two types of substrates for green roofs: one based on Lightweight Expanded Clay Aggregate (LECA) consisting of 60%

horticultural soil, 20% sand, and 20% LECA (4–8 mm in size), and the other based on perlite comprising 60% horticultural soil, 20% sand, 15% perlite, and 5% LECA (4–8 mm in size) [62]. The study investigated various parameters, including TN, NO_3^--N, NO_2^--N, NH_4^+-N, TP, PO_4^{3-}-P, as well as heavy metals such as copper (Cu), zinc (Zn), lead (Pb), and cadmium (Cd). The findings suggest that no substantial enhancement was observed in the water quality of green roofs. The concentrations of TN, copper (Cu), and zinc (Zn) in the LECA substrate and perlite substrate were found to be significantly higher than those in rainwater, with a notable increase observed specifically in the concentration of the perlite substrate.

The composition of green roof substrates (13 samples) and commonly used mineral compounds (29 samples of building aggregates) in terms of phosphorus (P), copper (Cu), nickel (Ni), cadmium (Cd), and zinc (Zn) concentration was investigated by Karczmarczyk et al. [63]. The concentration of P, Cu, Ni, Cd, and Zn in the runoff water from green roofs was also determined. The results revealed a positive correlation between the metal concentration in substrate composition and the quality of runoff water. In the study, with the exception of one group where substrate samples had previously received fertilization, disregarding the fertilizer factor, the findings suggest that potential pollution in green roofs is associated with the specific substrate chosen and its composition of compounds. Natural materials such as sand and gravel, as well as artificial materials like expanded clay and ash and crushed red brick, are considered to potentially contribute to higher levels of phosphorus pollution. Sand and crushed red brick may result in increased leaching of nickel concentration.

Rey et al. employed a combination of organic components, including compost (C), coco-peat (CP), rice husk (R), and humic soil (So), along with inorganic components such as expanded clay (ECl), perlite (P), coarse pumice (Pu), sand (S), and zeolite (Z) materials, to design various volume ratios of substrates [64]. These substrates were compared against commercially available extensive and intensive substrates. The findings revealed that the substrate (So20:ECl10:Pu40:S10:P10:Z10) and the substrate (CP20:ECl5:Pu60:S5:P5:Z5) exhibited favorable physical characteristics, including low bulk density and high water retention capacity, which supported normal plant growth. *Paepalanthus alpinus* plants demonstrated a 100% survival rate when grown in these two substrates. In comparison to the commercial extensive substrate, both mixed substrates showed significantly reduced concentrations of TKN, PO_4^{3-}-P, TSS, turbidity, COD, BOD, and coliforms; however, they still acted as sources for these pollutants when compared to rainwater inflow. No significant difference was observed in TP, NO_3^--N, and NO_2^--N between the effluent concentrations of the mixed substrates and influent concentrations. The modified mixed substrates displayed lower pollutant concentrations compared to the commercial extensive substrate, suggesting their effectiveness in reducing runoff pollution.

The leaching of nutrients and organic matter from the substrate can result in the eutrophication of water and an elevation in pollutant concentrations in the effluent, potentially surpassing those found in rainwater. Therefore, to mitigate the substrate's potential as a source of pollutants, careful selection of appropriate substrate materials and proportions is crucial to enhance both pollutant degradation and retention capacity within the substrate.

The substrate design by Vijayaraghavan and Raja involved varying proportions of vermiculite, perlite, crushed brick, sand, and coco-peat [65]. Out of the 18 designs tested, the mixture consisting of 20% vermiculite, 30% perlite, 20% crushed brick, 10% sand, and 20% coco-peat exhibited superior characteristics compared to other combinations and individual media groups. This particular composition demonstrated a bulk density of 431 kg m^{-3}, an air-filled porosity of 19.5%, a hydraulic conductivity reaching up to 4570 mm h^{-1}, and a water holding capacity of up to 39.4%. Moreover, Portulaca grandiflora plants cultivated in this mixed substrate displayed optimal growth performance with a biomass increase of approximately 380%. Furthermore, this specific mixture composition showcased an effective removal rate exceeding 97% for heavy metal ions (Al, Cd, Cr, Cu, Fe, Ni, Pb, and Zn).

Afterward, Vijayaraghavan and Badavane employed varying volume ratios of Purosil (a processed siliceous soil), vermiculite, Sand, LECA (Lightweight Expanded Clay Aggregate), coco-peat, and Sargassum wightii (commonly known as Sargassum seaweed) for the substrate design [66]. The findings revealed that out of the 13 different ratio designs tested, the optimal combination for the substrate mixture consisted of 20% Purosil, 30% vermiculite, 10% sand, 20% LECA, 10% coco-peat, and 10% S. wightii. This blend exhibited a bulk density of 495 kg m^{-3} with an air-filled porosity of 21%, a hydraulic conductivity reaching up to 5524 mm $hour^{-1}$, and a water holding capacity of up to 67.6%. Moreover, the Portulaca grandiflora plants cultivated on this substrate demonstrated robust growth performance, with an approximately 2.72-fold increase in biomass over a period of 40 days of operation. Furthermore, the mixed substrates composed of these six materials at different volume ratios also displayed remarkable binding capacities towards heavy metals (Al, Cd, Cr, Cu, Fe, Ni, Pb, and Zn), with removal rates surpassing 93.7%.

4.2.2. Substrate Thickness

When constructing green roof substrates, it is crucial to consider not only the composition and proportion of the substrate but also its optimal thickness. The thickness of the substrate plays a significant role in water quality improvement by influencing plant and microbial growth as well as functionality. A thicker substrate prolongs water passage, thereby increasing contact time between pollutants and the substrate, plant roots, and microorganisms. This extended contact enhances pollutant removal efficiency. Moreover, a thicker substrate with more pores and a larger surface area facilitates improved filtration and sedimentation of particulate matter. Additionally, it supports higher richness and diversity of microbial attachment growth, which efficiently transforms nutrients and decomposes organic matter for effective pollutant treatment. Furthermore, plants' roots can fully develop in a thicker substrate, enabling them to absorb more nutrients directly while providing robust physical filtering at root zones along with larger surfaces for microbial growth. Considering these comprehensive factors mentioned above, it is evident that a thicker substrate often yields superior treatment results; however, system operation costs and overall design should also be taken into account.

Chai et al. utilized perlite and recycled bricks to establish green roofs with substrate thicknesses of 10 cm and 20 cm, respectively [46]. The findings indicated that the adsorption capacity of suspended solids (SS) was influenced by the substrate thickness, with a thicker substrate exhibiting greater SS adsorption ability and reducing the concentration of SS in the effluent. Furthermore, substrate thickness had a significant impact on NH_4^+-N concentration in the effluent ($p < 0.05$). Increasing the substrate depth from 10 cm to 20 cm resulted in a decreased NH_4^+-N concentration, indicating that increasing the substrate depth can better regulate NH_4^+-N through adsorption, retention, and transformation processes. However, this study did not observe any significant effect of substrate thickness on TN, TP, and COD concentrations.

In their study, Gong et al. investigated the efficacy of green roofs with varying substrate thicknesses (10 cm, 15 cm, and 20 cm) and a module area of 0.5 square meters in treating runoff water quality [16]. The findings revealed a gradual reduction in average concentrations of TN, NO_3^--N, and NH_4^+-N in the effluent as the substrate thickness increased.

The study conducted by Thomaidi et al. employed perlite and vermiculite as substrates for the establishment of green roofs, with substrate thicknesses of 10 cm and 20 cm, respectively [49]. The findings revealed that when the substrate thickness was increased to 20 cm, both types of substrates demonstrated enhanced removal efficiencies for TSS, Turbidity, BOD, and COD, achieving optimal removal rates of 93%, 93%, 91%, and 91%, respectively. However, a notable decline in removal efficiency was observed when the substrate thickness decreased to 10 cm, resulting in removal rates ranging from 60% to 75% for the aforementioned indicators.

The study conducted by Park et al. investigated the impact of varying substrate thicknesses on effluent quality in green roofs, employing substrate thicknesses of 10 cm, 20 cm, and 40 cm [24]. The findings revealed a negative correlation between substrate thickness and heavy metal concentrations (Cu, Zn, Mn, and Cd) in roof runoff, particularly for Cu and Zn. Notably, an increase in substrate thickness was observed to significantly reduce the concentrations of these metals.

4.2.3. Substrate Amendment

Incorporating soil amendments into the substrate can enhance the development of aggregate structures (AS), which serve as fundamental components of soil. Well-developed AS possess excellent water retention capacity and mitigate nutrient leaching. Amendments are recognized as effective measures for controlling nutrient leaching, and their application in green roof substrates can contribute to controlling pollution.

When utilized as a soil amendment, biochar has the potential to enhance the absorption of both inorganic and organic pollutants while mitigating the leaching of nitrogen and phosphorus from the soil. In recent years, researchers have employed biochar on green roofs to investigate its impact on runoff quantity and quality. However, studies have yielded divergent conclusions regarding the effects of biochar on nutrient concentrations in green roof runoff. Some studies suggest that biochar can reduce nutrient concentrations in runoff [59,67], while others indicate that it may release nutrients during operation, resulting in elevated nutrient levels in the runoff. This phenomenon is attributed to the fact that biochar is derived from mineral-rich sewage sludge [68]. Researchers have conducted a comprehensive investigation on green roofs, examining the impact of three different sources of biochar, namely wood biochar, sewage sludge biochar, and food waste biochar [69]. The incorporation of wood biochar significantly influences the effluent quality of green roofs by reducing average concentrations of NH_4^+-N, NO_3^--N, TP, COD, and BOD_5 by 63%, 4%, 13.4%, 32.7%, and 4.7%, respectively, whereas the average concentrations of metals As, Ca, Cd, Cr, Mg, Ni, and Zn decrease within a range of 6.9–99%. However, TN concentration experiences an average increase of 527%. Average concentrations of metals Cu, Hg, K, and Pb exhibit an increase within a range between 8.3 and 325.5%. These findings suggest that while wood biochar demonstrates positive effects in mitigating specific pollutants, it also acts as a source of certain contaminants. In a separate investigation conducted by Xiong et al., maize straw biochar (MSB) and rice husk biochar (RHB) were employed as materials [70]. The findings demonstrated that in comparison to rice husk biochar, corn straw biochar exhibited superior efficacy in reducing TN, NO_3^--N, and DOC concentrations. Both types of biochar significantly elevated the levels of TP and PO_4^{3-}-P in runoff water; however, the impact on phosphorus elements was relatively less pronounced for rice husk biochar. Furthermore, this study investigated the effects of varying ratios of added biochar on runoff water with three ratios set at 10%, 15%, and 20%, respectively. The results revealed a significant decrease in TN and NO_3^--N concentrations in runoff water as the ratio of added biochar increased while TP and PO_4^{3-}-P concentrations gradually rose.

The granulation process and particle size of biochar have significant impacts on stormwater runoff from green roofs. Granulated biochar can enhance plant growth, leading to increased leaf area and final biomass [71]. Moreover, the use of granulated biochar reduces TSS concentration and improves the water quality index (WQI). Smaller particle-sized biochar is more effective in reducing particle loss and nutrient leaching compared to larger particles due to its ability to form water-stable aggregates and stronger water retention capacity [42].

In addition to utilizing biochar as an amendment, researchers are also investigating the utilization of alternative substances to enhance green roof performance. Zhang et al. conducted a study on the pollution control capability of polyaluminium chloride (PAC) and bentonite when introduced into runoff water from green roofs [72]. The findings revealed that both amendments compromised the ability of green roofs to regulate NH_4^+-N. Various concentrations of PAC and bentonite were examined, demonstrating that green roofs with PAC consistently exhibited superior pollutant control compared to those with bentonite. Specifically, the green roof incorporating 2.0% PAC displayed an increase in removal rates for NO_3^--N, TN, and TP by 204.50%, 148.36%, and 38.00%, respectively, in comparison to the group without any amendment. Expanding upon this research, Fei et al. further investigated the treatment performance of green roofs supplemented with additions of 2% polyaluminium chloride (PAC), polyferric sulfate (PFS), polyvinyl alcohol (PVA), methylcellulose (MC), carboxymethyl cellulose sodium (CMC), and hydroxypropyl methylcellulose (HPMC) [73]. The results indicated that PVA readily formed unstable aggregates, leading to the leaching of N and P nutrients. Additionally, the inclusion of MC, CMC, or HPMC did not significantly enhance pollutant interception capabilities either; thus, it is not recommended to employ these additives at a concentration level of 2%. Both PAC and PFS facilitated bridging adsorption and coagulation between particles while improving retention capacity for pollutants within the substrate layer. However, it should be noted that incorporating PAC may result in aluminum contamination, which inhibits plant growth and pollutes the environment; therefore, PFS appears to be a more suitable choice.

Table 2. Different green roof main substrate component and depth settings.

Main Substrate Component	Depth	References
Expanded clay, Spongilite, Peat, Brick rubble, Biochar from wood, Biochar from sewage sludge, Biochar from food waste, and Dried sewage sludge	150 mm	[69]
Expanded clay, Crushed marl, Peat, Recycled bricks, Biochar	100 mm	[74]
Peat soil, vermiculite	250 mm	[73]
Fractured tiles, Red lava, Fine pumice, Compost, Peat, Sand, Coconut fiber, Gravel	60 mm, 90 mm	[75]
Soil, Rice husk biochar, Maize stalk biochar, Perlite, Vermiculite	100 mm	[70]
Soil, Cocopeat, Loofah, Perlite	80 mm	[76]
Compost, Paper sludge, Pelletized paper sludge, Vulcaflor	100 mm	[77]
Unprocessed biochar, Granulated biochar	80 mm	[42]
Perlite, Vermiculite, LECA	150 mm, 250 mm	[49]
Biochar, Vermiculite, Porous aggregates, Composted organic matter, Fine sand	80 mm	[71]
Perlite, Peatmoss, Vermiculite	100 mm, 200 mm, 400 mm	[24]
Rural soil, Peat soil, Pine needle, Perlite, Vermiculite	50 mm, 100 mm, 150 mm	[45]
Peat soil, Vermiculite, Perlite, Biochar, Sawdust	50 mm, 100 mm, 150 mm, 200 mm	[78]
Wheat straw	>200 mm	[79]
Sand, Gravel, Limestone, Lightweight Aggregates, Expanded clay and ash, Crushed red brick	-	[63]
Sand, Gravel, Brick, Rubble, Bark, Peat, Compost, Polonite	120 mm	[80]
Humic soil, Compost, Coco-peat, Rice husk, Coarse pumice, Expanded clay, Sand, Zeolite, Perlite	100 mm	[64]
Horticultural soil, Sand, Expanded clay aggregate, Light expanded clay aggregate, Perlite	80 mm	[62]
Waste building material substrate, Local natural soil	200 mm, 250 mm, 300 mm	[81]
Stabilized sludge, Biochar, Pumice, Wood chips, Topsoil, Controlled release fertilizer	100 mm, 150 mm	[82]
Rural soil, Peat soil, Pine needle, Perlite, Vermiculite	50 mm, 100 mm, 150 mm	[34]
Pastoral soil, Turfy soil, Pine needles	50 mm, 100 mm, 150 mm	[16]
Peat, Vermiculite, Perlite, Sawdust, Biochar	100 mm	[59]
Rural soil, Peat soil, Pine needles, Perlite, Vermiculite	50 mm, 100 mm, 150 mm	[83]
Modified perlite, Modified recycled bricks,	100 mm, 200 mm	[46]
Perlite, Coal ash	200 mm	[84]
Local soil, Peat soil, Vermiculite, Perlite	50 mm, 100 mm	[85]
Pumice, Lava, Perlite, Activated charcoal, Zeolite	50 mm, 100 mm, 150 mm	[21]
Peat, Volcanic rock, Wood biochar, Olive husk biochar	200 mm	[86]
Expanded clay, Granulated cork, Organic matter from urban solid waste compost, Crushed egg shell	150 mm	[87]
Crushed bark, Sphagnum moss, Compost, Recycled, Crushed brick, Biochar from Birch Wood	30 mm, 40 mm	[68]
Crushed, Recycled brick, Compost, Crushed bark	50 mm	[88]
Purosil, Vermiculite, Sand, Lightweight expanded clay aggregates, Coco-peat, Sargassum wightii	100 mm	[66]
Expanded slate, Compost	10 mm	[89]
Peat soil, Vermiculite, Perlite, Sawdust,	150 mm	[90]
Vermiculite, Perlite, Crushed brick, Sand, Coco-peat	100 mm	[65]

4.3. Slope

Unlike artificial wetlands or vertical green walls, green roofs are often not positioned vertically due to their unique location and may have varying slopes. Therefore, when designing green roofs, it is crucial to consider the variable roof slope in order to accurately evaluate the performance of green roof treatments [91].

Beecham and Razzaghmanesh conducted an investigation on the water purification effects of green roofs with slopes of 1° and 25° [92]. The results revealed no significant variations in pH, turbidity, NO_3^--N, NO_2^--N, NH_4^+-N, potassium (K), sodium (Na), calcium (Ca), and magnesium (Mg) concentrations between the two slopes when considering the presence of vegetation cover as well as identical substrate type and thickness ($p > 0.05$). These findings suggest that factors related to vegetation and substrate may exert a more substantial influence on water quality compared to slope factors.

The study conducted by Castro et al. examined the impact of green roofs with slopes of 0° and 15° on runoff water quality [93]. The research findings revealed no significant differences in TN, NH_4^+-N, NO_3^--N, TP, pH, and turbidity concentration between the two slope treatments with vegetation cover. However, green roofs with a 15° slope exhibited slightly higher TSS, BOD_5, and COD concentrations compared to the group treated with a 0° slope. This observation can be attributed to the steeper incline of the 15° slope, which enhances rainwater flow and erosion capacity, facilitating the transport of solid particles and organic matter from the substrate.

In their study, Liu et al. conducted experiments on green roofs with varying slopes of 2%, 7%, and 12% [45]. The findings revealed a positive correlation between the slope of the green roof and the concentrations of F- and TP in water. Specifically, the F- concentration was significantly lower on the green roof with a slope of 2% compared to those with slopes of 7% and 12% ($p < 0.01$), while the TP concentration was significantly lower on the green roof with a slope of 7% compared to that on the green roof with a slope of 12% ($p < 0.05$). However, no significant differences were observed among the three slopes regarding TN, NH_4^+-N, NO_2^--N, Cl^-, SO_4^{2-}, pH, EC, ESP, and TSS concentration.

4.4. Operating Conditions

Different operating conditions, such as hydraulic load rate (HLR) and hydraulic retention time (HRT), exert specific influences on the treatment performance of green roofs. Excessively high HLR in the system can detrimentally impact filtration rates, while insufficient HRT diminishes water-plant and substrate contact time. Hence, when designing green roofs for water treatment purposes, it is imperative to establish reasonable and effective operating conditions.

4.4.1. Hydraulic Retention Time

The study conducted by Xu et al. investigated the treatment efficiency of a hydroponic green roof system under three different hydraulic retention times, namely 4 days, 6 days, and 8 days [27]. The findings revealed that at a hydraulic retention time of 4 days, the system exhibited an average turbidity removal rate of 67.4%, which further increased to 80.0% when the retention time was extended to 8 days. However, it is noteworthy that at a retention time of 6 days, the effluent turbidity surpassed the influent turbidity levels. In terms of chemical oxygen demand (COD) removal rates, values were determined as follows: for hydraulic retention times of 4, 6, and 8 days, respectively –69.5%, 66.0%, and 81%. Similarly, BOD_5 removal rates were observed to be 69.4%, 57.1%, and 97%. Regarding anionic surfactant (MBAS concentration indication), its elimination rates were recorded as 22.8%, 31.4%, and 88%. The extension of the appropriate hydraulic retention time leads to a significant reduction in turbidity, organic matter, and anionic surfactant concentration, thereby enhancing the quality of wastewater.

4.4.2. Hydraulic Load Rate

As the hydraulic load ratio increases, it will have a certain impact on the growth condition and survival rate of plants, as well as limit their functionality. Thi-Dieu-Hien et al. investigated the influence of wetland roofs on septic tank wastewater purification under two different hydraulic load rates (HLR1: 288 ± 19 m^3 ha^{-1} day^{-1}; HLR2: 394 ± 13 m^3 ha^{-1} day^{-1}) [26]. Under HLR1, all plants exhibited normal survival rates. However, with an increase in hydraulic load ratio from HLR1 to HLR2, plant growth was delayed, and some plants displayed yellowing leaves or even mortality. The COD removal efficiency at HLR1 ranged from 16 to 30%, with a removal rate of 67–86 kg ha^{-1} day^{-1}, whereas at HLR2, the COD removal efficiency ranged from 27 to 33%, with a removal rate of 61–79 kg ha^{-1} day^{-1}. The results revealed that although HLR1 exhibited a lower COD removal rate compared to HLR2, it demonstrated a higher removal rate than HLR2. This discrepancy can be attributed to the limited absorption and decomposition of organic matter by plants and microorganisms due to the relatively high hydraulic load rates, which result in shorter retention times. Bui Xuan et al. also conducted experiments using four different hydraulic load rates (HLR1: 137 ± 6 m^3 ha^{-1} day^{-1}; HLR2: 210 ± 7 m^3 ha^{-1} day^{-1}; HLR3: 338 ± 9 m^3 ha^{-1} day^{-1}; HLR4: 456 ± 6 m^3 ha^{-1} day^{-1}) to investigate the treatment performance of wetland roofs for septic tank effluent [94]. When the Hydraulic Load Ratio is HLR1, HLR2, HLR3, and HLR4, the ratios of nitrogen uptake by plants to nitrogen removal by the system are 0.09, 0.10, 0.13, and 0.11, respectively. The ratios of phosphorus uptake by plants to phosphorus removal by the system are 0.47, 0.50, 0.68, and 0.53, respectively. This indicates that when the hydraulic loading rate is too high, plants' ability to absorb nitrogen and phosphorus is also inhibited.

4.4.3. Water Feeding Patterns

The study conducted by Nguyen et al. investigated the impact of two water feeding patterns, namely continuous and intermittent, on the performance of wetland roof wastewater treatment [58]. Under identical substrate conditions and plant settings, the intermittent inflow method exhibited significantly higher efficiency in removing COD (62–64%) compared to the continuous inflow method (52–54%). Similarly, in terms of TN removal efficiency, the intermittent inflow method (80–87%) outperformed the continuous inflow method (73–80%). This can be attributed to enhanced oxygen diffusion into the system facilitated by the intermittent inflow method, which promotes nitrification and subsequently improves ammonium nitrogen removal. Consequently, it is evident that implementing an intermittent inflow approach can considerably enhance both COD and TN removal efficiencies in wetland roof systems; however, no significant impact was observed on TP removal.

4.4.4. Other Additional Conditions

The greywater treated by green roofs does not meet the standards for indoor non-potable use, and chlorination is considered a crucial step in the reuse of greywater. Petousi et al. integrated green roofs with chlorination technology to eliminate pathogens from greywater [95]. Their study demonstrates that when the storage period is less than 24 h, a chlorine dosage ranging from 3 to 7 mg L^{-1} can be added to the effluent to ensure water quality within microbial standards. However, if the storage period exceeds 24 h, there is a significant regeneration of pathogenic microorganisms. In such cases, adding a chlorine dosage of 7 mg L^{-1} can guarantee the complete inactivation of pathogenic microorganisms within three days for indoor non-potable use.

To address the issue of elevated total coliform concentrations, it is crucial to enhance filtration and disinfection treatment following green roof runoff management [96]. This strategic approach can effectively enhance the safety and accessibility of green roof runoff, thereby facilitating optimal utilization of rainwater resources.

4.5. Time

The newly established or installed green roofs may become a source of water pollutants, as the organic matter in these new green roofs is prone to decomposition or requires initial fertilization. Whether green roofs act as sinks or sources of pollutants depends not only on the substrate but also on their age [59].

Due to plant growth and substrate erosion, the water quality of green roofs may undergo temporal changes. Gong et al. conducted a two-year monitoring study on the water quality of a green roof from 2012 to 2013 [97]. The results demonstrated that after one year of operation, there was a significant decrease in nitrogen and phosphorus concentration in the outflow, both being lower than rainwater concentrations. Furthermore, it was observed that the green roof exhibited acid rain neutralization capabilities, pH stabilization, and reduction in turbidity and COD concentrations, leading to substantial enhancement of water quality.

The study conducted by Speak et al. compared the runoff water quality between a 43-year-old green roof and an adjacent traditional roof surface [98]. The findings demonstrated that the green roof functioned as a sink for PO_4^{3-}-P and NO_3^--N, effectively removing these pollutants from the runoff. However, monitoring revealed higher concentrations of plumbum (Pb) in the green roof runoff compared to rainwater, indicating that green roofs may contribute to Pb pollution in water bodies. Therefore, it is crucial to carefully consider plant and substrate selection while establishing green roofs for purification and restoration purposes, particularly considering the potential for aging green roofs to become sources of metal contaminants.

Harper et al. conducted a nine-month study investigating the impact of green roofs on runoff water quality [91]. As the duration of operation increased, there was a gradual decrease in the concentrations of TN and TP in the effluent from the green roof. The initial concentration of TN, which exceeded 60 mg L^{-1}, decreased to approximately 10 mg L^{-1}, while TP concentration reduced from an initial value exceeding 30 mg L^{-1} to approximately 5 mg L^{-1}. Furthermore, TOC levels in the system started at an initial concentration of 500 mg L^{-1} and gradually declined to 50 mg L^{-1} after several weeks, indicating a progressive reduction in organic matter dissolution and loss within the green roof with increasing operational time.

Akther et al. conducted laboratory and field investigations to examine nutrient leaching from green roofs [99]. The findings revealed an initial higher level of nutrient leaching, particularly nitrogen and phosphorus, during the early stages of green roof operation. However, with a prolonged duration of operation, there was a general decrease in the extent of nutrient leaching. This decline can be attributed to a reduction in available exchangeable nutrients within the substrate over time, resulting in diminished nutrient leaching. Consequently, long-term operation may contribute to an enhancement in water quality originating from green roofs.

4.6. Weather

Meteorological conditions, such as temperature, humidity, and precipitation, can all have an impact on the water quality of green roofs. Temperature and humidity play a crucial role in plant growth and physiological processes, as well as microbial activity, which subsequently affects water quality. The frequency and intensity of precipitation also influence both the runoff volume and water quality of green roofs; higher frequencies and intensities may result in increased pollutant transport from the substrate. Due to seasonal variations in temperature, humidity, and precipitation patterns, there is often a discernible seasonal fluctuation in the water quality of green roofs.

4.6.1. Temperature

Buffam et al. observed a positive correlation between temperature and nutrient leaching (N and P) from green roofs [100], potentially attributed to enhanced microbial mineralization and organic matter decomposition with increasing temperature, resulting in elevated N and P concentration in runoff water. However, contrary findings from other studies [101,102] suggest that temperature does not significantly influence net mineralization rates of N and P in soils. Akther et al. discovered a negative relationship between nutrient concentrations (N and P) on green roofs and substrate temperatures [103]. The results imply that as temperature rises, the nutrient concentration decreases within the system. This phenomenon may be due to plants thriving under higher temperatures with improved nutrient absorption capabilities, surpassing the effects of microbial mineralization and nutrient decomposition caused by elevated temperatures, ultimately leading to reduced nutrient leaching.

4.6.2. Humidity

The impact of humidity on the water quality of green roofs was also examined by Akther et al. from a humidity perspective [103]. It was observed that an increase in humidity led to an increase in nutrient leaching while metal element leaching decreased. These findings suggest that in high-humidity environments, accelerated organic matter decomposition results in greater leaching of nutrients (N and P), whereas metals are retained due to enhanced precipitation or adsorption.

4.6.3. Precipitation

Studies have indicated that green roofs exhibit lower concentrations of NO_3^--N in their runoff during large precipitation events, while TP and PO_4^{3-}-P concentrations are higher [104]. However, a separate study suggests that the concentration of TP on green roofs is not affected by precipitation events [105]. Buffam et al. investigated the impact of temperature, humidity, and precipitation conditions on the concentrations of dissolved nutrients, alkaline cations, and metals in runoff from green roofs [100]. The findings demonstrate relatively low levels of various element concentrations in green roof runoff during significant precipitation events. This study also highlights that temperature has a more substantial influence on water quality compared to humidity and precipitation for green roof runoff. Nevertheless, research also indicates that the magnitude of precipitation events significantly affects the water quality of green roof runoff [106]. The results reveal an increase in TN, NO_3^--N, PO_4^{3-}-P, SO_4^{2-}, and DOC concentrations in outflow with rising rainfall amounts as well. During larger rainfall events, TN and TP removal rates display a negative correlation with rainfall intensity [107], suggesting that nutrient removal efficiency by green roofs weakens with increased rainfall intensity.

4.6.4. Climate Zone

The meteorological conditions in the vicinity of green roof systems typically encompass factors such as temperature, humidity, and rainfall. It is imperative to particularly focus on the operational status and water treatment efficacy of green roofs within specific climatic regions.

Guo et al. conducted experiments in the Mediterranean climate region, where they planted 11 types of plants on green roofs, including five herbaceous plants, three subshrubs, and three shrubs [108]. Their research was carried out under the hot and dry conditions of the local summer and confirmed that shrubs and subshrubs exhibited a higher survival rate compared to herbaceous plants. This disparity may be attributed to the morphological characteristics possessed by shrubs and subshrubs that enable them to adapt to arid environments, such as thicker wax layers and more efficient leaf-shedding mechanisms. The impact of the Mediterranean climate on green roof vegetation is primarily manifested through factors like high temperatures and dryness during summer, which determine plant survival and growth performance on green roofs while indirectly influencing water purification effects. Rocha et al., also conducting related experiments in the Mediterranean region, observed better water purification effects during autumn and winter compared to spring and summer [57]. This discrepancy can be attributed to rainfall being predominantly concentrated in autumn and winter within the Mediterranean climate, whereas spring and summer are relatively drier seasons. During periods of increased rainfall, green roofs can effectively absorb and filter rainwater, thereby reducing runoff volume. During autumn/winter, when soil moisture levels are higher, plant growth remains robust with an increased root density that aids in retaining water infiltration capacity, consequently enhancing water purification effects.

Akther et al. investigated the impact of cold semi-arid climates on the water purification effectiveness of green roofs [103]. The region experiences prolonged and frigid winters with repeated freeze-thaw cycles, which may influence the mineralization and leaching of nutrients in green roof substrates, consequently affecting their water purification performance. Moreover, due to limited precipitation in the area, snowmelt plays a significant role in runoff and significantly influences the chemical leaching behavior of green roofs; indeed, similar chemical leaching behaviors are observed during rainfall events and snowmelt events. During rainfall events, higher nutrient leaching rates occur in spring due to increased soil moisture content and elevated growth medium temperatures that favor nutrient mineralization processes. However, lower nutrient leaching rates are observed during the summer and autumn seasons. It is important to note that these nutrient leaching and mineralization processes not only impact the water purification effectiveness of green roofs themselves but also pose potential risks to downstream water bodies. Therefore, a comprehensive understanding and effective control of these nutrient leaching and mineralization processes are crucial for optimizing design.

The study conducted by Sultana et al. involved a comprehensive assessment of water quality in rainwater collected from green roofs under tropical climate conditions, with a primary focus on indicators such as dissolved oxygen (DO), pH value, electrical conductivity, and temperature [109]. The findings revealed that the green roof system exhibited excellent water quality performance in tropical climates, enabling direct utilization of untreated rainwater for toilet flushing and garden irrigation purposes. Furthermore, all samples maintained temperatures within the standard range, which indicates effective heat regulation by the green roof system.

Rey and his team conducted research in a neotropical mountain climate characterized by a bimodal precipitation pattern, with two wet seasons (March–April and October–November) and two dry seasons (January–February and July–August) throughout the year [64]. The study categorized rainfall events into three groups based on their depth and duration: 'large events' (longer duration or greater depth), 'intermediate events' (falling between large and small events in terms of duration and depth), and 'small events' (shorter duration or lesser depth). The findings revealed that intermediate-sized rainfall events were linked to longer preceding drought periods, indicating that the dry substrate has the capacity to absorb more water during such events after experiencing an extended period of drought, thereby preventing rapid water loss. This phenomenon potentially enhances rainwater purification on green roofs. Additionally, following the end of a drought period, nutrients within the substrate may have accumulated to certain levels. Intermediate-sized rainfall events could transport these nutrients downstream through runoff, which might impact water quality. However, if both plants and substrate can efficiently utilize these nutrients, it may reduce nutrient losses via runoff while improving water quality from green roof outflows.

The experiment conducted by Ferrans et al. took place in the same location as Rey et al., which is characterized by a typical subtropical highland climate [14]. Bogota's rainfall pattern exhibits a bimodal distribution, with increased precipitation during the rainy season leading to higher runoff from green roofs, while reduced precipitation during the dry season impacts the water retention capacity of green roof systems. Seasonal variations were observed in pollutant concentrations within the outflow from the system, with elevated levels of BOD and TSS during the dry season and heightened levels of COD and total coliforms during the rainy season. These fluctuations may be attributed to rainfall patterns, temperature fluctuations, and plant growth activity.

4.7. Processing Objects

The management of rainwater runoff has been a primary focus in the investigation of green roof systems, with extensive research also conducted on harnessing these systems to enhance the water quality of rainwater. However, there is limited research on integrating green roofs into wastewater treatment processes. The utilization of conventional green roofs for wastewater treatment may yield suboptimal outcomes; nevertheless, wetland roofs provide an effective approach for treating and reusing such waste streams. By utilizing wastewater as an irrigation source for green roofs, irrigation costs can be minimized, and a secure and environmentally friendly solution for waste management can be attained. The current state of research on the treatment performance of green roof systems in wastewater treatment is presented in Table 3.

4.7.1. Greywater

Petreje et al. used recycled crushed building rubble containing a large proportion of brick as the substrate for wetland roofs to treat greywater [74]. The findings demonstrated that wetland roofs significantly mitigated the concentrations of total nitrogen and orthophosphate in the greywater. Ramprasad et al. employed the GROW (Green Roof-top Water Recycling System) wetland roof system for greywater treatment, incorporating eight different plant species into their design [110]. The GROW system exhibited remarkable efficacy in treating greywater with removal rates of 91.7% for TN, 83.6% for NO_3^--N, 87.9% for TP, 90.8% for BOD, 92.5% for COD, 91.6% for TSS, 93.4% for PG, 91.4% for FC, 88.9% for TMA, and 85.7% for SDS. Thomaidi et al. utilized green roofs to address greywater generated from buildings [49]. Among their systems, *Atriplex halimus* planted and filled with a layer of vermiculite at a depth of 20 cm achieved optimal treatment performance with removal rates reaching up to 93%, 93%, 91%, and 91% for TSS, turbidity, BOD, and COD, respectively.

4.7.2. Blackwater

The green roof exhibits not only a significant treatment effect on greywater but also effectively processes blackwater for potential reuse. Bui Xuan et al. (2014) developed a shallow subsurface flow wetland roof system planted with *Melampodium paludosum* to treat septic tank wastewater [94]. The findings demonstrated that the system achieved an average removal rate of 88–91% for TN, 77–78% for COD, and 72–78% for TP, thereby meeting local standards for water reuse and surface water discharge. Thi-Dieu-Hien et al. designed four distinct plant-based shallow wetland systems to treat septic tank wastewater, among which the average COD removal rates ranged from 61% to 79%, TN removal rates ranged from 54% to 81%, TP removal rates ranged from 62% to 83%, and suspended solids exhibited an average removal rate of 88 $\pm$ 3% [25]. Furthermore, the treated water complied with national standards for both discharge and reuse.

4.7.3. Other Types of Water

Green roofs are capable of effectively managing not only rainwater but also various types of wastewater, such as greywater and blackwater. A study conducted by researchers investigated the influence of seawater irrigation on green roofs planted with salt-tolerant plants [111]. The findings indicate that both partial and complete utilization of seawater for irrigation purposes are viable options. Particularly, when alternating between seawater and tap water for irrigation every four days, there is no negative impact on plant growth. Additionally, irrigating with seawater exclusively every four days still enables satisfactory plant development. This study presents novel insights for future irrigation water sources in green roofs, proposing the utilization of seawater in regions experiencing water scarcity. Simultaneously, considering the prevailing issue of severe water pollution, it is worthwhile to explore the potential use of green roofs for treating contaminated bodies of water such as lakes and seas. This approach not only conserves irrigation water but also exhibits a positive impact on water purification.

Table 3. Overview of main studies concerning wastewater treatment through green roofs.

Processing Objects	Influent (mg/L)		Effluent (mg/L)		References
	N and P Concentration	Organic Concentration	N and P Concentration	Organic Concentration	
Greywater	TN: 10.1 ± 2.7 TP: 7.6 ± 2.4	COD: 226 ± 60 BOD: 132 ± 36	TN: 4.9 ± 2.7 TP: 3.9 ± 2.1	COD: 20–36 BOD: 20 ± 11	[49]
Greywater	-	COD: 226 ± 60 BOD_5: 132 ± 36	-	COD: 25 ± 17 BOD_5: 14 ± 10	[95]
Greywater	NH_4^+-N: 1.9–3.3 TN: 2.3–4.6 TP: 0.34–0.36	COD: 234–313 BOD_5: 121–149	NH_4^+-N < 3 TN < 4 TP < 0.4	COD: 53.6 BOD_5: 4.1	[27]
Greywater	TN: 16.3 (HLR1) TN: 16.7 (HLR2) TP: 1.2 (HLR1) TP: 2.6 (HLR2)	COD: 635.2 (HLR1) COD: 1115 (HLR2) BOD_5: 393.3 (HLR1) BOD_5: 407.7 (HLR2)	TN: 1.2 (HLR1) TN: 1.1 (HLR2) TP: 0.4 (HLR1) TP: 0.3 (HLR2)	COD: 69.1 (HLR1) COD: 61.3 (HLR2) BOD_5: 10.6 (HLR1) BOD_5: 5.9 (HLR2)	[112]
Greywater	NH_4^+-N: 10.28 –14.56 NO_3^--N: 12.32–7.84 TP: 2.934–3.84	COD: 216–320 BOD5: 68–120	NH_4^+-N: 0.67–0.95 NO_3^--N: 1.2–3.5 TP: 0.8–1.4	COD < 10BOD_5 < 20	[110]
Greywater	NH_4^+-N: 1.2 ± 0.3 NO_3^--N: 1.6 ± 0.3 TP: 0.7 ± 0.1	COD: 81.9 ± 4.1 BOD5: 19.0 ± 0.9	No clear change	BOD_5 < 10	[113]
Greywater	-	COD: 87 (low level) COD: 495 (high level) BOD_5: 20 (low level) BOD_5: 164 (high level)	-	COD: 19 (low level) COD: 159 (high level) BOD_5: 2 (low level) BOD_5: 80 (high level)	[114]
Blackwater	TP: 5.8 ± 0.6	COD: 176 ± 43	TP: 1.3–7	COD: 25–65	[94]
Blackwater	TKN: 42 ± 7 NH_4^+-N: 38 ± 2 NO_3^--N: 0.5 ± 0.3 TP: 1.5 ± 0.7	COD: 108 ± 53	TN: 14 ± 3 TP: 0.4 ± 0.3	COD: 32 ± 26,	[25]
Blackwater	TKN: 42 ± 7 NH_4^+-N: 38 ± 2 NO_3^--N: 0.5 ± 0.3 TP: 1.5 ± 0.7	COD: 108 ± 53	TN: 10 ± 4 (HLR1) TN: 10 ± 2 (HLR2) TP: 0.7 ± 0.3 (HLR1) TP: 0.4 ± 0.3 (HLR2)	COD: 29 ± 16 (HLR1) COD: 34 ± 23 (HLR2)	[26]

5. Conclusions

The selection of appropriate plant species is a crucial factor in establishing green roof systems, as plants have the potential to impact the functionality, stability, and long-term viability of such roofs. Plants play a significant role in maintaining the physical, chemical properties, and microbial conditions of green roof substrates and can effectively reduce nutrient concentrations in runoff post-planting. It is advisable to avoid deciduous plants due to their leaf litter decomposition, which may elevate nutrient levels in runoff. The limited range of plant species suitable for rooftop growth is primarily attributed to specific environmental constraints. When utilized for wastewater treatment purposes, selecting plants with well-developed root systems, large leaf areas, and high growth rates can yield favorable treatment outcomes. Additionally, considering interactions between different plant species and substrates is essential for optimizing treatment performance from green roofs. Substrates also hold critical importance in green roof system design; varying materials and proportions may result in nutrient leaching during rainfall runoff, which could classify green roofs as sources of pollution. Therefore, it is imperative to carefully select materials that do not contribute to pollution while designing substrate mixtures appropriately for optimal performance. Thicker substrates exhibit the ability to decrease pollutant concentrations in effluent while providing effective water purification effects. Furthermore, incorporating suitable amendments can enhance pollutant interception capacity by green roofs. Consequently, when designing compositionally or materially diverse green roof substrates based on site conditions and actual requirements, it becomes necessary to determine substrate proportions, thicknesses, and amendment application methods.

The slope of green roofs does not significantly affect the retention and removal efficiency of most pollutants. This may be due to the fact that the actual roof slopes are usually relatively gentle. Furthermore, during the design phase of green roofs, operational conditions such as hydraulic loading rate (HLR), hydraulic retention time (HRT), and water feeding patterns should be taken into consideration. With the prolonged operation of green roofs, there is a potential for improved pollutant removal efficiency. Enhanced treatment performance can be achieved through mature plant growth, stable substrate composition, and a diverse microbial community. When establishing long-term existing green roofs with purification and restoration functions, careful selection of suitable plants and substrates is crucial to avoid the accumulation of pollutants from components and materials used. Local climatic conditions, including temperature, humidity, and precipitation, also influence the performance of green roofs. When designing green roof systems, priority should be given to the environmental adaptability of the plants and the stability of the substrate to cope with seasonal changes, extreme temperatures, and precipitation conditions in different climate zones.

6. Prospective

In general, while some studies have considered green roofs as a potential source of pollutants, these findings can be attributed to substrate and plant factors. However, specific research has demonstrated that green roofs, particularly wetland green roofs, exhibit significant potential in the treatment of wastewater. It has been discovered that properly designed green roofs in urban areas can effectively treat both greywater and blackwater. Consequently, they serve as promising on-site rainwater treatment technology and an integral component of sustainable urban design and green building practices in the future.

The article critically examined green roofs and investigated their efficacy in nutrient removal, organic pollutant degradation, and heavy metal contamination mitigation, thereby showcasing the immense potential of green roofs in promoting sustainable water resource management and enhancing water quality purification in future urban areas. Drawing upon existing research findings, the following avenues for further development are suggested:

(1) Future studies could concentrate on selecting plant species with superior pollutant removal capabilities to enhance the efficiency of water quality purification by green roofs. Additionally, attention should be given to mixed planting strategies aimed at augmenting biodiversity within green roof systems.

(2) To address concerns regarding substrate composition and leaching of pollutants from materials, it is imperative to conduct research focused on developing novel materials characterized by reduced nutrient leaching rates so as to minimize pollution risks associated with green roofs.

(3) While most investigations have primarily centered around plants and substrates within green roof systems, insufficient emphasis has been placed on comprehending the impact of microorganisms on water quality purification processes.

(4) Green roofs have demonstrated effectiveness in treating rainwater, greywater, and blackwater; henceforth, additional research endeavors can be undertaken to explore their applicability for remediating other polluted bodies of water, such as lakes or seawater.

(5) Limited attention has been devoted to investigating the treatment efficacy of pathogens, antibiotics, and recalcitrant pharmaceuticals present in wastewater using green roof technologies. Therefore, future studies should prioritize strengthening research efforts in this area.

Author Contributions: Conceptualization, J.Y. and P.Y.; methodology, J.Y. and P.Y.; software, J.Y. and P.Y.; validation, Z.W.; formal analysis, J.Y. and P.Y.; investigation, J.Y. and P.Y.; resources, J.Y. and P.Y.; data curation, J.Y. and P.Y.; writing—original draft preparation, J.Y., P.Y., B.W. and S.W.; writing—review and editing, S.W., M.Z., Y.Z. and C.F.; visualization, Z.W.; supervision, Y.Z. and C.F.;

project administration, J.Y. and P.Y.; funding acquisition, X.Z. and C.F. All authors have read and agreed to the published version of the manuscript.

Funding: This research was funded by the Wenzhou Ecological Park Research Project (grant number SY2022ZD-1002-07) and the Wenzhou Science and Technology Project for Basic Society Development (grant number S20220015).

Data Availability Statement: The data presented in this study are available on request from the corresponding author. The data are not publicly available due to funder restrictions.

Acknowledgments: The authors express their sincere gratitude for the work of the editor and the anonymous reviewers.

Conflicts of Interest: The authors declare no conflicts of interest.

Nomenclature

TN	Total nitrogen	SS	Suspended solids
NO_3^--N	Nitrate nitrogen	EC	Electrical conductivity
NO_2^--N	Nitrite nitrogen	TDS	Total dissolved solids
NH_4^+-N	Ammonium nitrogen	ESP	Exchangeable sodium percentage
TKN	Total kjeldahl nitrogen	MBAS	Methylene blue active substance
TP	Total phosphorus	DOC	Dissolved organic carbon
PO_4^{3-}-P	Orthophosphate	FC	Fecal coliform
COD	Chemical oxygen demand	TMA	Trimethyl amine
BOD	Biochemical oxygen demand	SDS	Sodium do-decyl sulphate
TSS	Total soluble solid	PG	Propylene glycol

References

1. Oberndorfer, E.; Lundholm, J.; Bass, B.; Coffman, R.R.; Doshi, H.; Dunnett, N.; Gaffin, S.; Koehler, M.; Liu, K.K.Y.; Rowe, B. Green roofs as urban ecosystems:: Ecological structures, functions, and services. *Biosci.* **2007**, *57*, 823–833. [CrossRef]
2. Rowe, D.B.; Getter, K.L.; Durhman, A.K. Effect of green roof media depth on Crassulacean plant succession over seven years. *Landsc. Urban Plan.* **2012**, *104*, 310–319. [CrossRef]
3. Rafael, S.; Correia, L.P.; Ascenso, A.; Augusto, B.; Lopes, D.; Miranda, A.I. Are green roofs the path to clean air and low carbon cities? *Sci. Total. Environ.* **2021**, *798*, 149313. [CrossRef] [PubMed]
4. Motlagh, S.H.B.; Pons, O.; Hosseini, S.M.A. Sustainability model to assess the suitability of green roof alternatives for urban air pollution reduction applied in Tehran. *J. Affect. Disord.* **2021**, *194*, 107683. [CrossRef]
5. Feitosa, R.C.; Wilkinson, S.J. Attenuating heat stress through green roof and green wall retrofit. *J. Affect. Disord.* **2018**, *140*, 11–22. [CrossRef]
6. Wooster, E.; Fleck, R.; Torpy, F.; Ramp, D.; Irga, P. Urban green roofs promote metropolitan biodiversity: A comparative case study. *J. Affect. Disord.* **2022**, *207*. [CrossRef]
7. Friedler, E.; Hadari, M. Economic feasibility of on-site greywater reuse in multi-storey buildings. *Desalination* **2006**, *190*, 221–234. [CrossRef]
8. Ghaitidak, D.M.; Yadav, K.D. Characteristics and treatment of greywater—a review. *Environ. Sci. Pollut. Res.* **2013**, *20*, 2795–2809. [CrossRef] [PubMed]
9. Leigh, N.G.; Lee, H. Sustainable and Resilient Urban Water Systems: The Role of Decentralization and Planning. *Sustainability* **2019**, *11*, 918. [CrossRef]
10. Capodaglio, A.G. Integrated, Decentralized Wastewater Management for Resource Recovery in Rural and Peri-Urban Areas. *Resources* **2017**, *6*, 22. [CrossRef]
11. Lam, L.; Kurisu, K.; Hanaki, K. Comparative environmental impacts of source-separation systems for domestic wastewater management in rural China. *J. Clean. Prod.* **2015**, *104*, 185–198. [CrossRef]
12. Capodaglio, A.G.; Callegari, A.; Cecconet, D.; Molognoni, D. Sustainability of decentralized wastewater treatment technologies. *Water Pr. Technol.* **2017**, *12*, 463–477. [CrossRef]
13. Santos, A.V.; Lin, A.R.A.; Amaral, M.C.S.; Oliveira, S.M.A.C. Improving control of membrane fouling on membrane bioreactors: A data-driven approach. *Chem. Eng. J.* **2021**, *426*, 131291. [CrossRef]
14. Ferrans, P.; Rey, C.V.; Pérez, G.; Rodríguez, J.P.; Díaz-Granados, M. Effect of Green Roof Configuration and Hydrological Variables on Runoff Water Quantity and Quality. *Water* **2018**, *10*, 960. [CrossRef]
15. Liu, R.; Stanford, R.L.; Deng, Y.; Liu, D.; Liu, Y.; Yu, S.L. The influence of extensive green roofs on rainwater runoff quality: A field-scale study in southwest China. *Environ. Sci. Pollut. Res.* **2019**, *27*, 12932–12941. [CrossRef] [PubMed]

16. Gong, Y.; Yin, D.; Li, J.; Zhang, X.; Wang, W.; Fang, X.; Shi, H.; Wang, Q. Performance assessment of extensive green roof runoff flow and quality control capacity based on pilot experiments. *Sci. Total. Environ.* **2019**, *687*, 505–515. [CrossRef] [PubMed]
17. Lim, H.S.; Segovia, E.; Ziegler, A.D. Water quality impacts of young green roofs in a tropical city: A case study from Singapore. *Blue-Green Syst.* **2021**, *3*, 145–163. [CrossRef]
18. Chen, C.-F. Performance evaluation and development strategies for green roofs in Taiwan: A review. *Ecol. Eng.* **2013**, *52*, 51–58. [CrossRef]
19. Alsup, S.; Ebbs, S.; Retzlaff, W. The exchangeability and leachability of metals from select green roof growth substrates. *Urban Ecosyst.* **2009**, *13*, 91–111. [CrossRef]
20. Berndtsson, J.C.; Bengtsson, L.; Jinno, K. Runoff water quality from intensive and extensive vegetated roofs. *Ecol. Eng.* **2008**, *35*, 369–380. [CrossRef]
21. Wang, X.; Tian, Y.; Zhao, X. The influence of dual-substrate-layer extensive green roofs on rainwater runoff quantity and quality. *Sci. Total. Environ.* **2017**, *592*, 465–476. [CrossRef]
22. Hashemi, S.S.G.; Bin Mahmud, H.; Ashraf, M.A. Performance of green roofs with respect to water quality and reduction of energy consumption in tropics: A review. *Renew. Sustain. Energy Rev.* **2015**, *52*, 669–679. [CrossRef]
23. Pearlmutter, D.; Pucher, B.; Calheiros, C.S.C.; Hoffmann, K.A.; Aicher, A.; Pinho, P.; Stracqualursi, A.; Korolova, A.; Pobric, A.; Galvão, A.; et al. Closing Water Cycles in the Built Environment through Nature-Based Solutions: The Contribution of Vertical Greening Systems and Green Roofs. *Water* **2021**, *13*, 2165. [CrossRef]
24. Park, S.-Y.; Oh, D.-K.; Lee, S.-Y.; Yeum, K.-J.; Yoon, Y.-H.; Ju, J.-H. Combined Effects of Substrate Depth and Vegetation of Green Roofs on Runoff and Phytoremediation under Heavy Rain. *Water* **2022**, *14*, 2792. [CrossRef]
25. Thi-Dieu-Hien, V.; Thi-Bich-Ngoc, D.; Xuan-Thanh, B.; Van-Truc, N.; Dinh-Duc, N.; Sthiannopkao, S.; Lin, C. Improvement of septic tank effluent and green coverage by shallow bed wetland roof system. *Int. Biodeter. Biodegr.* **2017**, *124*, 138–145.
26. Thi-Dieu-Hien, V.; Xuan-Thanh, B.; Dinh-Duc, N.; Van-Truc, N.; Ngo, H.-H.; Guo, W.; Phuoc-Dan, N.; Cong-Nguyen, N.; Lin, C. Wastewater treatment and biomass growth of eight plants for shallow bed wetland roofs. *Bioresource. Technol.* **2018**, *247*, 992–998.
27. Xu, L.; Yang, S.; Zhang, Y.; Jin, Z.; Huang, X.; Bei, K.; Zhao, M.; Kong, H.; Zheng, X. A hydroponic green roof system for rainwater collection and greywater treatment. *J. Clean. Prod.* **2020**, *261*. [CrossRef]
28. Thi-Dieu-Hien, V.; Xuan-Thanh, B.; Lin, C.; Van-Truc, N.; Thi-Khanh-Dieu, H.; Hong-Hai, N.; Phuoc-Dan, N.; Ngo, H.H.; Guo, W. A mini-review on shallow-bed constructed wetlands: A promising innovative green roof. *Curr. Opin. Environ. Sci. Health.* **2019**, *12*, 38–47.
29. Mahmoudi, A.; Mousavi, S.A.; Darvishi, P. Greywater as a sustainable source for development of green roofs: Characteristics, treatment technologies, reuse, case studies and future developments. *J. Environ. Manag.* **2021**, *295*, 112991. [CrossRef]
30. Arden, S.; Ma, X. Constructed wetlands for greywater recycle and reuse: A review. *Sci. Total. Environ.* **2018**, *630*, 587–599. [CrossRef]
31. Boano, F.; Caruso, A.; Costamagna, E.; Ridolfi, L.; Fiore, S.; Demichelis, F.; Galvão, A.; Pisoeiro, J.; Rizzo, A.; Masi, F. A review of nature-based solutions for greywater treatment: Applications, hydraulic design, and environmental benefits. *Sci. Total. Environ.* **2019**, *711*, 134731. [CrossRef] [PubMed]
32. Khan, A.U.; Khan, A.N.; Waris, A.; Ilyas, M.; Zamel, D. Phytoremediation of pollutants from wastewater: A concise review. *Open Life Sci.* **2022**, *17*, 488–496. [CrossRef] [PubMed]
33. 3Chaudhary, V.B.; Sandall, E.L.; Lazarski, M.V. Urban mycorrhizas: Predicting arbuscular mycorrhizal abundance in green roofs. *Fungal Ecol.* **2019**, *40*, 12–19. [CrossRef]
34. Liu, W.; Wei, W.; Chen, W.; Deo, R.C.; Si, J.; Xi, H.; Li, B.; Feng, Q. The impacts of substrate and vegetation on stormwater runoff quality from extensive green roofs. *J. Hydrol.* **2019**, *576*, 575–582. [CrossRef]
35. Han, W.; Li, T.; Cheng, L.; Liu, L.; Yu, L.; Peng, Z. Effect of adding microorganism and carbon source to substrate on nitrogen removal treating the drainage of WWTP. *Water. Sci. Technol.* **2019**, *79*, 1947–1955.
36. John, J.; Kernaghan, G.; Lundholm, J. The potential for mycorrhizae to improve green roof function. *Urban Ecosyst.* **2016**, *20*, 113–127. [CrossRef]
37. Pradhan, S.; Al-Ghamdi, S.G.; Mackey, H.R. Greywater recycling in buildings using living walls and green roofs: A review of the applicability and challenges. *Sci. Total. Environ.* **2019**, *652*, 330–344. [CrossRef]
38. Kader, S.; Chadalavada, S.; Jaufer, L.; Spalevic, V.; Dudic, B. Green roof substrates—A literature review. *Front. Built Environ.* **2022**, *8*. [CrossRef]
39. Hachoumi, I.; Pucher, B.; De Vito-Francesco, E.; Prenner, F.; Ertl, T.; Langergraber, G.; Fürhacker, M.; Allabashi, R. Impact of Green Roofs and Vertical Greenery Systems on Surface Runoff Quality. *Water* **2021**, *13*, 2609. [CrossRef]
40. Xiao, M.; Lin, Y.; Han, J.; Zhang, G. A review of green roof research and development in China. *Renew. Sustain. Energy Rev.* **2014**, *40*, 633–648. [CrossRef]
41. Manso, M.; Teotónio, I.; Silva, C.M.; Cruz, C.O. Green roof and green wall benefits and costs: A review of the quantitative evidence. *Renew. Sustain. Energy Rev.* **2020**, *135*, 110111. [CrossRef]
42. Liao, W.; Drake, J.; Thomas, S.C. Biochar granulation, particle size, and vegetation effects on leachate water quality from a green roof substrate. *J. Environ. Manag.* **2022**, *318*, 115506. [CrossRef] [PubMed]
43. Vijayaraghavan, K.; Reddy, D.H.K.; Yun, Y.-S. Improving the quality of runoff from green roofs through synergistic biosorption and phytoremediation techniques: A review. *Sustain. Cities Soc.* **2018**, *46*, 101381. [CrossRef]

44. Liu, H.; Yin, H.; Kong, F.; Middel, A.; Zheng, X.; Huang, J.; Sun, T.; Wang, D.; Lensky, I.M. Change of nutrients, microorganisms, and physical properties of exposed extensive green roof substrate. *Sci. Total. Environ.* **2021**, *805*, 150344. [CrossRef] [PubMed]
45. Liu, W.; Engel, B.A.; Chen, W.; Wei, W.; Wang, Y.; Feng, Q. Quantifying the contributions of structural factors on runoff water quality from green roofs and optimizing assembled combinations using Taguchi method. *J. Hydrol.* **2020**, *593*, 125864. [CrossRef]
46. Chai, H.; Tang, Y.; Su, X.; Wang, W.; Lu, H.; Shao, Z.; He, Q. Annual variation patterns of the effluent water quality from a green roof and the overall impacts of its structure. *Environ. Sci. Pollut. Res.* **2018**, *25*, 30170–30179. [CrossRef] [PubMed]
47. Liu, L.; Cao, J.; Ali, M.; Zhang, J.; Wang, Z. Impact of green roof plant species on domestic wastewater treatment. *Environ. Adv.* **2021**, *4*, 100059. [CrossRef]
48. Gong, Y.; Zhang, X.; Li, H.; Zhang, X.; He, S.; Miao, Y. A comparison of the growth status, rainfall retention and purification effects of four green roof plant species. *J. Environ. Manag.* **2021**, *278*, 111451. [CrossRef] [PubMed]
49. Thomaidi, V.; Petousi, I.; Kotsia, D.; Kalogerakis, N.; Fountoulakis, M. Use of green roofs for greywater treatment: Role of substrate, depth, plants, and recirculation. *Sci. Total. Environ.* **2021**, *807*, 151004. [CrossRef]
50. Vymazal, J. Long-term performance of constructed wetlands with horizontal sub-surface flow: Ten case studies from the Czech Republic. *Ecol. Eng.* **2011**, *37*, 54–63. [CrossRef]
51. Tran, S.; Lundholm, J.T.; Staniec, M.; Robinson, C.E.; Smart, C.C.; Voogt, J.A.; O'Carroll, D.M. Plant survival and growth on extensive green roofs: A distributed experiment in three climate regions. *Ecol. Eng.* **2018**, *127*, 494–503. [CrossRef]
52. Cáceres, N.; Robbiati, F.O.; Suárez, M.; Hick, E.C.; Matoff, E.; Jim, C.Y.; Galetto, L.; Imhof, L. Growth performance of multi-species plant mixtures on an extensive vegetated roof: A two-year experimental study. *Urban Ecosyst.* **2024**, 1–17. [CrossRef]
53. Johnson, C.; Schweinhart, S.; Buffam, I. Plant species richness enhances nitrogen retention in green roof plots. *Ecol. Appl.* **2016**, *26*, 2130–2144. [CrossRef] [PubMed]
54. Liao, W.; Sidhu, V.; Sifton, M.A.; Margolis, L.; Drake, J.A.; Thomas, S.C. Biochar and vegetation effects on discharge water quality from organic-substrate green roofs. *Sci. Total. Environ.* **2024**, *922*, 171302. [CrossRef] [PubMed]
55. Wright, I.J.; Reich, P.B.; Westoby, M.; Ackerly, D.D.; Baruch, Z.; Bongers, F.; Cavender-Bares, J.; Chapin, T.; Cornelissen, J.H.C.; Diemer, M.; et al. The worldwide leaf economics spectrum. *Nature* **2004**, *428*, 821–827. [CrossRef] [PubMed]
56. Hu, X.; Song, X.; Song, S.; Li, X.; Li, C.; Zhang, Z.; Li, G.; Chen, Q. Effect of Combinations of Green Roof Substrate with Vegetation Coverage on Rainwater Quality Improvement. *Pol. J. Environ. Stud.* **2023**, *33*, 1149–1158. [CrossRef] [PubMed]
57. Rocha, B.; Paço, T.A.; Luz, A.C.; Palha, P.; Milliken, S.; Kotzen, B.; Branquinho, C.; Pinho, P.; de Carvalho, R.C. Are Biocrusts and Xerophytic Vegetation a Viable Green Roof Typology in a Mediterranean Climate? A Comparison between Differently Vegetated Green Roofs in Water Runoff and Water Quality. *Water* **2021**, *13*, 94. [CrossRef]
58. Nguyen, V.-T.; Bui, X.-T.; Lin, C.; Nguyen, H.-H.; Vo, T.-D.; Tran, L.-L.; Nguyen, T.-B.; Bui, M.-H.; Nguyen, D.-T.; Nguyen, D.-D.; et al. Influence of plant types, bed media and feeding patterns on wastewater treatment performance of wetland roofs. *J. Water Process. Eng.* **2021**, *40*, 101972. [CrossRef]
59. Zhang, Q.; Miao, L.; Wang, H.; Wang, L. Analysis of the effect of green roof substrate amended with biochar on water quality and quantity of rainfall runoff. Environ. *Monit. Assess.* **2019**, *191*, 304.
60. Guo, J.; Zhang, Y.; Che, S. Performance analysis and experimental study on rainfall water purification with an extensive green roof matrix layer in Shanghai, China. *Water Sci. Technol.* **2018**, *77*, 670–681. [CrossRef]
61. Leite, F.R.; Antunes, M.L.P. Green roof recent designs to runoff control: A review of building materials and plant species used in studies. *Ecol. Eng.* **2023**, *189*. [CrossRef]
62. Pęczkowski, G.; Szawernoga, K.; Kowalczyk, T.; Orzepowski, W.; Pokładek, R. Runoff and Water Quality in the Aspect of Environmental Impact Assessment of Experimental Area of Green Roofs in Lower Silesia. *Sustainability* **2020**, *12*, 4793. [CrossRef]
63. Karczmarczyk, A.; Baryła, A.; Fronczyk, J.; Bus, A.; Mosiej, J. Phosphorus and Metals Leaching from Green Roof Substrates and Aggregates Used in Their Composition. *Minerals* **2020**, *10*, 112. [CrossRef]
64. Rey, C.V.; Franco, N.; Peyre, G.; Rodríguez, J.P. Green Roof Design with Engineered Extensive Substrates and Native Species to Evaluate Stormwater Runoff and Plant Establishment in a Neotropical Mountain Climate. *Sustainability* **2020**, *12*, 6534. [CrossRef]
65. Vijayaraghavan, K.; Raja, F.D. Design and development of green roof substrate to improve runoff water quality: Plant growth experiments and adsorption. *Water Res.* **2014**, *63*, 94–101. [CrossRef] [PubMed]
66. Vijayaraghavan, K.; Badavane, A. Preparation of growth substrate to improve runoff quality from green roofs: Physico-chemical characterization, sorption and plant-support experiments. *Urban Water J.* **2016**, *14*, 804–810. [CrossRef]
67. Beck, D.A.; Johnson, G.R.; Spolek, G.A. Amending greenroof soil with biochar to affect runoff water quantity and quality. *Environ. Pollut.* **2011**, *159*, 2111–2118. [CrossRef] [PubMed]
68. Kuoppamäki, K.; Hagner, M.; Lehvävirta, S.; Setälä, H. Biochar amendment in the green roof substrate affects runoff quality and quantity. *Ecol. Eng.* **2016**, *88*, 1–9. [CrossRef]
69. Novotný, M.; Šipka, M.; Miino, M.C.; Raček, J.; Chorazy, T.; Petreje, M.; Tošić, I.; Hlavínek, P.; Marković, M. Influence of different alternative organic substrates as fillings for green roofs on the quality of rainfall runoff. *Sustain. Chem. Pharm.* **2024**, *38*. [CrossRef]
70. Xiong, W.; Li, J.; Wang, H.; Wu, Y.; Li, D.; Xue, J. Biochar Addition and the Runoff Quality of Newly Constructed Green Roofs: A Field Study. *Sustainability* **2023**, *15*, 4081. [CrossRef]
71. Liao, W.; Drake, J.; Thomas, S.C. Biochar granulation enhances plant performance on a green roof substrate. *Sci. Total. Environ.* **2022**, *813*, 152638. [CrossRef] [PubMed]

72. Zhang, J.; Xu, C.; Fu, D.; Liu, W. Improvement in pollutant controlling performance by modified aggregate structure in extensive green roof substrate layer. *Environ. Sci. Water Res. Technol.* **2022**, *8*, 1709–1718. [CrossRef]
73. Fei, Y.; Xu, C.; Miao, S.; Fu, D.; Zhang, J. Comparison of rainwater management performance of modified extensive green roof substrate layer with different additives in rainstorm events. *Environ. Sci. Water Res. Technol.* **2022**, *9*, 3243–3256. [CrossRef]
74. Petreje, M.; Sněhota, M.; Chorazy, T.; Novotný, M.; Rybová, B.; Hečková, P. Performance study of an innovative concept of hybrid constructed wetland-extensive green roof with growing media amended with recycled materials. *J. Environ. Manag.* **2023**, *331*, 117151. [CrossRef] [PubMed]
75. van der Kolk, H.J.; van den Berg, P.; van Veen, T.; Bezemer, M. Substrate composition impacts long-term vegetation development on blue-green roofs: Insights from an experimental roof and greenhouse study. *Ecol. Eng.* **2023**, *186*, 106847. [CrossRef]
76. Cakar, H.; Saracoglu, O.A.; Akat, H.; Kilic, C.C.; Adanacioglu, H. THE POTENTIAL FOR USING DIFFERENT SUBSTRATES IN GREEN ROOFS. *J. Environ. Eng. Landsc. Manag.* **2023**, *31*, 44–51. [CrossRef]
77. Vannucchi, F.; Bibbiani, C.; Caudai, C.; Bretzel, F. Mediterranean Extensive Green Roof Self-Sustainability Mediated by Substrate Composition and Plant Strategy. *Horticulturae* **2023**, *9*, 1117. [CrossRef]
78. Meng, R.; Zhang, Q.; Li, D.; Wang, H. Influence of Substrate Layer Thickness and Biochar on the Green Roof Capacity to Intercept Rainfall and Reduce Pollution in Runoff. *Pol. J. Environ. Stud.* **2021**, *30*, 4085–4103. [CrossRef]
79. Xu, C.; Yuan, Q.; Zhao, S.; He, T.; Song, N. Effects of pretreatments on physical and chemical characteristics of wheat straw used as a maintenance-free compressed green roof substrate material. *J. Clean. Prod.* **2020**, *277*, 123381. [CrossRef]
80. Bus, A.; Karczmarczyk, A.; Baryła, A. The use of reactive material for limiting P-leaching from green roof substrate. *Water Sci. Technol.* **2016**, *73*, 3027–3032. [CrossRef]
81. Fan, L.; Wang, J.; Liu, X.; Luo, H.; Zhang, K.; Fu, X.; Li, M.; Li, X.; Jiang, B.; Chen, J.; et al. Whether the carbon emission from green roofs can be effectively mitigated by recycling waste building material as green roof substrate during five-year operation? Environ. *Sci. Pollut. R.* **2020**, *27*, 40893–40906. [CrossRef]
82. Qiu, D.; Peng, H.; Li, T.; Qi, Y. Application of stabilized sludge to extensive green roofs in Shanghai: Feasibility and nitrogen leaching control. *Sci. Total. Environ.* **2020**, *732*, 138898. [CrossRef] [PubMed]
83. Liu, W.; Feng, Q.; Chen, W.; Wei, W.; Si, J.; Xi, H. Runoff retention assessment for extensive green roofs and prioritization of structural factors at runoff plot scale using the Taguchi method. *Ecol. Eng.* **2019**, *138*, 281–288. [CrossRef]
84. Agra, H.; Solodar, A.; Bawab, O.; Levy, S.; Kadas, G.J.; Blaustein, L.; Greenbaum, N. Comparing grey water versus tap water and coal ash versus perlite on growth of two plant species on green roofs. *Sci. Total. Environ.* **2018**, *633*, 1272–1279. [CrossRef]
85. Zhang, W.; Zhong, X.; Che, W. Nutrient leaching from extensive green roofs with different substrate compositions: A laboratory study. *Water Sci. Technol.* **2017**, *77*, 1007–1014. [CrossRef]
86. Piscitelli, L.; Rivier, P.-A.; Mondelli, D.; Miano, T.; Joner, E.J. Assessment of addition of biochar to filtering mixtures for potential water pollutant removal. *Environ. Sci. Pollut. Res.* **2018**, *25*, 2167–2174. [CrossRef]
87. Monteiro, C.; Calheiros, C.; Martins, J.P.; Costa, F.M.; Palha, P.; de Freitas, S.S.; Ramos, N.M.M.; Castro, P. Substrate influence on aromatic plant growth in extensive green roofs in a Mediterranean climate. *Urban Ecosyst.* **2017**, *20*, 1347–1357. [CrossRef]
88. Kuoppamäki, K.; Lehvävirta, S. Mitigating nutrient leaching from green roofs with biochar. *Landsc. Urban Plan.* **2016**, *152*, 39–48. [CrossRef]
89. Malcolm, E.G.; Reese, M.L.; Schaus, M.H.; Ozmon, I.M.; Tran, L.M. Measurements of nutrients and mercury in green roof and gravel roof runoff. *Ecol. Eng.* **2014**, *73*, 705–712. [CrossRef]
90. Zhang, Q.; Wang, X.; Hou, P.; Wan, W.; Li, R.; Ren, Y.; Ouyang, Z. Quality and seasonal variation of rainwater harvested from concrete, asphalt, ceramic tile and green roofs in Chongqing, China. *J. Environ. Manag.* **2014**, *132*, 178–187. [CrossRef]
91. Harper, G.E.; Limmer, M.A.; Showalter, W.E.; Burken, J.G. Nine-month evaluation of runoff quality and quantity from an experiential green roof in Missouri, USA. *Ecol. Eng.* **2015**, *78*, 127–133. [CrossRef]
92. Beecham, S.; Razzaghmanesh, M. Water quality and quantity investigation of green roofs in a dry climate. *Water Res.* **2015**, *70*, 370–384. [CrossRef] [PubMed]
93. Castro, A.S.; Goldenfum, J.A.; da Silveira, A.L.; Bertani DallAgnol, A.L.; Loebens, L.; Demarco, C.F.; Leandro, D.; Nadaleti, W.C.; Quadro, M.S. The analysis of green roof's runoff volumes and its water quality in an experimental study in Porto Alegre, Southern Brazil. *Environ. Sci. Pollut. R.* **2020**, *27*, 9520–9534. [CrossRef] [PubMed]
94. Bui Xuan, T.; Phan Thi Hai, V.; Nguyen Thanh, T.; Vo Thi Dieu, H.; Nguyen Phuoc, D.; Koottatep, T. Performance of wetland roof with Melampodium paludosum treating septic tank effluent. *Desalin. Water. Treat.* **2014**, *52*, 1070–1076.
95. Petousi, I.; Thomaidi, V.; Kalogerakis, N.; Fountoulakis, M.S. Removal of pathogens from greywater using green roofs combined with chlorination. *Environ. Sci. Pollut. Res.* **2022**, *30*, 22560–22569. [CrossRef] [PubMed]
96. Sá, T.S.W.; Najjar, M.K.; Hammad, A.W.A.; Vazquez, E.; Haddad, A. Assessing rainwater quality treated via a green roof system. *Clean Technol. Environ. Policy* **2021**, *24*, 645–660. [CrossRef]
97. Gong, K.; Wu, Q.; Peng, S.; Zhao, X.; Wang, X. Research on the characteristics of the water quality of rainwater runoff from green roofs. *Water Sci. Technol.* **2014**, *70*, 1205–1210. [CrossRef]
98. Speak, A.; Rothwell, J.; Lindley, S.; Smith, C. Metal and nutrient dynamics on an aged intensive green roof. *Environ. Pollut.* **2014**, *184*, 33–43. [CrossRef]
99. Akther, M.; He, J.; Chu, A.; van Duin, B. Nutrient leaching behavior of green roofs: Laboratory and field investigations. *Sci. Total. Environ.* **2020**, *754*, 141841. [CrossRef]

100. Buffam, I.; Mitchell, M.E.; Durtsche, R.D. Environmental drivers of seasonal variation in green roof runoff water quality. *Ecol. Eng.* **2016**, *91*, 506–514. [CrossRef]
101. Menge, D.N.L.; Field, C.B. Simulated global changes alter phosphorus demand in annual grassland. *Glob. Chang. Biol.* **2007**, *13*, 2582–2591. [CrossRef]
102. Rinnan, R.; Michelsen, A.; Bååth, E.; Jonasson, S. Mineralization and carbon turnover in subarctic heath soil as affected by warming and additional litter. *Soil Biol. Biochem.* **2007**, *39*, 3014–3023. [CrossRef]
103. Akther, M.; He, J.; Chu, A.; van Duin, B. Chemical leaching behaviour of a full-scale green roof in a cold and semi-arid climate. *Ecol. Eng.* **2020**, *147*, 105768. [CrossRef]
104. Teemusk, A.; Teemusk, Ü. Rainwater runoff quantity and quality performance from a greenroof: The effects of short-term events. *Ecol. Eng.* **2007**, *30*, 271–277. [CrossRef]
105. Gregoire, B.G.; Clausen, J.C. Effect of a modular extensive green roof on stormwater runoff and water quality. *Ecol. Eng.* **2011**, *37*, 963–969. [CrossRef]
106. Todorov, D.; Driscoll, C.T.; Todorova, S.; Montesdeoca, M. Water quality function of an extensive vegetated roof. *Sci. Total. Environ.* **2018**, *625*, 928–939. [CrossRef]
107. Gong, Y.; Zhang, X.; Li, J.; Fang, X.; Yin, D.; Xie, P.; Nie, L. Factors affecting the ability of extensive green roofs to reduce nutrient pollutants in rainfall runoff. *Sci. Total. Environ.* **2020**, *732*, 139248. [CrossRef]
108. Guo, B.; Arndt, S.; Miller, R.; Lu, N.; Farrell, C. Are succulence or trait combinations related to plant survival on hot and dry green roofs? *Urban. For. Urban. Gree.* **2021**, *64*, 127248. [CrossRef]
109. Sultana, N.; Akib, S.; Ashraf, M.A.; Abidin, M.R.Z. Quality assessment of harvested rainwater from green roofs under tropical climate. *Desalination Water Treat.* **2015**, 1–8. [CrossRef]
110. Ramprasad, C.; Smith, C.S.; Memon, F.A.; Philip, L. Removal of chemical and microbial contaminants from greywater using a novel constructed wetland: GROW. *Ecol. Eng.* **2017**, *106*, 55–65. [CrossRef]
111. Paraskevopoulou, A.T.; Ntoulas, N.; Bourtsoukli, D.; Bertsouklis, K. Effect of Seawater Irrigation on *Arthrocnemum macrostachyum* Growing in Extensive Green Roof Systems under Semi-Arid Mediterranean Climatic Conditions. *Agronomy* **2023**, *13*, 1198. [CrossRef]
112. Zehnsdorf, A.; Blumberg, M.; Müller, R.A. Helophyte mats (wetland roofs) with high evapotranspiration rates as a tool for decentralised rainwater management—process stability improved by simultaneous greywater treatment. *Water Supply* **2018**, *19*, 808–814. [CrossRef]
113. Avery, L.M.; Frazer-Williams, R.A.; Winward, G.; Shirley-Smith, C.; Liu, S.; Memon, F.A.; Jefferson, B. Constructed wetlands for grey water treatment. *Ecohydrol. Hydrobiol.* **2007**, *7*, 191–200. [CrossRef]
114. Winward, G.P.; Avery, L.M.; Frazer-Williams, R.; Pidou, M.; Jeffrey, P.; Stephenson, T.; Jefferson, B. A study of the microbial quality of grey water and an evaluation of treatment technologies for reuse. *Ecol. Eng.* **2007**, *32*, 187–197. [CrossRef]

water

Article

Impact of an Integral Management System with Constructed Wetlands in Pig Slurry Traceability and GHG/NH_3 Emissions

Melisa Gómez-Garrido *, Martire Angélica Terrero Turbí, Oumaima El bied and Ángel Faz Cano

Sustainable Use, Management and Reclamation of Soil and Water Research Group, Agronomic Engineering Department, Technical University of Cartagena, Paseo Alfonso XIII, 48, 30203 Cartagena, Spain; angelica.terrero@upct.es (M.A.T.T.); oumaima.elbied@upct.es (O.E.b.); angel.fazcano@upct.es (Á.F.C.)
* Correspondence: melisa.gomez@upct.es

Abstract: The sustainable management of pig slurry (PS) in intensive farms is essential to reduce adverse environmental impacts and reduce the ecological footprint. If not managed properly, PS can release GHG/NH_3 gases into the atmosphere and contaminate waters. This study evaluates the impact of an integral management system with physical and biological stages to mitigate the impact of PS. The system resulted in effective PS traceability, studying its physicochemical properties. The synergism in the whole system allowed a decrease in the most analyzed parameters during the autumn, spring, and summer. The pretreatment contributed significantly to obtaining an appreciable percentage of reduction in the constructed wetlands of SS (99–100%), COD (56–87%), TN (50–57%), and PO_4^{3-} (88–100%). The emission values (g/m^2/day) were 0–2.14 (CH_4), 0–473.76 (CO_2), 0–179.77 (H_2O), 0–0.265 (N_2O), and 0–0.195 (NH_3), highlighting the raw, separated and manure fractions with the highest values. It is concluded that the system proves to be a practical, low-cost, and efficient technique for the treatment of PS. It significantly reduces the concentration of nutrients, and the intercepted byproducts can be valuable for application to the soil. In addition, the system effectively reduces GHG/NH_3 emissions in decanted, purified, and wetland PS fractions.

Keywords: pig slurry; manure; integral management system; constructed wetlands; GHG emissions; ammonia; solid-liquid phase separation; decantation; *Phragmites australis*

1. Introduction

The need to preserve natural resources and protect the environment, in addition to preventing potential negative effects that could result from pig farming, makes it essential to align the wealth of resources from this important livestock activity with the regulatory requirements of this sector at the worldwide level.

The development of modeling techniques for the sustainable and cost-effective treatment of agricultural wastewater is a widespread problem, especially across the European Union, where solutions and processes that contribute to nutrient recycling within a circular economy are increasingly required [1]. These solutions aim to prevent the contamination of groundwater and surface water, in addition to protecting the environment from greenhouse gases (GHG) and other pollutants that can result from livestock manure management.

Around Europe, pig slurry management is a significant challenge for farmers. Often, the environmental harm caused by pig farm effluents results from the high density of animals in confined areas and poor waste management practices. The EU aims to address this issue through the "Nitrate Directive" (91/676/EEC), which seeks to minimize environmental problems triggered by water pollution generated by nitrates from agricultural sources. This Directive requires EU Member States to identify vulnerable zones where action plans must be implemented to reduce nitrate leaching into the mass surface and/or subsurface water [2,3].

The properties of pig slurry can vary greatly from farm to farm, even though nitrogen is the component of main concern due to the environmental risk that its transformation and management may entail. In particular, untreated pig slurry contains considerable amounts of non-stabilized organic matter and high concentrations of ammonium, it depends on the farm characteristics. Separation of slurry generates a solid fraction with a high concentration of dry matter (DM) and phosphorus (P), and a liquid fraction with low DM content and a relatively high concentration of total ammoniacal nitrogen [4].

Pig slurry management has several options to be applied from more simple/economical to more sophisticated/expensive methods such as phase separation, drying, use of additives biological or chemical, nitrification-denitrification, composting, incineration of solid fractions, membrane filtration, and others [3–5]. Although, the combination of different technologies could result in suitable and effective legislation suggested for pig slurry treatment, resulting in a reduction in contaminants concentrations as well as emissions. Therefore, considering these guidelines, this study was focused on using an integrated system composed of phase separation, decanters, and constructed wetlands.

The separation process is well known and widely used firstly to obtain two fractions of slurries and as a pretreatment for other techniques. The different pathways to separate the solid and liquid fractions from slurry can be physical, mechanical, or chemical methods. Physical separation could reach over than 80% of the total solids [6,7]. In this study, physical and mechanical separation was used throughout a sieve separator plus screw followed by a decantation unit.

On the other hand, constructed wetlands (CW) are considered tertiary [7–9], capable of removing a wide range of contaminants, including pathogenic microorganisms [3,10,11].

Treatment with CW became a very attractive option for farmers because this system has been demonstrated to be effective, low-cost, low-maintenance, and environmentally friendly, further beneficial to pig slurry treatment by being a viable, sustainable, and cost-effective alternative to other traditional treatments (for instance, anaerobic digestion or bio-membranes) [2,9,12]. Constructed wetlands can be used for several treatments like agricultural wastewater [7,13,14], industrial dairy wastewater, industrial tannery, acid mine drainage wastewater [15,16] pulp and paper industry wastewater [17,18], industrial textile wastewater, etc. According to Vymazal [16,17] there are three basic concepts that constructed wetlands could be categorized, (1) hydrology (open water-surface flow and sub-surface flow), (2) type of macrophytic growth (emergent, submerged, free-floating, and floating-leaved) and (3) flow path in sub-surface wetlands (horizontal and vertical). It is possible to combine the different types of CW depending on the purpose of the design and the specific objective to achieve [9,16,19].

In accordance with the previous comment about the CW flow path, surface flow constructed wetlands closely resemble natural environments and are typically more suitable for wetland species due to the presence of permanent standing water; in contrast, subsurface flow wetlands direct water laterally through a porous medium, such as sand and gravel, supporting fewer macrophyte species and generally lacking standing water. Subsurface flow is categorized into vertical flow (VF) CW, horizontal flow (HF) CW, french vertical flow (FVF) CW, and hybrid type CW [20,21].

There is literature with practical evidence of physical and biological techniques for treating slurry, such as solid-liquid separation and phytoremediation with CW. However, there are hardly any publications with comprehensive results of slurry treatment systems that combine both techniques and also carry out analytical monitoring of slurry properties and GHG and NH_3 emissions at all stages of processing and recycling. In this study, a Horizontal Flow Subsurface Constructed Wetland (was used for the integrated treatment of pig slurry. This type usually has predominant anoxic/anaerobic mechanisms and thus provides suitable conditions for the denitrification process if nitrate is present [20]. Conversely, it very much limited the nitrification process because of the lack of oxygen in the water-saturated filtration bed, and for this reason, ammonia reduction tends to be low.

Consequently, it is essential to address sustainable solutions for pig manure treatment with respect to nutrient removal like biodegradable organic matter, suspended solids, phosphorus, and nitrogen with special attention towards environmental and agricultural benefits, always in line with environmental legislation and European regulations.

The monitored parameters of pig slurry during the study were T, pH, CE, SS, COD, TN, NH_4^+, NO_3^-, PO_4^{3-}, K^+, Cu, and Zn, as well as the measured gases were CH_4, CO_2, N_2O, and NH_3. The objective of this study was to evaluate the effects of three stages of treatment in an integrated management system on pig slurry, how it influences its composition, enhancing its properties, and which mechanisms are involved in mitigating the emissions of CH_4, CO_2, N_2O, and NH_3. Stages:

Stage 1: Physical stage of solid–liquid phase separation with phase separator with sieving and press filter.

Stage 2: Physical stage of solid–liquid phase separation with gravity decantation.

Stage 3: Biological stage of purification in artificial wetlands or biofilters.

2. Materials and Methods

2.1. Operation

In 2018, an integral slurry treatment system was implemented in the southeast of Spain in a maternity farm with a census of 2750 places for breeding sows and 232 replacement places. This farm generated a total of 14,605 m^3/year of liquid and semi-liquid manure (Spanish Royal Decree 306/2020). Initially, 50% of the slurry production was processed, but currently, 100% of the slurry production is processed, being agronomic recycling is the final destination of all the slurry fractions generated (manure and pig slurry). A summary of the operation parameters is shown in Table 1 and Figure 1. The integral system consisted of different slurry treatment stages:

Table 1. Main design characteristics for the integral system of pig slurries with constructed wetlands.

Stage	Work Units	Processed Volume	Characteristic *
Phase separator (Segalés, Kompact 1-100)	1	10–12 m^3/h	Mesh: 500 µm
Decanters (bricklaying and plumbing work)	6 units in series	160 m^3/week	3.20 × 36.25 × 0.42 m
Constructed wetlands (bricklaying and plumbing work)	25 independent units	50–100 m^3/week	Cell size: 25 × 1.7 × 1.2 m Filling substrates (from below): 30 cm fine gravel (ø = 2–20 mm) 50 cm coarse gravel (ø = 20–40 mm) 10 cm fine gravel 30 cm washed sand HRT: 3-5-20 days *Phragmites australis* (5 plants/m^2)

Notes: (*) HRT: hydraulic retention time. The gravels are composed of hydrated carbonates of alkaline and alkaline earth metals.

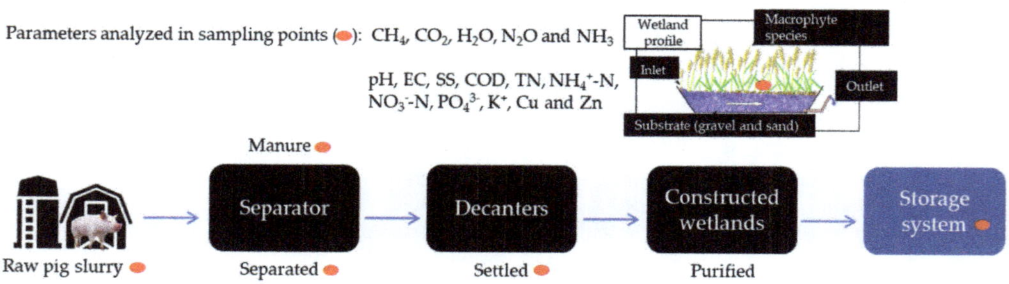

Figure 1. Operational design diagram of the treatment system with artificial wetlands.

2.2. Pig Slurry Parameters and Methodology

The pig slurry samples were taken in triplicates when the measurements of gas emissions were carried out. The selected parameters for the characterization of the pig slurry were pH, electrical conductivity (EC), settleable solids (SS), chemical oxygen demand (COD), total nitrogen (TN), ammoniacal nitrogen (NH_4^+-N), nitrogen as nitrates (NO_3^--N), phosphate ion (PO_4^{3-}) and potassium ion (K^+), Copper (Cu) and Zinc (Zn).

The standardized methodology used to analyze the pig slurry was the following: the pH and EC were measured in situ using a HANNA multiparameter equipment (ref. HI98194). Settleable solids (SS) were measured in situ by natural sedimentation in an Inhoff vessel, after 60 min [22]. The COD was determined via photometric analysis of the chromium (III)concentration after 2 h of oxidation with potassium dichromate/sulfuric acid and silver sulfate at 148 C (Macherey–Nagel GmbH & Co., KG, Nanocolor Test; ref. 985 028/29, Weilheim, Germany) according to American standard methods, APHA, [22].

Total nitrogen was calculated from the sum of Kjeldahl N, N-NO_3^- and N-NO_2^-; the Kjeldahl N content was measured using a modified Kjeldahl method [23], 1 mL of pig slurry was used for digestion and the form NH_4^+-N was determined via steam distillation, followed by titration with HCl 0.1 N. Kjeldahl N comprised Org.-N and NH_4^+-N. NH_4^+-N was determined with the previous methodology but did not include digestion. N-NO_3^-, N-NO_2^-, PO_4^{3-}, and K^+ were determined by ionic chromatography technique (Methrom, 861 Advanced Compact IC) after sample preparation. Copper (Cu) and zinc (Zn) were determined by inductively coupled plasma mass spectrometry (ICP-MS).

2.3. Experimental Design and Methodology for Measuring Emissions in Pig Slurry Storage Systems

Floating dynamic chambers are one of the most used systems to capture and measure GHG and contaminant gases (CH_4, CO_2, N_2O, and NH_3) in ponds or slurry storage systems. The principle of this technique is to isolate a part of the surface where the slurry is stored and measure the change in concentration of the gases in the chamber over time. The results are expressed per unit area of slurry and per unit volume. This method uses PVC plastic chambers with certain dimensions to isolate part of the surface from which emissions are to be determined. For its correct operation, an air pump of known flow brings air to the dynamic chamber (gas inlet), while another second pump of also known flow is placed at the other end (gas outlet). To measure GHG and NH_3 emissions (F = flow measured with dynamic chambers), the analyzer determines the emission concentration of the gases at the inlet (Ce) and at the outlet (Cs) in mg/m^3 and multiplies by the airflow (Qa) of the dynamic chamber (m^3 air/h) using the following relationship for each of the gases: $F = (Cs - Ce) \times Qa$.

The analyzer used to measure gas concentrations (CH_4, CO_2, H_2O, N_2O, and NH_3) at both the inlet and outlet quantifies the concentrations in parts per million (ppm) by infrared spectrometry. The gas analyzer equipment allows continuous measurement of gases. The gases are introduced into the analyzer through a tube, the internal pump extracts the gas sample through the instrument displaying the measurements on the device. The analyzer measures and analyzes an infrared spectrum of gas samples using a photoacoustic sensor based on an optical microphone. To carry out this methodology, the principle described in the protocol "Vera of Environmental Technologies for Agricultural Production Test Protocol for Covers and other Mitigation Technologies for Reduction of Gaseous Emissions from Stored Manure" and the design according to "Reference procedures" have been considered for the measurement of gaseous emissions from livestock houses and storages of animal manure." It is an international protocol used as a reference and recommended by the Ministry of Agriculture, Fisheries and Food of Spain. GHG and NH_3 emissions in manure piles have been quantified using the same measurement equipment. However, a static gas chamber similar to that used to measure soil emissions has been used, being a cylindrical steel chamber that is inserted 5 cm into the soil or contact surface. For this study, measurements of GHG and NH_3 emissions have been carried out for five consecutive weeks in autumn 2020, spring 2021, and summer 2021 seasons in five storage systems that correspond to the following systems of slurry fractions:

- Raw slurry (RAW, measurements made in the storage tank that receives raw pig slurry).
- Slurry after the phase separator (SEP, measurements made in the first settling pond).
- Settled slurry (DEC, measurements made between the fifth and sixth settling pond).
- Purified slurry (PUR, measurements made in the purified slurry storage pond subjected to drying conditions).
- Manure (MAN, measurements made on a pile of fresh manure resulting from the phase separator).
- Wetland surface without vegetation (WC).
- Wetland surface planted with *Phragmites australis* (WV).

Emissions of the WC and WV fractions were only recorded in the spring 2021 season due to the availability of the measurement chambers. The emissions were taken after 3–4 h of filling the wetlands.

3. Results

3.1. Pig Slurry and Manure

The results presented in Tables 2–4 represent the average of the 3 replicates for each sampling at each stage of the integrated treatment system (RAW-SEP-DEC-PUR) and in each season of the study (autumn, spring, and summer). The same procedure was followed for manure. The physical-chemical and biological characterization from the analytical results were reported showing the traceability of the pig slurry quality throughout the integrated management system.

Table 2. Mean and standard deviation values, percentage of reduction, agronomic dose of application and macronutrients (N-P-K) content per year during autumn (n = 3).

Season	Autumn								
Sample type **	RAW		SEP		DEC		PUR		Red (%) ***
Parameter *	Mean ± SD		Mean ± SD		Mean ± SD		Mean ± SD		
pH	7.37 ± 0.02	a	7.55 ± 0.01	b	7.92 ± 0.03	c	7.90 ± 0.06	c	-
EC ($ds\ m^{-1}$)	15.34 ± 0.13	c	15.62 ± 0.22	c	12.77 ± 0.20	b	12.05 ± 0.26	a	21
SS ($mg\ L^{-1}$)	483.3 ± 28.9	c	366.7 ± 28.9	b	4.2 ± 3.4	a	0.0 ± 0.0	a	100
COD ($g\ L^{-1}$)	25.67 ± 3.79	c	17.67 ± 3.21	b	5.47 ± 0.40	a	5.15 ± 1.24	a	80
TN ($g\ L^{-1}$)	2.13 ± 0.51	b	1.97 ± 0.41	b	1.29 ± 0.05	ab	1.05 ± 0.06	a	50
NH_4^+-N ($g\ L^{-1}$)	1.52 ± 0.04	b	1.65 ± 0.36	b	1.00 ± 0.02	a	0.70 ± 0.05	a	54
NO_3^--N ($mg\ L^{-1}$)	5.87 ± 0.19	b	6.55 ± 0.11	b	5.74 ± 0.56	b	4.26 ± 0.34	a	27
PO_4^{3-} ($mg\ L^{-1}$)	323.7 ± 302.3	ab	553.2 ± 3.1	b	78.2 ± 11.4	a	0.0 ± 0.0	a	100
K^+ ($mg\ L^{-1}$)	1348.8 ± 51.2	d	1279.7 ± 3.6	c	1021.7 ± 1.8	b	917.2 ± 11.1	a	32
Cu ($mg\ L^{-1}$)	0.05 ± 0.01	b	0.05 ± 0.01	b	0.04 ± 0.00	b	0.00 ± 0.00	a	-
Zn ($mg\ L^{-1}$)	0.07 ± 0.12	a	0.08 ± 0.14	a	0.03 ± 0.05	a	0.00 ± 0.00	a	-
[1] Agronomic dosage ($L\ ha^{-1}\ yr^{-1}$)	79,982		86,364		131,734		161,250		
[1] N ($kg\ ha^{-1}$)	170.0		170.0		170.0		170.0		
[1] P_2O_5 ($kg\ ha^{-1}$)	19.35		35.71		7.70		0.00		
[1] K_2O ($kg\ ha^{-1}$)	129.9		133.1		162.1		178.2		

Notes: (*) EC: electrical conductivity; SS: settleable solids; COD: chemical oxygen demand; TN: total nitrogen; NH_4^+-N: ammoniacal nitrogen; NO_3^--N: nitrogen as nitrates; PO_4^{3-}: phosphates; K^+: potassium ion; Cu: copper; Zn: Zinc. (**) RAW: raw pig slurry, SEP: separated pig slurry, DEC: decanted pig slurry, PUR: purified pig slurry. (***) Percentage reduction = 100 − ((PUR/RAW) × 100)); (-) indicates not reduction. Different letters indicate significant differences ($p < 0.05$) between phase of treatment. [1] Considering the ceiling of 170 kg N $ha^{-1}\ yr^{-1}$ according to Nitrates Directive (91/676/EEC) for vulnerable areas.

Table 3. Mean and standard deviation values, percentage of reduction, agronomic dose of application and macronutrients (N-P-K) content per year during spring (n = 3).

Season	Spring																
Sample type **	RAW				SEP				DEC				PUR				Red (%) ***
Parameter *	Mean	±	SD		Mean	±	SD		Mean	±	SD		Mean	±	SD		
pH	7.39	±	0.31	a	7.30	±	0.31	a	7.50	±	0.32	a	7.86	±	0.33	a	−6
EC (ds m^{-1})	8.18	±	0.35	a	8.34	±	0.35	a	8.27	±	0.35	a	7.58	±	0.35	a	7
SS (mg L^{-1})	172.7	±	7.3	b	276.3	±	11.7	c	168.7	±	7.1	b	1.0	±	0.5	a	99
COD (g L^{-1})	10.56	±	0.45	c	19.73	±	0.83	d	7.70	±	0.32	b	4.64	±	0.20	a	56
TN (g L^{-1})	1.37	±	0.06	c	2.23	±	0.09	d	1.08	±	0.05	b	0.59	±	0.02	a	57
NH_4^+-N (g L^{-1})	0.90	±	0.04	b	1.21	±	0.05	c	0.86	±	0.04	b	0.41	±	0.02	a	54
NO_3^--N (mg L^{-1})	0.00	±	0.00	a	0.00	±	0.00	a	0.00	±	0.00	a	0.19	±	0.01	b	-
PO_4^{3-} (mg L^{-1})	399.5	±	16.9	c	624.5	±	26.4	d	132.1	±	5.6	b	31.9	±	1.3	a	92
K^+ (mg L^{-1})	831.2	±	35.1	a	831.8	±	35.1	a	868.1	±	36.6	a	830.9	±	35.1	a	0
Cu (mg L^{-1})	0.04	±	0.00	a	0.04	±	0.00	a	0.05	±	0.00	b	0.04	±	0.00	ab	-
Zn (mg L^{-1})	0.16	±	0.01	a	0.14	±	0.01	a	0.19	±	0.01	b	0.15	±	0.01	a	-
[1] Agronomic dosage (L ha^{-1} yr^{-1})	124,369				76,328				157,330				287,452				
[1] N (kg ha^{-1})	170.0				170.0				170.0				170.0				
[1] P_2O_5 (kg ha^{-1})	37.14				35.63				15.54				6.86				
[1] K_2O (kg ha^{-1})	124.5				76.5				164.5				287.7				

Notes: (*) EC: electrical conductivity; SS: settleable solids; COD: chemical oxygen demand; TN: total nitrogen; NH_4^+-N: ammoniacal nitrogen; NO_3^--N: nitrogen as nitrates; PO_4^{3-}: phosphates; K^+: potassium ion; Cu: copper; Zn: Zinc. (**) RAW: raw pig slurry, SEP: separated pig slurry, DEC: decanted pig slurry, PUR: purified pig slurry. (***) Percentage reduction = 100 − ((PUR/RAW) × 100)); (-) indicates not reduction. Different letters indicate significant differences ($p < 0.05$) between phase of treatment. [1] Considering the ceiling of 170 kg N ha^{-1} yr^{-1} according to Nitrates Directive (91/676/EEC) for vulnerable areas.

Table 4. Mean and standard deviation values, percentage of reduction, agronomic dose of application and macronutrients (N-P-K) content per year during summer (n = 3).

Season	Summer																
Sample type **	RAW				SEP				DEC				PUR				Red (%) ***
Parameter *	Mean	±	SD		Mean	±	SD		Mean	±	SD		Mean	±	SD		
pH	7.29	±	0.31	a	7.30	±	0.31	a	7.83	±	0.04	a	7.86	±	0.33	a	-
EC (ds m^{-1})	9.85	±	0.42	b	9.22	±	0.39	b	9.36	±	0.06	b	7.92	±	0.73	a	20
SS (mg L^{-1})	197.3	±	8.3	b	444.0	±	18.7	c	1.4	±	2.3	a	1.0	±	0.5	a	99
COD (g L^{-1})	10.75	±	0.45	b	29.60	±	1.25	c	5.93	±	1.01	a	4.64	±	0.20	a	57
TN (g L^{-1})	1.71	±	0.07	b	2.26	±	0.10	c	1.47	±	0.18	ab	1.33	±	0.06	a	22
NH_4^+ -N (g L^{-1})	1.10	±	0.05	ab	1.26	±	0.05	b	0.91	±	0.16	a	1.00	±	0.04	a	9
NO_3^- -N (mg L^{-1})	2.05	±	0.09	b	2.09	±	0.09	b	2.11	±	0.13	b	0.19	±	0.01	a	91
PO_4^{3-} (mg L^{-1})	271.2	±	11.4	b	620.8	±	26.2	c	81.2	±	35.3	a	31.9	±	1.3	a	88
K^+ (mg L^{-1})	910.0	±	38.4	a	954.5	±	40.3	ab	1077.8	±	69.5	b	830.9	±	35.1	a	9
Cu (mg L^{-1})	0.16	±	0.05	ab	0.08	±	0.08	ab	0.05	±	0.05	a	0.20	±	0.01	b	-
Zn (mg L^{-1})	0.51	±	0.10	a	0.29	±	0.22	a	0.25	±	0.21	a	0.54	±	0.02	a	-
[1] Agronomic dosage (L ha^{-1} yr^{-1})	99,532				75,065				115,811				127,609				
[1] N (kg ha^{-1})	170.0				170.0				170.0				170.0				
[1] P_2O_5 (kg ha^{-1})	20.17				34.83				7.03				3.05				
[1] K_2O (kg ha^{-1})	109.1				86.3				150.4				127.7				

Notes: (*) EC: electrical conductivity; SS: settleable solids; COD: chemical oxygen demand; TN: total nitrogen; NH_4^+-N: ammoniacal nitrogen; NO_3^--N: nitrogen as nitrates; PO_4^{3-}: phosphates; K^+: potassium ion; Cu: copper; Zn: Zinc. (**) RAW: raw pig slurry, SEP: separated pig slurry, DEC: decanted pig slurry, PUR: purified pig slurry. (***) Percentage reduction = 100 − ((PUR/RAW) × 100); (-) indicates not reduction. Different letters indicate significant differences ($p < 0.05$) between phase of treatment. [1] Considering the ceiling of 170 kg N ha^{-1} yr^{-1} according to Nitrates Directive (91/676/EEC) for vulnerable areas.

As can be observed there is a clear tendency to decrease the values for most parameters when pig slurry passed through each phase of the treatment for purification in all periods of the research, in autumn, spring, and summer.

Non-significant differences ($p < 0.05$) among phases of treatments were found for pH and EC, for the three periods of study, except the EC where differences were found between RAW and PUR during summer. Conversely, parameters like SS, COD, and TN were significantly different ($p < 0.05$) when comparing RAW to PUR in all seasons.

In general terms, the parameters Cu and Zn varied barely according to the tendency from season to season throughout the phases of the integrated management system.

Regarding to dose of application, it was calculated respecting the ceiling of 170 kg N ha^{-1} $year^{-1}$ according European Normative prescribed in Annex III of the Nitrates Directive (91/676/EEC). Tables 2–4 exhibit that it is possible to achieve a greater volume of PUR for application purposes on land when pig slurry is treated with an integrated treatment system, therefore the following pattern was detected PUR > DEC > SEP > RAW.

In addition, the dose of macronutrients (N-P-K) that can be applied per hectare during a year is calculated. Obviously, nitrogen will be 170 kg N ha^{-1} $year^{-1}$ according to the European Normative prescribed in Annex III of the Nitrates Directive (91/676/EEC), but P_2O_5 follows the pattern SEP > RAW > DEC > PUR in autumn and summer and in spring RAW > SEP > DEC > PUR; meanwhile, the behavior of K_2O was PUR > DEC > SEP > RAW in autumn, DEC > PUR > RAW > SEP during the summer and PUR > DEC > RAW > SEP during spring.

Regarding manure properties, Table 5 presents the obtained results for the solid phase of pig slurry after separation (MAN) concerning media values and standard deviation (DS). As can be seen, in DM significant differences ($p < 0.05$) between seasons were detected. The parameters pH and EC showed the same behavior with no significant differences ($p < 0.05$) between autumn and spring, but there were differences in summer with respect to the previous seasons. According to TN, in autumn was observed the greatest mean values were significantly different ($p < 0.05$) when compared to spring and summer, and on the other hand, ammoniacal nitrogen as well as nitrates presented significant differences ($p < 0.05$) between seasons. Total organic carbon presented the highest mean values during summer, conversely phosphates and potassium exhibited the lowest in this season with significant differences ($p < 0.05$) with respect to previous seasons.

Table 5. Mean and Standard Deviation of manure (n = 3).

Sample Type	**Manure**											
Season	**Autumn**				**Spring**				**Summer**			
*** Parameter**	**Mean**		**DS**		**Mean**		**DS**		**Mean**		**DS**	
DM (%)	25.80	±	0.01	a	30.71	±	0.10	b	43.66	±	1.54	c
pH	7.57	±	0.02	a	7.45	±	0.05	a	8.62	±	0.30	b
EC ($dS\ m^{-1}$)	1.18	±	0.01	a	1.31	±	0.16	a	2.00	±	0.07	b
TN ($g\ kg^{-1}$)	3.95	±	0.01	b	3.56	±	0.01	a	3.72	±	0.13	a
N-NH_4^+ ($g\ kg^{-1}$)	1.37	±	0.00	b	1.05	±	0.02	a	2.54	±	0.09	c
NO_3^- -N ($mg\ kg^{-1}$)	4.72	±	0.13	c	0.00	±	0.00	a	0.28	±	0.04	b
TOC (%)	10.56	±	0.02	a	10.56	±	0.26	a	15.81	±	0.15	b
PO_4^{3-} ($mg\ kg^{-1}$)	451.4	±	2.8	b	939.4	±	0.0	c	84.5	±	3.0	a
K^+ ($mg\ kg^{-1}$)	741.2	±	4.5	c	701.0	±	0.0	b	0.1	±	0.0	a
Cu ($mg\ kg^{-1}$)	1.20	±	0.02	b	0.15	±	0.01	a	1.86	±	0.07	c
Zn ($mg\ kg^{-1}$)	1.47	±	0.05	b	0.42	±	0.02	a	0.51	±	0.02	a

Notes: * DM: dry matter; EC: electrical conductivity; TN: total nitrogen; NH_4^+-N: ammoniacal nitrogen; NO_3^--N: nitrogen as nitrates; TOC: total organic carbon; PO_4^{3-}: phosphates; K^+: potassium ion; Cu: copper; Zn: Zinc. Different letters indicate significant differences ($p < 0.05$) between seasons.

3.2. Gas Emissions during Storage

Figures 2–6 show the results of atmospheric humidity, atmospheric T, and emissions of CH_4, CO_2, H_2O, N_2O, and NH_3 during autumn 2020, spring 2021, and summer 2021. Atmospheric humidity values ranged between 25.0% (S11, summer) and 55.0% (S6, spring).

The general trend is decreasing from autumn to summer, presenting values within a narrower range in summer. At the atmospheric level, T values varied from 19.5 °C (S6, spring) to 38.3 °C (S11, summer).

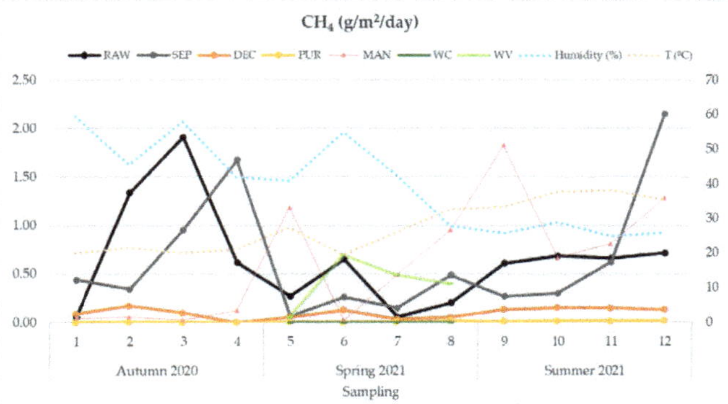

	1	2	3	4	5	6	7	8	9	10	11	12
RAW	A-a	B-c	C-c	A-ab	A-ab	B-b	A-a	A-ab	AB-ab	C-b	B-b	AB-b
SEP	B-ab	A-ab	B-d	C-d	A-a	A-ab	A-ab	AB-abc	A -ab	B-ab	B-bc	C-d
DEC	A-cd	A-g	A-de	A-a	A-bc	A-def	A-ab	A-bc	A-efg	AB-fg	A-fg	A-efg
PUR	A-a	A-a	A-a	A-a	A-a	A-a	A-a	A-b	A-b	A-b	A-b	A-b
MAN	A-a	A-a	A-a	AB-ab	B-abc	A-a	B-abc	B-abc	B-c	C- abc	B-abc	BC-bc
WC	-	-	-	-	A	A	A	A	-	-	-	-
WV	-	-	-	-	A-a	B-b	A-ab	AB-ab	-	-	-	-

Figure 2. Methane (CH_4) emissions evolution of pig slurry fractions. Different letters indicate significant differences between the fractions and samplings ($p < 0.05$). RAW: raw pig slurry; SEP: separated pig slurry; DEC: decanted pig slurry; PUR: purified pig slurry; WC: wetland control; WV: wetland vegetation; S: sampling.

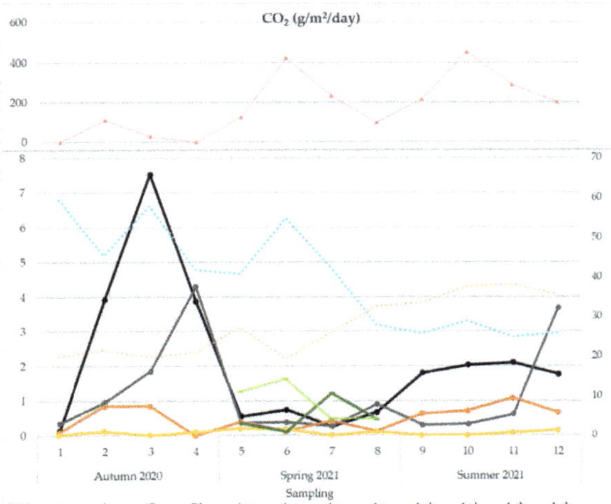

	1	2	3	4	5	6	7	8	9	10	11	12
RAW	a	b	AB-c	B-b	A-a	A-a	A-a	A-a	A-ab	A-ab	A-ab	A-ab
SEP	a	a	A-a	B-b	A-a	A-a	A-a	A-a	A-a	A-a	A-a	A-b
DEC	a	cd	A-cd	A-a	A-abc	A-ab	A-abc	A-ab	A-bcd	A-cd	A-d	A-cd
PUR	ab	bcd	A-ab	A-abcd	A-d	A-cd	A-ab	A-abcd	A-a	A-a	A-abc	A-cd
MAN	ab	ab	B-ab	A-a	B-ab	B-ab	B-ab	B-ab	B-ab	B-b	B-ab	B-ab
WC	-	-	-	-	A-b	A-a	A-c	A-b	-	-	-	-
WV	-	-	-	-	A	A	A	A	-	-	-	-

Figure 3. Carbon dioxide (CO_2) emissions evolution with stacked upper scale of pig slurry fractions. Different letters indicate significant differences between the fractions and samplings ($p < 0.05$). RAW: raw pig slurry; SEP: separated pig slurry; DEC: decanted pig slurry; PUR: purified pig slurry; WC: wetland control; WV: wetland vegetation; S: sampling.

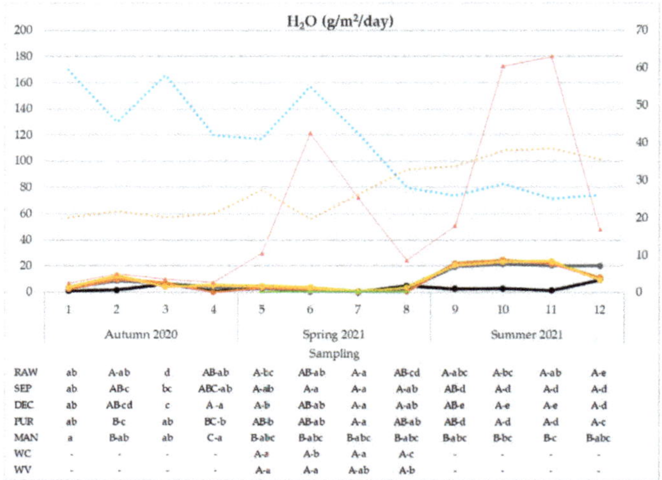

Figure 4. Water (H_2O) emissions evolution with stacked upper scale of pig slurry fractions. Different letters indicate significant differences between the fractions and samplings ($p < 0.05$). RAW: raw pig slurry; SEP: separated pig slurry; DEC: decanted pig slurry; PUR: purified pig slurry; WC: wetland control; WV: wetland vegetation; S: sampling.

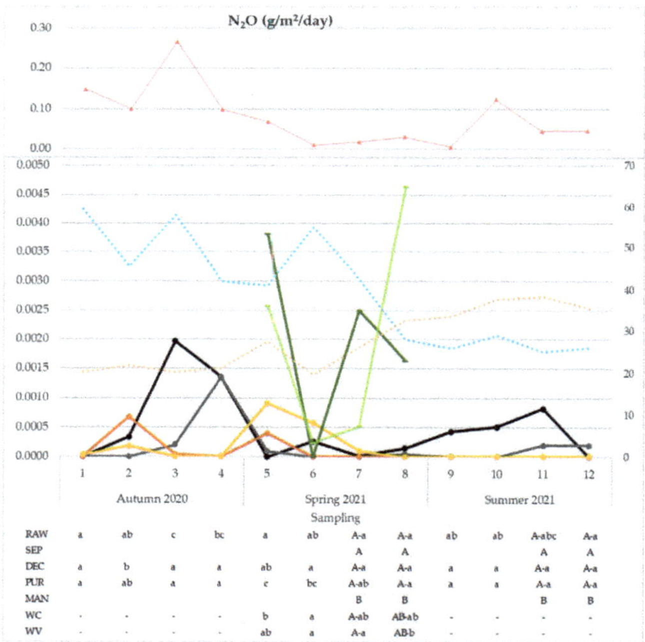

Figure 5. Nitrous oxide (N_2O) emissions evolution with stacked upper scale of pig slurry fractions. Different letters indicate significant differences between the fractions and samplings ($p < 0.05$). RAW: raw pig slurry; SEP: separated pig slurry; DEC: decanted pig slurry; PUR: purified pig slurry; WC: wetland control; WV: wetland vegetation; S: sampling.

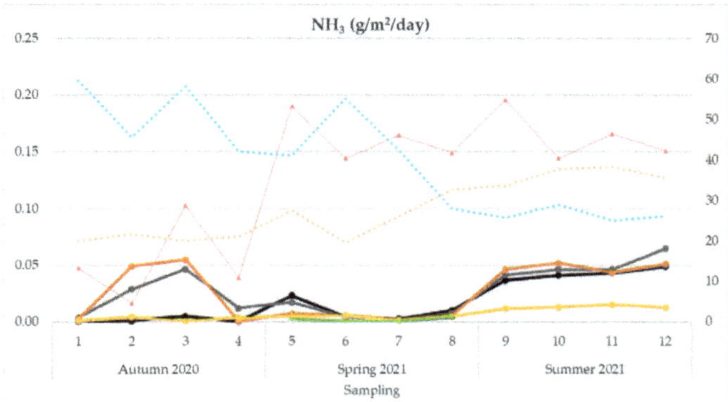

	1	2	3	4	5	6	7	8	9	10	11	12
RAW	a	A-a	A-ab	A-a	A-c	A-ab	A-a	A-b	A-d	de	A-e	f
SEP	a	B-d	A-e	AB-bc	A-c	A-a	A-a	A-ab	A-e	e	A-e	f
DEC	ab	C-cde	AB-e	A-a	A-b	A-ab	A-ab	A-ab	AB-cd	de	A-c	de
PUR	a	A-b	A-a	AB-ab	A-b	A-b	A-a	A-b	A-c	cd	A-d	cd
MAN		AB	B	B	B	B	B	B	B		B	
WC	-	-	-	-	A-b	A-a	A-a	A-c	-	-	-	-
WV	-	-	-	-	A	A	A	A	-	-	-	-

Figure 6. Ammonia (NH_3) emissions evolution with stacked upper scale of pig slurry fractions. Different letters indicate significant differences between the fractions and samplings ($p < 0.05$). RAW: raw pig slurry; SEP: separated pig slurry; DEC: decanted pig slurry; PUR: purified pig slurry; WC: wetland control; WV: wetland vegetation; S: sampling.

Figure 2 shows the results of CH_4 emissions for all the slurry fractions. Values are between 0 g/m^2/day (DEC (S4), PUR (S4), and WC (S5; S6; S7; S8)) and 2.14 g/m^2/day (SEP (S12)). Over time, no well-defined trend is observed in the results. It can be seen that CH_4 emissions present statistically higher values at the beginning of the autumn season and then decrease, later in spring display an increase (not as notable as in autumn) and then decrease (RAW, SEP, and DEC). In summer, CH_4 emissions show an increasing trend for RAW, SEP, and MAN and a slightly increasing trend for DEC and PUR. Between treatments, significantly higher emissions are recorded for the RAW (0.05–1.91 g/m^2/day), SEP (0.06–2.14 g/m^2/day) and MAN (0.02–1.82 g/m^2/day) treatments compared to DEC (0–0.17 g/m^2/day), PUR (0–0.02 g/m^2/day), WC (0 g/m^2/day) and WV (0.05–0.70 g/m^2/day). This fact reveals, on the one hand, a notable reduction in CH_4 emissions in the slurry fractions generated in the decantation and phytopurification stages, and on the other hand, a peak in CH_4 emissions from WV due to the recent filling, taking into account that the measurement of emissions was carried out 3–4 h after completing the filling of the wetland.

With respect to CO_2 emissions (Figure 3) the range of values recorded ranges between 0 g/m^2/day (DEC (S4) and 453.76 g/m^2/day (MAN (S10). The concentration ranges recorded for the fractions are: RAW (0.15–7.53 g/m^2/day), SEP (0.26–4.30 g/m^2/day), DEC (0–1.08 g/m^2/day), PUR (0.01–0.21 g/m^2/day). day), MAN (0.50–453.76 g/m^2/day), WC (0.12–1.22 g/m^2/day) and WV (0.49–1.63 g/m^2/day). MAN emissions are statistically the most outstanding, except for S4, which decreases drastically to a value of 0.50 g/m^2/day, and RAW (3.87 g/m^2/day) and SEP (4.30 g/m^2/day) stand out. The emissions of the RAW and SEP fractions present in a certain way a similar trend line because they present a similar analytical characterization. The results recorded for all fractions display a fluctuating trend with peaks and decreases, as occurred with CH_4 emissions. In this sense, the correlation analysis has shown a statistically significant correlation between CH_4 and CO_2 emissions (R = 0.626**).

The recorded emissions of H_2O in slurry fractions range between 0 g/m^2/day (DEC (S4) and 179.77 g/m^2/day (MAN (S11)). For each fraction, the following ranges are detected: RAW (0.71–9.39 g/m^2/day), SEP (0.47–21.60 g/m^2/day), DEC (0–24.45 g/m^2/day), PUR (0.49–23.38 g/m^2/day), MAN (6.61–179.77 g/m^2/day), WC (0.03–0.14 g/m^2/day) and

WV (0.01–0.12 g/m^2/day), being in the summer season the highest values for all fractions. Statistically, the results of MAN stand out with respect to the rest of the fractions, possibly because it is not a liquid fraction and the water present in the manure tends to evaporate into the atmosphere. The WC and WV fractions present lower H_2O emissions than the rest of the fractions. Although WC and WV do not stand out statistically, these values could be influenced by the fact that the slurry is not in aerobic conditions, with a layer of wetland fill substrate existing between the emissions-emitting surface and the emissions measurement chamber. Correlations are recorded with values of R = 0.32** (H_2O emission/T) and R = 0.68** (H_2O emission/NH_3 emission). Although the H_2O emission/T correlation presents a low regression coefficient (R = 0.32**), this proportionality can be notably seen in the trend recorded in the summer season, as the T increases (33.5–38.3 °C) H_2O emissions increase.

For N_2O gas, the global warming potential factor for a given time of 100 years is 298, being 1 for CO_2 and 25 for CH_4. N_2O emissions are the lowest of all recorded emissions, ranging between 0 g/m^2/day (RAW (S1; S5; S12), SEP (S1; S2; S6; S7; S9; S10) DEC (S1, S4, S6; S7, S9; S11; S12), PUR (S3; S4, S9; S10; S11) and 0.265 g/m^2/day (MAN (S3)). The highest value recorded for each fraction is 0.002 g/m^2/day (RAW), 0.001 g/m^2/day (SEP), 0.001 g/m^2/day (DEC), 0.001 g/m^2/day (PUR), 0.265 g/m^2/day (MAN), 0.003 g/m^2/day (WC) and 0.004 g/m^2/day (WV). During the weekly measurements of all stations and fractions, peaks of rise and fall in N_2O emissions are recorded, being more notable in the autumn season for RAW, SEP, and MAN, and in spring for DEC and PUR. In the summer season, a drop in N_2O emissions is recorded, which can be partially justified based on the low correlation recorded N_2O emission/atmospheric humidity (R = −0.269**). Other correlations are also recorded: N_2O emission/CO_2 emission (R = 0.443**) and N_2O emission/CH_4 emission (R = 0.217**).

NH_3 is a polluting and toxic gas at certain concentrations. Its concentrations in the atmosphere are combated to be mitigated in pig farms through nutritional and technical strategies. In this study, a range of NH_3 concentrations is recorded between 0 g/m^2/day (RAW (S1; S4), DEC (S4), WC (S6; S7) and WV (S6; S7) and 0.195 g/m^2/day (MAN (S9)). In Figure 6 it can be seen how in most samples the emissions in the MAN fraction stand out. The slurry fractions resulting from the decanting and phytopurification stages present NH_3 values closer to 0 g/m^2/day, especially the WC and WV fractions, although these fractions do not stand out statistically. The NH_3 emission/H_2O emission correlation (R = 0.679**) stands out, influenced by the humidity/T correlation (R = −0.575**). In fact, similar trend lines are observed for both gases, influenced by parameters such as T (NH_3 emission/T, R = 0.282**) and atmospheric humidity (NH_3 emission/humidity, R = −0.172**). Also noteworthy are the correlations NH_3 emission/CH_4 emission (R = 0.387**) and NH_3 emission/CO_2 emission (R = 0.419**).

4. Discussion

4.1. Investigation Facilities and Schedule of the Integral Treatment with Wetlands

Physical stage of solid–liquid phase separation with phase separator with sieving and press filter (stage 1): the application of the solid–liquid phase separation technique is justified based on the BREF-MTD19/Group 12 on the In Situ Processing of Manure (Guide to the Best Available Techniques) [24]. This BAT indicates that the application of mechanical slurry separation reduces emissions of nitrogen, phosphorus, odors, and pathogenic microorganisms to the atmosphere and water, and facilitates the storage and/or application of manure to the field. It specifies that mechanical separation can be performed using a screw press separator, a centrifugal decanter, coagulation-flocculation, sieving, and filter presses. In this study, a static separator separates the solid and liquid fraction of the raw slurry by filtration through a sieve (500 μm) and pressing with an endless screw, with a working performance of 10–12 m^3/h. The raw pig slurry initially passes through a sieve consisting of a mesh whose pore must allow the retention of solid particles (≥500 μm). The raw pig slurry that does not filter through the sieve is introduced into a cylinder with a filter-shaped wall (thickness 0.5–1 mm) inside which a helical screw is located. The raw slurry is introduced into the lower part of the cylinder, passing the separated liquid fraction through a filter and being drained into a separate container. Meanwhile, the solids in suspension are subjected to pressure by rotating the screw about its central axis. The solid phase is compacted by the loss of liquid and the result will be a solid fraction with a high dry matter content that exits the cylinder from the opposite end (manure). Increasing the applied pressure will increase the dry matter content. The sieved liquid fraction (separated pig slurry) and filtered is drained to a collector or storage system.

Physical stage of solid–liquid phase separation with gravity decantation (stage 2): the technique of settling solid-liquid phases by gravity is not recognized in the Guide to BAT [23] or the Implementing Decision (EU) 2017/302 of the Commission of February 15, 2017, which establishes the conclusions on BAT within the framework of Directive 2010/75/EU of the European Parliament and of the Council regarding the intensive farming of poultry or pigs. According to the BAT Guide, it is recognized as BAT when the decanter used is a centrifugal type, which is not the one used in the integral system of this study, given that decantation occurs with the sedimentation of the slurry as it moves through several interconnected ponds. The present stage consists of a separation of the solid and liquid fractions of the slurry separated by natural sedimentation by gravity in several decanters. It consists of decanting the slurry by gravity into a container shaped like an orthogonal rectangular prism with adequate dimensions. To improve sedimentation, it works with at least three decanters connected in series where the slurry is added constantly. At the present farm, the system works with six decanters in series, and the decanted liquid fraction overflows and drains to a storage system. The most solid fraction settles at the bottom of the decanters and is returned from time to time to the raw slurry storage system or is incorporated with the manure.

Biological stage of purification in artificial wetlands or biofilters: the biological purification of slurry with artificial wetlands or biofilters is considered in the BREF documents "Intensive Rearing of Poultry or Pigs" as a technique for wastewater purification [24]. It is a low-cost system, with high environmental integration and greater resistance to load variations than conventional systems [8,17] included in any BAT; however, it is included as a technique that reduces atmospheric emissions of wastewater according to the technical document Evaluation of Manure Management Techniques in Livestock (bovine, pork, poultry, and meat sectors) issued by the Ministry of Agriculture, Food and Environment in 2015 and also in Commission Implementing Decision (EU) 2017/302 of 15 February 2017 in which the conclusions on the BAT are established within the framework of Directive 2010/75/EU of the European Parliament and of the Council regarding the intensive farming of poultry or pigs. Specifically, it is indicated that in the purified slurry storage pond (compared to untreated raw slurry) the volatilization and rapid emission of ammonia is reduced by up to 50%. Furthermore, at an experimental level, a reduction in the concentration of NO_3^- of up to 22% has been recorded between the samples of the input effluent and the output effluent of the artificial wetlands. This fact shows that the treatment of manure with this biological technique also involves nitrification-denitrification and could also be included within the BREF/MTD19. At a conceptual level, artificial wetlands consist of a mono or polycrop of macrophyte plants arranged in lagoons, tanks, or shallow, waterproofed channels filled with different substrates (sand, fine gravel, medium gravel, and coarse gravel). The treated wastewater is filtered through the filter media and collected through a drainage system at the bottom.

Plants are the center of wastewater treatment, being an integral and indispensable part of these systems [25]. General requirements for selecting the appropriate plant in constructed wetlands for wastewater treatment include [26]: (a) ecological acceptability, (b) tolerant to local climatic conditions, pests and diseases, (c) tolerance to contaminants and submerged hypertrophic conditions, (d) easy propagation, rapid establishment, spreading and growth (e) high capacity for the removal of contaminants, either by direct assimilation and storage or indirectly by increasing microbial transformation such as nitrification (through the release of oxygen in the root zone) or denitrification (through the production of carbon in the substrates). The selection of plants in wetlands is a factor of great importance, due to their purification capacity and tolerance to contaminants. Plants are attributed to a high capacity for removing nitrogen and phosphorus. Root exudate positively influences microbial transformation, the amount of denitrifying bacteria, root biodegradation, and the purification capacity of wetlands [27]. In addition, wetlands provide greater oxygen transfer because when they are emptied, air penetrates the wetland, leading to the nitrification process. Subsequently, when filled and with the appropriate anaerobiosis conditions (influenced by hydraulic retention times among other factors), the biological process of denitrification could be completed to produce the conversion to forms of gaseous nitrogen that is released into the atmosphere [28,29]. In the case of artificial wetlands, the amount of sludge generated is not appreciable, so when the wetland cells are emptied, N_2 escapes into the atmosphere and the output effluent is directed to a final storage pond. The purifying species *Phagmites australis* is the most used phytopurifying species in wetland treatment [3,27] in semiarid climates such as that of the Murcia Region, having adequate effectiveness for the elimination of physical-chemical and microbiological parameters [7,28,29]. This species has been planted in the present study (5 plants/m^2).

It should be mentioned that in the slurry-purified ponds spontaneously certain types of microalgae overgrow. Those microalgae continue to promote the purification of the slurry [30]. Purification occurs through the photosynthesis of microalgae with solar irradiation, and among the processes that occur, denitrification also takes place. These types of microalgae grow mainly with the consumption of soluble phosphorus and NO_3^-, and even have the capacity to take atmospheric nitrogen when the medium in which they are found lacks other nitrogen sources in oxidized form (being NO_3^- the most oxidized form of nitrogen [30]. Within this type of microalgae, benthic algae, and *Scenedesmus* sp. stand out. In a doctoral thesis on artificial wetlands with fattening pig slurry at the Integrated Center for Training and Agricultural Experiences (Lorca, Spain) it was recorded a certain degree of bioremediation in the purified slurry storage pond by the action of microalgae (*Scenedesmus* sp.); specifically, reductions were recorded in the concentrations of NO_3^-, N-NH_4^+, total N, and soluble Cu and Zn [28]. Pig slurry presents a great variability in its composition, and for that reason is necessary to carry out an analysis before/after treatment with artificial wetlands. In this way, the purification efficiency of the system and the agronomic value of the purified effluent, which can be used as fertilizer, are known. Thus, the application dose adjusted to the type of crop and the characteristics of the soil and irrigation water can be calculated.

Agronomic recycling stage of water and nutrients: the agronomic recycling of pig slurry involves several recommendations and application techniques that are included in Group 13 Application of manure to the field MTD20-MTD22 [24] such as carrying out soil analysis, application recommendations on land with runoff, preparation of fertilization plans according to the demand of the soil–water–plant system and application techniques. Among the techniques for applying slurry to the soil, the application of slurry with deep injection (>15 cm) is the most practiced, which reduces ammonia emissions by up to 90%, and also acidification with a view to reducing ammonia emissions. At an environmental level, it is highly recommended to carry out a fractional agronomic application when the agronomic doses are between 50,000 L/ha–100,000 L/ha, as well as carrying out environmental control of the receiving soil through periodic annual analyses. In addition, the Spanish Royal Decree 1051 on sustainable nutrition in agricultural soils will be taken into consideration in order to mitigate the environmental impact of the application of manures on agricultural soils. All this with the aim of achieving a sustainable supply of nutrients in agricultural soils and reducing greenhouse gas emissions and other polluting gases. Similarly, with the recommendations of the JRC SAFEMANURE working group created by the European Union to develop criteria and agricultural resources for the safe use of processed manure in areas vulnerable to contamination by nitrates at doses above the limits established in Directive 91/ 676/EEC. Such resources are known by the acronym

'RENURE' for "recovered nitrogen from manure" and are defined as "any substance containing nitrogen wholly or partially derived from livestock manure by means of a treatment that can be used in areas with contamination of the water for nitrogen." For RENURE resources, similar provisions apply to nitrogen-containing chemical fertilizers as defined in the Nitrate Regulation (Directive (91/676/EEC), as long as compliance with the nitrate directive is ensured and adequate agronomic benefits are provided to achieve good productivity. Regarding the criteria for a purified pig slurry to be considered RENURE: mineral N/total N ratio $\geq$ 90% or TOC/TN ratio $\leq$ 3 and not exceeding the limits of Cu $\leq$ 300 mg/kg dry matter and Zn $\leq$ 800 mg/kg dry matter dry. The experimental analytical results of slurry on farms with integral slurry management systems approach or meet the aforementioned criteria. Analytical results can even be optimized by promoting certain techniques in some of the aforementioned stages, such as aeration, microfiltration (120–140 μm) of the slurry entering the wetlands and even working with longer hydraulic retention times. All actions aimed at transforming liquid manure or slurry into RENURE could be effective manure management strategies to protect waters from nitrate leaching and ensure adequate agronomic benefits. In this way, having an estimate of nitrogen emissions in RENURE fractions or agricultural soil from recycled RENURE fractions (or possible RENURE fractions) would be very useful from a bibliographic point of view according to the JRC since there is hardly any data at an international level.

4.2. *Pig Slurry Traceability*

As can be observed EC did not experience large variation after separation or decanter modules in any season (Tables 2–4) showing no significant differences ($p < 0.05$) when compared SEP and DEC with respect to RAW. Electrical conductivity is proportional to the content of dissolved salts and, therefore, is directly related to the sum of cations or anions that are determined chemically [31] and, in general, presents a close relation with the total dissolved solids.

As expected, PUR slurry resulted in highly effective SS removal. In this study, the system reached up to 99% reduction in autumn, spring, and summer. This finding agreed with previous studies [8,32,33]. Several authors [32–34] have demonstrated that planted CW is more effective in a reduction in SS, reporting high percentages of 100% 99%, and 98%.

The internal slope of the Horizontal subsurface constructed wetland could contribute to sedimentation, in addition to it, filtration phenomena because of the small space among particles also triggered the reduction in SS [8,35]. It is important to highlight that within the CW the processes by which wastewater is purified include a wide range of interacting biological, physical, and chemical mechanisms, as well as plant uptake, which may contribute to a synergism for the system. Nutrients are absorbed by plants from the water column through the roots, which serve as an ideal support medium for bacterial growth and the filtration/adsorption of suspended solids [9].

It should be noted that pretreatment of the pig slurry (separation and decantation), cooperates significantly on one hand to avoid media clogging, on the other hand with at least sedimentation of settleable solids in the CW beds [35]; therefore, these effects promoted the reduction in solids.

A range of 56–80% reduction in COD was achieved in the integral treatment system during the assessment for the three studied seasons. The main phenomena associated with COD reduction in pig slurry are volatilization, photochemical oxidation, sedimentation, adsorption, and biological degradation [8,19,20]. Scholz [33] reported 95% of removal for COD calculated in the outflow respect to the inflow; Caballero-Lajarín [32] observed an efficiency of 68% in a study using CW combined with a pretreatment composed by a separator, decanter, and sedimentation tanks, similar to the one used in this study for the comprehensive treatment of slurry. A study carried out by Haddis [20] highlighted CW as a natural solution to remove organic pollutants, reporting 65% and 62% of removal in planted systems.

Phosphate concentration reductions were generally greater with 100% in autumn, 92% in spring, and 88% in summer. Previous researchers reported the potential efficiency of the CW to remove this nutrient from the influent [36].

The solubility and reactivity of various forms of phosphorus are influenced by the physical, chemical, and biological properties of a wetland system. Previous authors suggest that the most important mechanisms phosphorus retention pathway in wetlands are via physical sedimentation [37], adsorption, and chemical precipitation associated with long-term storage in CW [32].

In a review concerning to treatment of industrial wastewater with CW, Vymazal [17] found in different studies that 5 days resulted be the most effective HRT for percentages of reduction. For instance, 84.4% for aquaculture wastewater, 35% for mixed wastewater, 80% for potato processing wastewater, and 85% for treating diluted olive mill wastewater. In an integral treatment system used in our study, Terrero [8] reported 95% of TP removal with an HRT of 7 days treating pig slurry, and Caballero-Lajarín [32] reported a significant reduction of 90% after 4 weeks of HRT.

Plants absorb nutrients to sustain their metabolism, they can also take in trace chemicals from the root zone, which may be stored or, in some cases, expelled as gases. This uptake primarily occurs through the roots, typically located in wetland soils, although fortuitous roots can sometimes extend into the water column. Submerged plants may also absorb nutrients and metals directly from the water into their stems and leaves. While plant uptake is a key removal mechanism for certain pollutants, it plays a principal role only in lightly loaded systems. Nevertheless, plants are essential for maintaining high-quality water treatment performance in most wetland systems [9]. This phenomenon can also explain the minimization of the TP content in the effluent, the purified slurry with the HSFCW treatment system. Likewise, TP reduction could be related to coprecipitation with Ca and Mg due to a limestone gravel bed as explained by Terrero [8]. Schulz [38] obtained a 49% removal in TP with a very short HRT of 7.5, 2.5, and 1.5 h, treating rainbow trout farm effluents in HSFCW with emergent plants.

Concerning adsorption via substrate, Vymazal [17] pointed out that to improve phosphorus removal, it is important to choose materials with high phosphorus adsorption capacity, which is determined by their chemical and physical properties. These materials may include minerals with reactive iron or aluminum hydroxide or oxide groups on their surfaces, or calcareous materials that can encourage the precipitation of calcium phosphate. Thus, previous studies observed that the decrease in TP could be related to adsorption and chemical precipitation with Ca^{+2} coupled with iron, aluminum, and organic matter fixed in the used substrate [8,32,35].

Results concerning nitrogen concentrations presented the highest percentage reduction of 57% during spring (Table 3) when compared to PUR to RAW slurry, followed by 50% during autumn (Table 2) and 22% in summer (Table 4). The different mechanisms that occur to reduce nitrogen forms throughout integrated treatment systems like those used in this study could be very wide, including separation, sedimentation, nitrification and denitrification, microbial transformation during storage [39,40], and other processes like adsorption that could take place during the whole treatment. Kadlec [9] exposed that when the wastewater moves through the wetland, it undergoes treatment through processes such as sedimentation, filtration, oxidation, reduction, adsorption, and precipitation. Additionally, the wetland nitrogen cycle includes a number of pathways like atmospheric nitrogen inputs, ammonia adsorption, and ammonia volatilization, affecting nitrogen compounds. Although nitrite and nitrate, the oxidized nitrogen forms do not adhere to solid substrates, but ammonia is capable of sorption to both organic and inorganic substrates, due to the positive charge of the ammonium ion, it is susceptible to cation exchange.

Wetlands plants need to assimilate nitrogen, especially in the forms of ammonia and nitrate nitrogen. In this process, plants absorb nitrogen primarily through their root systems, which are mostly situated in the wetland soil [9]; therefore, this is the main pathway to reduce the nitrogen in CW thanks to plant uptake. Our results are slightly below those experienced by Huang [41] with 73–61% TN reduction in a study concerning the effects of plants in a horizontal subsurface flow pilot-scale constructed wetlands, or Caballero-Lajarín [32] where a decrease of 63% was achieved for TN after wetland. A study presented by Hjorth [39] pointed out that around 25% of the nitrogen and phosphorous is retained in the solid fraction using a screw separator.

Taking into account the importance of the Spanish livestock subsector, it is considered necessary to accomplish the regulations for the dosage and application of manure to soils that ensure the protection of human health and the environment; therefore, the proper management of manure is crucial, with farmers being responsible for it within the scope of their respective obligations manifested in RD 306/2020 [42]. In this way, the calculation of the agronomic dosage is useful to estimate the agricultural land necessary for better application.

Nitrates Directive has established a maximum of 170 kg N ha^{-1} year $^{-1}$ as application dosage; therefore, countries must adopt techniques in order to avoid unnecessarily high application levels of nitrogen per hectare of land. According to Bref documents [24], Best Available Techniques can be applied to pig slurry aiming to facilitate the manure's agricultural use (better dosing).

Although in many cases the use of certain techniques is limited for technical and/or economic reasons, agricultural valorization as the final destination of slurry should be considered the main and most favorable option. But it should always be considered that when the agricultural application is not performed correctly and the capacity of the receiving agrosystem is exceeded, due to risks of contamination and alteration the environment may occur.

As can be observed in Tables 2–4, purified pig slurry allowed a greater volume of application compared with the previous phase a CW, and furthermore raw slurry. Regarding the volume of application respecting the limited agronomic dosage of 170 kg N ha^{-1} $year^{-1}$, in the three studied seasons, the pattern followed by the modules of the integrated management system was RAW < DEC < PUR. In all cases of study, SEP resulted higher in TN concentrations compared to RAW.

The solid manure is a product of the separation of the RAW slurry and as well-known is composed basically of dried matter and phosphorous. The characterization of MAN in this study is in accordance with values reported by Møller [43] in terms of dried matter and TN. Those authors reported mean values of 21.9–31.7% of DM and 0.4–0.48% of TN, comparable with our findings of 43.7–25.8% of DM and 0.4–0.37% of TN.

Table 5 verifies that phosphate content in autumn and spring with mean values of 451.1 mg kg^{-1} and 939.1 mg kg^{-1} were higher than mean values found in RAW 323.7 mg kg^{-1} and 399.5 mg kg^{-1} during autumn and spring, respectively. Our findings are within the range 264.0–501.6 kg^{-1} presented in research carried out by Kowalski [44] during the same season.

4.3. GHG/NH_3 Emissions in Pig Slurry Fractions

The emissions results are consistent with previous studies that have investigated CH_4 emissions in slurry and animal waste management systems. In a study conducted by Dinuccio [45], CH_4 emissions were evaluated in different slurry fractions and significant variability in emissions was observed between the different slurry fractions. Furthermore, the authors found that CH_4 emissions were higher during certain seasons of the year and in liquid fractions (not in manure), which is consistent with the findings reported in this study, confirming an increase in CH_4 emissions with the increase in T [46]. Another relevant study is the one carried out by Veillete [47], in which CH_4 emissions in slurry treatment systems were investigated. The results showed that the decantation and phytopurification stages significantly reduced CH_4 emissions compared to other treatment stages. These findings support the observation of a reduction in CH_4 emissions in the slurry fractions generated in the settling and phytopurification stages reported in this study. Furthermore, a study conducted by Zhou [48] examined CH_4 emissions in wetlands with the presence of vegetation. The results showed that newly filled wetlands with the presence of vegetation can experience methane emission peaks due to the decomposition of organic matter and associated microbial activity. These findings support the observation of a CH_4 emission peak in WV in this study.

In manure, CO_2 originates from three sources: (1) the rapid hydrolysis of urea into NH_3 and CO_2 catalyzed by the enzyme urease; (2) the anaerobic fermentation of organic matter into intermediate volatile fatty acids (VFA), CH_4, and CO_2; (3) the aerobic degradation of organic matter [49–51]. Based on the above, the positive correlation between the recorded emissions of CO_2 and CH_4 in waste and slurry management systems is justified. A study conducted by Dinuccio [45] examined CO_2 emissions with temperature and at different stages of slurry treatment and found that the RAW and MAN fractions showed the highest CO_2 emissions. These findings support the results recorded in this study, where MAN, RAW, and SEP also present notable CO_2 emissions. Furthermore, a study carried out by Philippe [52] investigated CO_2 emissions in agricultural waste management systems and found a significant correlation between CH_4 and CO_2 emissions. The authors highlighted that the decomposition and fermentation processes of organic matter in agricultural waste can generate both CO_2 and CH_4. These results support the statistically significant correlation between CH_4 and CO_2 emissions (R = 0.626**) reported in this study. Regarding H_2O, little data concerning H_2O emissions during storage of both liquid and solid fractions are currently available.

Appreciable N_2O concentrations are only recorded for the MAN fraction, coinciding with other studies. Several studies support these findings for liquid slurry fractions in terms of the low N_2O emissions recorded. For example, Dinuccio [45] conducted research on greenhouse gas emissions in agricultural systems and found that N_2O emissions were generally lower compared to CO_2 and CH_4 emissions. These results can be attributed to the lower production and release of N_2O compared to other gases, as well as the lower atmospheric persistence of N_2O. The small N_2O fluxes from cattle and pig slurry storage can be explained by the absence of crust during most of the storage period. N_2O may be emitted during the storage of manure either as a byproduct of incomplete ammonium oxidation or as a by-product of incomplete denitrification [53]. Under aerobic conditions, NH_4^+ will be oxidized to NO_2^-; as an intermediate, the diffusion of N_2O from the nitrification reaction system to the atmosphere results in the emission of N_2O [54]. Furthermore, previous research has indicated that the formation of N_2O in CWs is mainly caused by a nitrification process [55,56], deducing that the process is favored under the aerobic conditions of cell filling. In this sense, the fact that the PUR fraction presents lower N_2O emissions than WC and WV could be due to alterations in the structure of the microbial community involved in the transformation of nitrogen in the wetlands or in the final storage pond, particularly in denitrifying microbial species [57].

Gases such as hydrogen, hydrogen sulfide (H_2S), NH_3, and volatile organic compounds are also generated in slurry fractions [58]; however, they do not have a direct effect on global warming. The results confirm a positive relationship between NH_3 emission and parameters like temperature, pH, and NH_4^+-N found by other studies [45,59,60]. A study conducted by Dinuccio [45] examined NH_3 emissions in different pig manure fractions. The results showed that raw slurry had the highest ammonia emissions, followed by separated slurry and decanted slurry. These fractions, which contain a higher concentration of nitrogen, are more likely to release ammonia due to microbial decomposition and volatilization of ammoniacal nitrogen. Another study by Osada [61] investigated the emissions of CH_4, N_2O, and NH_3 in pig manure treated with constructed wetlands. The study found that this slurry treatment method can significantly reduce ammonia emissions. Constructed wetlands act as biological filters and promote the transformation of ammonia into less volatile forms, such as nitrate. This helps reduce the release of ammonia into the atmosphere and reduce environmental impact. Furthermore, a study by Zhou [48] evaluated NH_3 emissions in different stages of pig manure management. The study found that manure, as a solid slurry fraction, can have significant ammonia emissions due to its nitrogen-rich composition. However, slurry fractions settled and treated with artificial wetlands showed a considerable reduction in ammonia emissions due to the separation and biological transformation of nitrogen.

5. Conclusions

All efforts to convert liquid manure or slurry into RENURE can serve as effective manure management strategies, safeguarding water sources from nitrate leaching while also providing essential agronomic benefits. By adopting these practices, we can enhance nutrient utilization in agriculture, reducing environmental impact, and contributing to more sustainable farming systems. Therefore, the integral management system of this study has demonstrated: (1) to be a practical, low cost and efficient technique to pig slurry treatment that offers a successful opportunity to decrease the concentration of nutrients in pig slurry fractions in line with the European normative, (2) the interception of nitrogen, phosphorus, other nutrients, and organic matter could provide a valuable subproduct that subsequently can be useful to be applied to the soil for its nutritional and water value, (3) the potential of removal up to 95% of SS, 56% of COD, 52% of TN and 80% of PO_4^{3-} in CW promoted by the pretreatment linked to the phytoextraction and several biological and physico-chemical processes in the system, (4) the results of emissions support the importance of the physical separation and phytopurification stages in reducing emissions of CH_4, CO_2, N_2O, and NH_3 and highlight the practical potential of artificial wetlands to treat slurry and reduce the impact of emissions derived from the pig sector and the related environmental and analytical factors, (5) with respect to the liquid fractions, the MAN fraction presents higher emissions of CO_2 and N_2O, this aspect could be the subject of future research due to its great contribution to the global warming potential and (6) the substrate surface of constructed wetlands (WC and WV) has an effect similar to that of a rigid coverage and stands out for displaying lower NH_3 emissions compared to the rest of the fractions (RAW, SEP, DEC, and PUR).

Author Contributions: Conceptualization, M.G.-G. and M.A.T.T.; methodology, M.G.-G. and M.A.T.T.; validation, M.G.-G., M.A.T.T. and O.E.b.; formal analysis, M.G.-G.; investigation, M.G.-G., M.A.T.T., O.E.b. and Á.F.C.; resources, M.G.-G., M.A.T.T., O.E.b. and Á.F.C.; data curation, M.G.-G. and M.A.T.T.; writing—original draft preparation, M.G.-G., M.A.T.T. and O.E.b.; writing—review and editing, M.G.-G., M.A.T.T., O.E.b. and Á.F.C.; visualization, M.G.-G., M.A.T.T. and O.E.b.; supervision, Á.F.C. All authors have read and agreed to the published version of the manuscript.

Funding: This research received no external funding.

Data Availability Statement: The data presented in this study are confidential and belong to a finished project. This publication is the first associated with the project. Access to the information related to the complete project data is restricted to authorized personnel only due to privacy and confidentiality considerations. For further inquiries or potential data access requests, please contact the corresponding author.

Conflicts of Interest: The authors declare no conflicts of interest. The funders had no role in the design of the study; in the collection, analyses, or interpretation of data; in the writing of the manuscript; or in the decision to publish the results.

References

1. Huygens, D.; Orveillon, G.; Lugato, E.; Tavazzi, S.; Comero, S.; Jones, A.; Gawlik, B.; Saveyn, H. *Technical Proposals for the Safe Use of Processed Manure above the Threshold Established for Nitrate Vulnerable Zones by the Nitrates Directive (91/676/EEC)*, EUR 30363 EN, 2020th ed.; Publications Office of the European Union: Luxembourg, 2020. [CrossRef]
2. Torrens, A.; Folch, M.; Salgot, M. Design and Performance of an Innovative Hybrid Constructed Wetland for Sustainable Pig Slurry Treatment in Small Farms. *Front. Environ. Sci.* **2021**, *8*, 577186. [CrossRef]
3. Scholz, M. Piggery Wastewater Treatment with Integrated Constructed Wetlands. In *Wetlands for Water Pollution Control*; Elsevier: Amsterdam, The Netherlands, 2016; pp. 419–432. [CrossRef]
4. Cocolo, G.; Hjorth, M.; Zarebska, A.; Provolo, G. Effect of acidification on solid–liquid separation of pig slurry. *Biosyst. Eng.* **2016**, *143*, 20–27. [CrossRef]

5. Karakashev, D.; Schmidt, J.E.; Angelidaki, I. Innovative process scheme for removal of organic matter, phosphorus and nitrogen from pig manure. *Water Res.* **2008**, *42*, 4083–4090. [CrossRef]
6. Burton, C.H. Manure management: Treatment strategies for sustainable agriculture, 2nd edition. *Livest. Sci.* **2006**, *102*, 256–257. [CrossRef]
7. Burton, C.H. The potential contribution of separation technologies to the management of livestock manure. *Livest. Sci.* **2007**, *112*, 208–216. [CrossRef]
8. Terrero, M.A.; Muñoz, M.Á.; Faz, Á.; Gómez-López, M.D.; Acosta, J.A. Efficiency of an integrated purification system for pig slurry treatment under mediterranean climate. *Agronomy* **2020**, *10*, 208. [CrossRef]
9. Kadlec, H.R.; Wallace, D.S. *Treatment Wetlands*, 2nd ed.; Taylor & Francis Group: Boca Ratón, FL, USA, 2009.
10. Parde, D.; Patwa, A.; Shukla, A.; Vijay, R.; Killedar, D.J.; Kumar, R. A review of constructed wetland on type, treatment and technology of wastewater. *Environ. Technol. Innov.* **2021**, *21*, 101261. [CrossRef]
11. Ottová, V.; Balcarová, J.; Vymazal, J. Microbial characteristics of constructed wetlands. *Water Sci. Technol.* **1997**, *35*, 117. [CrossRef]
12. Harrington, C.; Scholz, M.; Culleton, N.; Lawlor, P.G. The use of integrated constructed wetlands (ICW) for the treatment of separated swine wastewaters. *Hydrobiologia* **2012**, *692*, 111–119. [CrossRef]
13. Scholz, M. *Wetlands for Water Pollution Control*; Elsevier: Amsterdam, The Netherlands, 2016; pp. 137–155. [CrossRef]
14. Knight, R.L.; Payne, V.W., Jr.; Borer, R.E.; Clarke, R.A., Jr.; Pries, J.H. Constructed wetlands for livestock wastewater management. *Ecol. Eng.* **2000**, *15*, 41–55. [CrossRef]
15. Hsu, C.-B.; Hsieh, H.-L.; Yang, L.; Wu, S.-H.; Chang, J.-S.; Hsiao, S.-C.; Su, H.-C.; Yeh, C.-H.; Ho, Y.-S.; Lin, H.-J. Biodiversity of constructed wetlands for wastewater treatment. *Ecol. Eng.* **2011**, *37*, 1533–1545. [CrossRef]
16. García-Valero, A.; Martínez-Martínez, S.; Faz, Á.; Terrero, M.A.; Muñoz, M.Á.; Gómez-López, M.D.; Acosta, J.A. Treatment of wastewater from the tannery industry in a constructed wetland planted with Phragmites australis. *Agronomy* **2020**, *10*, 176. [CrossRef]
17. Vymazal, J. Constructed wetlands for treatment of industrial wastewaters: A review. *Ecol. Eng.* **2014**, *73*, 724–751. [CrossRef]
18. Vymazal, J. The use constructed wetlands with horizontal sub-surface flow for various types of wastewater. *Ecol. Eng.* **2009**, *35*, 1–17. [CrossRef]
19. Kadlec, R.H.; Knight, R.L. *Treatment Wetland*; Lewis Publishers: Boca Ratón, FL, USA, 1997.
20. Haddis, A.; Van der Bruggen, B.; Smets, I. Constructed wetlands as nature based solutions in removing organic pollutants from wastewater under irregular flow conditions in a tropical climate. *Ecohydrol. Hydrobiol.* **2020**, *20*, 38–47. [CrossRef]
21. Vymazal, J.; Kröpfelová, L. Multistage hybrid constructed wetland for enhanced removal of nitrogen. *Ecol. Eng.* **2015**, *84*, 202–208. [CrossRef]
22. American Public Health Association; American Water Works Association; Water Environment Federation. *Standard Methods for the Examination of Water and Wastewater*, 23rd ed.; American Public Health Association: Washington, DC, USA, 2017.
23. Duchaufour, P. *Precis de Pedologie*; Masson: Paris, France, 1970.
24. Santonja, G.G.; Georgitzikis, K.; Scalet, B.M.; Montobbio, P.; Roudier, S.; Sancho, L.D. *Best Available Techniques (BAT) Reference Document for the Intensive Rearing of Poultry or Pigs*; Publications Office of the European Union: Luxembourg, 2017. [CrossRef]
25. Brix, H. Do macrophytes play a role in constructed treatment wetlands? *Water Sci. Technol.* **1997**, *35*, 11–17. [CrossRef]
26. Tanner, C.C. Plants for constructed wetland treatment systems—A comparison of the growth and nutrient uptake of eight emergent species. *Ecol. Eng.* **1996**, *7*, 59–83. [CrossRef]
27. Jiang, X.; Tian, Y.; Ji, X.; Lu, C.; Zhang, J. Influences of plant species and radial oxygen loss on nitrous oxide fluxes in constructed wetlands. *Ecol. Eng.* **2020**, *142*, 105644. [CrossRef]
28. Vymazal, J. Constructed Wetlands for Wastewater Treatment: Five Decades of Experience. *Environ. Sci. Technol.* **2011**, *45*, 61–69. [CrossRef] [PubMed]
29. Wu, H.; Zhang, J.; Ngo, H.H.; Guo, W.; Hu, Z.; Liang, S.; Fan, J.; Liu, H. A review on the sustainability of constructed wetlands for wastewater treatment: Design and operation. *Bioresour. Technol.* **2015**, *175*, 594–601. [CrossRef] [PubMed]
30. Guzmán, F.R.; de Ursinos, J.A.F.R.; Escobar, F.C.; Mateos, P.A.; Barrantes, M.M.D. Proceso ecológico-fotosintético para la depuración de purines en granjas. *Tecnol. Agua* **2000**, *201*, 61–66.
31. Caballero-Lajarín, A. Sistema de Depuración de Aguas Residuales de Origen Ganadero: Humedales artificiales. Ph.D. Thesis, Universidad Politécnica de Cartagena, Cartagena, Spain, 2013.
32. Caballero-Lajarín, A.; Zornoza, R.; Faz, A.; Lobera, J.B.; Muñoz, M.A.; Domínguez-Oliver, S.G. Combination of Low-Cost Technologies for Pig Slurry Purification under Semiarid Mediterranean Conditions. *Water Air Soil Pollut.* **2015**, *226*, 1–15. [CrossRef]
33. Scholz, M.; Harrington, R.; Carroll, P.; Mustafa, A. The Integrated Constructed Wetlands (ICW) concept. *Wetlands* **2007**, *27*, 337–354. [CrossRef]
34. Mashauri, D.A.; Mulungu, D.M.M.; Abdulhussein, B.S. Constructed wetland at the University of Dar es Salaam. *Water Res.* **2000**, *34*, 1135–1144. [CrossRef]
35. Polprasert, C.; Kittipongvises, S. Constructed Wetlands and Waste Stabilization Ponds. *Treatise Water Sci.* **2011**, *4*, 277. [CrossRef]
36. Christensen, M.L.; Hjorth, M.; Keiding, K. Characterization of pig slurry with reference to flocculation and separation. *Water Res.* **2009**, *43*, 773–783. [CrossRef] [PubMed]

37. Wang, N.; Mitsch, W.J. A detailed ecosystem model of phosphorus dynamics in created riparian wetlands. *Ecol. Model.* **2000**, *126*, 101–130. [CrossRef]
38. Schulz, C.; Gelbrecht, J.; Rennert, B. Treatment of rainbow trout farm effluents in constructed wetland with emergent plants and subsurface horizontal water flow. *Aquaculture* **2003**, *217*, 207–221. [CrossRef]
39. Hjorth, M.; Christensen, K.V.; Christensen, M.L.; Sommer, S.G. Soli-liquid separation of animal slurry in therory and practice. *A review. Agron. Sustain. Dev.* **2010**, *30*, 153.
40. Stottmeister, U.; Wießner, A.; Kuschk, P.; Kappelmeyer, U.; Kästner, M.; Bederski, O.; Müller, R.A.; Moormann, H. Effects of plants and microorganisms in constructed wetlands for wastewater treatment. *Biotechnol. Adv.* **2003**, *22*, 93–117. [CrossRef] [PubMed]
41. Huang, X.-F.; Ye, G.-Y.; Yi, N.-K.; Lu, L.-J.; Zhang, L.; Yang, L.-Y.; Xiao, L.; Liu, J. Effect of plant physiological characteristics on the removal of conventional and emerging pollutants from aquaculture wastewater by constructed wetlands. *Ecol. Eng.* **2019**, *135*, 45–53. [CrossRef]
42. Real Decreto 306, de 11 de Febrero, por el que se Establecen Normas Básicas de Ordenación de las Granjas Porcinas Intensivas, y se Modifica la Normativa Básica de Ordenación de las Explotaciones de Ganado Porcino Extensivo, Boletín Oficial Del Estado (2020) 13761–13791. Available online: https://www.boe.es/buscar/doc.php?id=BOE-A-2020-2110 (accessed on 21 August 2020).
43. Møller, H.B.; Lund, I.; Sommer, S.G. Solid–liquid separation of livestock slurry: Efficiency and cost. *Bioresour. Technol.* **2000**, *74*, 223–229. [CrossRef]
44. Kowalski, Z.; Makara, A.; Fijorek, K. Changes in the properties of pig manure slurry. *Acta Biochim. Pol.* **1970**, *60*, 845–850. [CrossRef]
45. Dinuccio, E.; Berg, W.; Balsari, P. Gaseous emissions from the storage of untreated slurries and the fractions obtained after mechanical separation. *Atmos. Environ.* **2008**, *42*, 2448–2459. [CrossRef]
46. Husted, S. Seasonal Variation in Methane Emission from Stored Slurry and Solid Manures. *J. Environ. Qual.* **1994**, *23*, 585–592. [CrossRef]
47. Veillette, M.; Girard, M.; Viens, P.; Brzezinski, R.; Heitz, M. Function and limits of biofilters for the removal of methane in exhaust gases from the pig industry. *Appl. Microbiol. Biotechnol.* **2012**, *94*, 601–611. [CrossRef] [PubMed]
48. Zhou, S.; Wang, C.; Liu, C.; Sun, H.; Zhang, J.; Zhang, X.; Xin, L. Nutrient removal, methane and nitrous oxide emissions in a hybrid constructed wetland treating anaerobic digestate. *Sci. Total Environ.* **2020**, *733*, 138338. [CrossRef]
49. Wolter, M.; Prayitno, S.; Schuchardt, F. Greenhouse gas emission during storage of pig manure on a pilot scale. *Bioresour. Technol.* **2004**, *95*, 235–244. [CrossRef]
50. Jeppsson, K.-H. SE—Structure and Environment: Carbon Dioxide Emission and Water Evaporation from Deep Litter Systems. *J. Agric. Eng. Res.* **2000**, *77*, 429–440. [CrossRef]
51. Møller, H.B.; Sommer, S.G.; Ahring, B.K. Biological Degradation and Greenhouse Gas Emissions during Pre-Storage of Liquid Animal Manure. *J. Environ. Qual.* **2004**, *33*, 27–36. [CrossRef]
52. Philippe, F.-X.; Nicks, B. Review on greenhouse gas emissions from pig houses: Production of carbon dioxide, methane and nitrous oxide by animals and manure. *Agric. Ecosyst. Environ.* **2015**, *199*, 10–25. [CrossRef]
53. Oenema, O.; Bannink, A.; Sommer, S.G.; Van Groenigen, J.W.; Velthof, G.L. Chapter 12—Gaseous Nitrogen Emissions from Livestock Farming Systems. In *Nitrogen in the Environment*, 2nd ed.; Hatfield, J.L., Follett, R.F., Eds.; Academic Press: San Diego, CA, USA, 2008; pp. 395–441. [CrossRef]
54. Wrage, N.; Velthof, G.L.; Van Beusichem, M.L.; Oenema, O. Role of nitrifier denitrification in the production of nitrous oxide. *Soil Biol. Biochem.* **2001**, *33*, 1723–1732. [CrossRef]
55. Inamori, R.; Wang, Y.; Yamamoto, T.; Zhang, J.; Kong, H.; Xu, K.; Inamori, Y. Seasonal effect on N2O formation in nitrification in constructed wetlands. *Chemosphere* **2008**, *73*, 1071–1077. [CrossRef] [PubMed]
56. Søvik, A.K.; Kløve, B. Emission of N2O and CH4 from a constructed wetland in southeastern Norway. *Sci. Total Environ.* **2007**, *380*, 28–37. [CrossRef] [PubMed]
57. Mander, Ü.; Dotro, G.; Ebie, Y.; Towprayoon, S.; Chiemchaisri, C.; Nogueira, S.F.; Jamsranjav, B.; Kasak, K.; Truu, J.; Tournebize, J.; et al. Greenhouse gas emission in constructed wetlands for wastewater treatment: A review. *Ecol. Eng.* **2014**, *66*, 19–35. [CrossRef]
58. Zahn, J.A.; Hatfield, J.L.; Laird, D.A.; Hart, T.T.; Do, Y.S.; DiSpirito, A.A. Functional Classification of Swine Manure Management Systems Based on Effluent and Gas Emission Characteristics. *J. Environ. Qual.* **2001**, *30*, 635–647. [CrossRef] [PubMed]
59. Dewes, T. Effect of pH, temperature, amount of litter and storage density on ammonia emissions from stable manure. *J. Agric. Sci.* **1996**, *127*, 501–509. [CrossRef]
60. Ni, J. Mechanistic Models of Ammonia Release from Liquid Manure: A Review. *J. Agric. Eng. Res.* **1999**, *72*, 1–17. [CrossRef]
61. Osada, T.; Shiraishi, M.; Hasegawa, T.; Kawahara, H. Methane, nitrous oxide and ammonia generation in full-scale swine wastewater purification facilities. *Front. Environ. Sci. Eng.* **2017**, *11*, 10. [CrossRef]

water

MDPI

Article

Migration and Transformation of Greenhouse Gases in Constructed Wetlands: A Bibliometric Analysis and Trend Forecast

Ruiyao Qi [1], Jiahao Dong [1], Yan Kang [2], Huijun Xie [3], Haiming Wu [1], Zhen Hu [1] and Zizhang Guo [1,*]

1 Shandong Key Laboratory of Water Pollution Control and Resource Reuse, School of Environmental Science and Engineering, Shandong University, Qingdao 266237, China; qiruiyao117@126.com (R.Q.); dongjiahao624@126.com (J.D.); haimingwu20@sdu.edu.cn (H.W.); huzhen885@sdu.edu.cn (Z.H.)
2 College of Environment and Safety Engineering, Qingdao University of Science and Technology, Qingdao 266042, China; kangyan@qust.edu.cn
3 Environment Research Institute, Shandong University, Qingdao 266237, China; xiehuij@sdu.edu.cn
* Correspondence: guozizhang@sdu.edu.cn

Abstract: Constructed wetlands (CWs), serving as an advanced wastewater treatment system, play a vital role in both the emission and sequestration of diverse GHGs. However, there are few papers reviewing and analyzing developments in the field. In this study, bibliometrics were used as an essential tool for identifying and establishing connections among key elements within a discipline, as well as for analyzing the research status and developmental trends of the research fields. CiteSpace 6.3.1 was utilized to conduct an analysis of the references from the Web of Science Core Collection pertaining to GHG emissions from CWs over the period from 1993 to 2023. This study showed the following conclusions. (1) Organic nitrogen conversion produces N_2O, which is eventually transformed into N_2 and released from CWs. Anammox represents an attractive route for nitrogen removal. (2) The CO_2 is the final product of the oxidation of organic matter in the influent of CWs and can be fixed by plant photosynthesis. Anaerobic fermentation and CO_2 reduction produce CH_4. The two are emitted through aerenchyma transport, bubble diffusion, and other forms. (3) In the past 30 years, the number of publications and citation frequency shows an increasing trend. China and the United States published more papers. The top ten authors contributed to 20.607% of the total 1019, and the cooperation between different author groups needs to be strengthened. (4) The emerging burst keywords following 2020 are "microbial fuel cell" and "microbial community", which highlights the current hotspots in research related to GHG emissions from CWs. (5) There is still a lack of long-term and applied discussion on the role of CWs in promoting GHG emission reduction. The relevant reaction conditions and mechanisms need to be explored and the possible research directions can be genetic regulation and information technology.

Keywords: constructed wetland; greenhouse gas; bibliometrics; CiteSpace; microorganism

1. Introduction

Greenhouse gases (GHGs) are constituents of the atmosphere and play a vital role in the greenhouse effect, contributing to the warming of the planet. Climate change and global warming, driven by GHG emissions, have consistently been significant subjects of international discourse [1–5]. Since the onset of the industrial revolution, human activities have profoundly disrupted the equilibrium of these gases, resulting in intensified greenhouse effects and global climate change. GHGs mainly include nitrous oxide (N_2O), carbon dioxide (CO_2), methane (CH_4), and various fluorinated gases. The Greenhouse Gas Bulletin (2022) published by the World Meteorological Organization (WMO) indicated that the global average surface molar fraction of CO_2, CH_4 and N_2O hit a record high in 2021. This climatic instability threatens agricultural productivity, water resources, and biodiversity, leading to food and water insecurity for millions [6]. Therefore, the mitigation of GHG emissions is essential [7]. This necessitates a comprehensive understanding of the dynamics of GHGs to develop effective strategies for mitigating climate change.

Constructed wetlands (CWs) are engineering systems extensively used in wastewater treatment. They treat wastewater by using the synergistic effect of wetland plants, substrates and microorganisms [8]. CWs serve as a vital link between the initial treatment of municipal wastewater and its final discharge into natural environments. Due to the great economic advantages and effectiveness in wastewater purification, CWs are extensively utilized in countries such as China [9], the United States [10], Canada [11], Spain [11], Germany [8], and others. With the widespread application of CWs, the environmental impacts of their GHG emissions must be seriously considered. Nevertheless, the role of CWs as either a carbon source or a carbon sink in wastewater treatment processes remains undetermined [12]. According to different estimates, wetlands cover only 5–8% of the global land area but account for 20–30% (2500 Pg) of the world's carbon pool [13]. The microbial transformations involved in wetlands produce several GHGs, such as N_2O, CO_2, and CH_4, which are crucial to climate change [14–16]. Studies show that the world's wetlands serve as significant carbon sinks, with an estimated capacity of approximately 830 Tg/year [13]. Most wetlands are net carbon sinks rather than sources contributing to climate change [13]. Some studies indicate that wetlands can function as both sources and sinks of carbon, depending on factors such as their age, management practices, and environmental boundary conditions including climate and location [17]. The specific effect of wetlands is still controversial [18]. Although their GHG emissions can be 2 to 10 times higher than those of natural wetlands [19], the relationship between GHG emissions and CWs is complex and multifaceted. This is attributed to the significantly greater microbial biomass and influent pollutant load present in CWs compared to natural wetlands [20,21]. Therefore, how to ensure the purification effect of CWs while minimizing GHG emissions is the key to achieving sustainable development of CWs. A comprehensive understanding of the GHG generation mechanism in CWs will help to assess its dual effects as a climate change mitigation measure.

In recent years, there has been an increasing interest in research concerning GHG emissions from CWs. However, there is a scarcity of papers that have an overall analysis and summary of the field. It is necessary to comprehend the development of relevant research to provide a reference for further work. To address this shortfall, bibliometrics approaches are critical to reviewing and synthesizing the literature to fully understand the complexity of GHG emissions in CWs [22]. Bibliometrics is utilized to analyze the research progress and developmental trends within a specific study field. It can be employed as a tool for identifying and establishing connections among key elements pertinent to a particular subject. It offers valuable insights into the growth of literature and the flow of knowledge within a particular field over time by analyzing data collected from databases, such as citations, authors, keywords, and the variety of journals referenced [23]. CiteSpace 6.3.1, a Java-based application, is designed for the visualization of bibliometric results through metrology, co-occurrence analysis, and cluster analysis [24,25]. As a scientometric tool, it serves several functions: evaluating the current state of research, mapping subject areas, delineating interdisciplinary connections, identifying trending research topics, and forecasting research trends [25]. Since its inception, the software has been widely used in bibliometrics research.

In view of the above points, this study analyzes the references of the Web of Science Core Collection in the field of GHG emissions in CWs from 1993 to 2023 through bibliometric analysis and visualization. Specifically, this study is driven by four primary objectives: (1) to analyze and summarize the process of GHG generation and emission in CWs; (2) to understand the rise and development of the field of GHG emissions in CWs through the number of papers published and the frequency of citations per year, and identify countries and authors with more research; (3) to explore research hotspots through keywords in the past three decades years; and (4) to identify the shortcomings of current research and consider the possible research directions in the future.

2. Methodology

2.1. Data Sources

The reference data for the paper were sourced from the literature database of the Web of Science Core Collection. The keywords for the literature search were determined as the synonyms of CWs and GHGs, with the Boolean operation formula being TS = ("constructed wetland*" OR "artificial wetland*" OR "treated wetland*") AND TS = ("greenhouse gas*" OR "house gas*" OR "GHG*" OR "carbon emission*" OR "carbon dioxide*" OR "CO_2" OR "methane*" OR "CH_4" OR "nitrous oxide*" OR "N_2O"). Document types were selected as "article" and "review article". While early studies in this field were limited, a noticeable uptick in publications has been observed since 1993. This growth reflects the evolution and milestones within the field. Consequently, the time span for this analysis was established as 1993 to 2023, resulting in 1019 documents. The search results encompassed a range of details pertaining to each document, such as title, year, citations, country, source (journal title), author(s), and keywords. Complete records were downloaded for subsequent analysis.

2.2. Analysis Methods

This paper summarizes the recent research on the GHG emission mechanism of CWs. Specifically, it focuses on N_2O, CO_2, and CH_4 to briefly elucidate the mechanisms of production and processes of release for these GHGs. CiteSpace 6.3.1 was used to analyze the fundamental information of the literature, including countries, authors, and keywords. This study followed the general procedure of visual analysis of CiteSpace 6.3.1. The 1019 articles were exported in plain text files for data preprocessing. For countries and authors, a collaborative network analysis was performed. In the CiteSpace 6.3.1 user interface, the "Years Per Slice" parameter was used to partition the time period. This function organizes the literature into chronological segments to facilitate a better understanding of research topics, trends, and developments over time. The "Years Per Slice" parameter was set to 1. In the resulting visualization, nodes (represented by circles) were labeled and sized according to their significance. Node color indicates the time sequence, progressing from earlier (center) to more recent (edge) studies.

Different selections and configurations of parameters influence the credibility level of the results. For keyword co-occurrence network analysis, "Keyword" was selected in "Node Types" and the selection criteria g-index (k) for selecting the appropriate number of nodes in each time slice was set to 4. The visualization was pruned utilizing "Minimum Spanning Tree" and "pruning sliced networks". On this basis, keyword clustering was conducted, and two important indicators offered insights into the overall structural characteristics of the network. Modularity Q, a community detection algorithm, indicates the extent to which the author or organization of authors of literature is divided into numerous independent modules and recombined together [26,27]. The value range of Q is from 0 to 1. If Q is > 0.3, it can be considered that the structure of the network community is rational and obvious. The silhouette metric is commonly employed as an index in cluster analysis, serving to assess the quality of clustering outcomes [28]. The closer the value approaches 1, the more effective the clustering outcome becomes. When S is > 0.5, the outcome of clustering is regarded as reasonable. By analyzing the timeline and visualizing the evolution in time, a timeline view of the keyword clusters was obtained. CiteSpace 6.3.1 calculates the occurrence frequency of each keyword in different time periods. When the frequency of a keyword experiences a significant increase over a specified period, it is considered a burst keyword. This article ultimately obtained 25 burst keywords, and the length of the red line segment displayed in the visualization represents the duration of the burst keywords. The specific start and end times are also listed accordingly.

3. Results and Discussion

3.1. The Generation and Release of N_2O, CO_2, and CH_4 in CWs

3.1.1. The Production and Release of N_2O

Comprehensively considering the migration and transformation of nitrogen in CWs, the pathways for N_2O production and release in CWs are obtained (Figure 1). The upper layer of CWs forms an aerobic layer due to atmosphere reaeration and radial oxygen loss (ROL) of plant root systems [29], while the lower layer forms an anaerobic layer due to the lack of a dissolved oxygen source and the consumption of dissolved oxygen by aerobic microorganisms. Organic nitrogen is converted to ammonia nitrogen by biological ammoniation, which can be carried out under aerobic or anaerobic conditions. The next step is nitrifying-denitrifying microbial nitrogen removal. Nitrification refers to the process by which ammonia nitrogen is initially oxidized to nitrite nitrogen by ammonia-oxidizing bacteria, followed by its further oxidation to nitrate nitrogen by nitrifying bacteria [30]. Microorganisms with ammonia oxidation activity include ammonia-oxidizing bacteria (AOB) and ammonia-oxidizing archaea (AOA) [30,31]. AOB is mainly concentrated in three genera: *Nitrosococcus*, *Nitrosomonas*, and *Nitrosospira* [32]. AOA generally outperforms AOB in terms of quantity and diversity, with known major species including *Nitrosopumilus*, *Nitrosophaera*, *Nitrosokaldus*, *Nitrosotalea*, etc. [31]. In this process, the possible production of N_2O arises from chemical oxidation with NO_2^- as an electron acceptor [33], or due to the chemical decomposition of hydroxylamine [34] and the intermediates of biological hydroxylamine oxidation [35]. The nitrifier denitrification process—a reduction of NO_2^- by AOB in combination with electron donors (e.g., pyruvate, hydrogen, or ammonia) under oxygen-limiting conditions or raised nitrite concentrations, also produces N_2O [34,36,37]. Denitrification is the process in which denitrifying bacteria convert nitrate nitrogen into nitrogen through a multi-step reaction [38]. Denitrifying bacteria are mostly facultative anaerobic and chemical heterotrophic bacteria [39]. The resulting intermediate product N_2O is eventually converted into N_2, which is the only way to remove N_2O inside the CWs. In addition, under anaerobic conditions, anaerobic ammonia-oxidizing bacteria use nitrite as an electron acceptor and directly convert ammonia nitrogen to nitrogen, which is called Anammox [40]. Anaerobic ammonia-oxidizing bacteria do not produce N_2O during the process. It is an attractive route for microbial nitrogen removal.

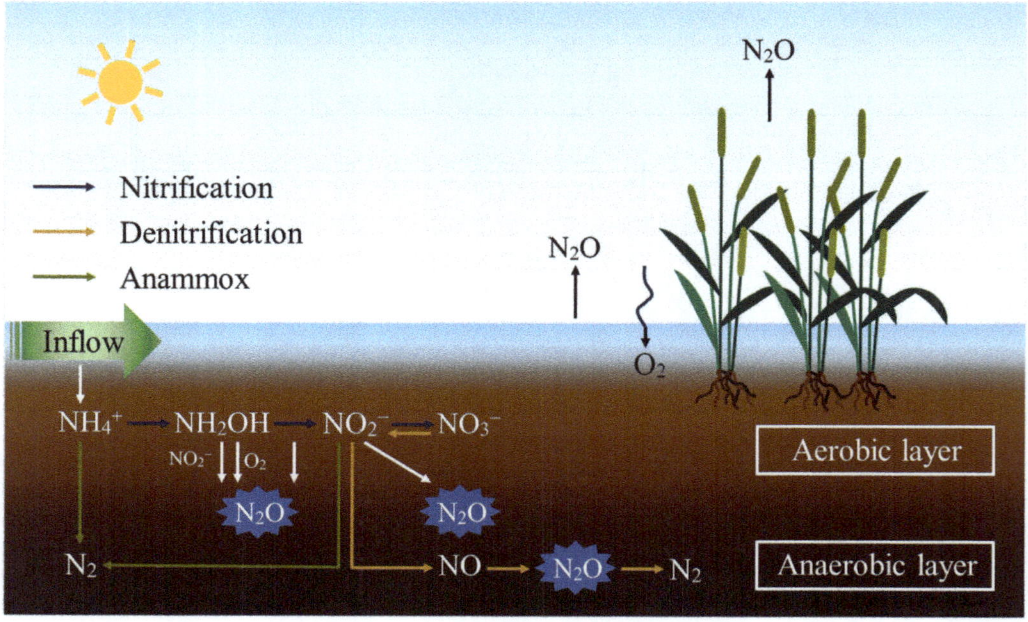

Figure 1. The generation and release of N_2O in CWs.

3.1.2. The Production and Release of CO_2 and CH_4

The production and release of CO_2 and CH_4 in CWs have a complex relationship with the carbon cycle [41]. CWs can fix CO_2 through plant photosynthesis, and emit CO_2 to the atmosphere through the respiration of plants and the oxidation and decomposition of organics in wastewater and substrate by microorganisms [42]. The release of CH_4 in CWs is the result of anaerobic fermentation and CO_2 reduction [43,44]. It is concluded that the CO_2 released from CWs is the natural destination of organic matter and should not be included in the GHG emission catalog. Therefore, the ratio of organic matter conversion to CH_4 is an important indicator to determine the final emission effect of carbon-based GHGs. Taking into account the migration and transformation of carbon in CWs, a diagram for the generation and release of CO_2 and CH_4 is obtained (Figure 2). Some organics in the influent of CWs are oxidized due to the activity of upper aerobic bacteria, and the final product contains CO_2. There are two main ways to produce CH_4—anaerobic fermentation and CO_2 reduction [43,44]. Complex organic matter is hydrolyzed under the action of hydrolytic microflora (including aerobic bacteria, anaerobic bacteria, and facultative bacteria), and undergoes anaerobic fermentation to produce monomers such as fatty acids and alcohols. Then, under the action of acid-producing bacteria, acid and hydrogen are produced and CO_2 is generated. Finally, methanogenic bacteria use acetic acid or CO_2/H_2 to produce CH_4 under anaerobic conditions [45]. The generated CH_4 may have oxidation in two forms: aerobic oxidation and anaerobic oxidation. Aerobic methane-oxidizing bacteria (MOB) oxidized CH_4 to CO_2 under aerobic conditions [46]. Anaerobic oxidation occurs in the denitrification process, where CH_4 is oxidized to CO_2 by denitrifying anaerobic methane-oxidizing bacteria (DamoB) and denitrifying anaerobic methane-oxidizing archaea (DamoA) [47,48]. The CO_2 and CH_4 generated and not eliminated in CWs are emitted through plant aeration tissue transport, bubble diffusion, and other forms [49]. Then, CO_2 is fixed by plant photosynthesis.

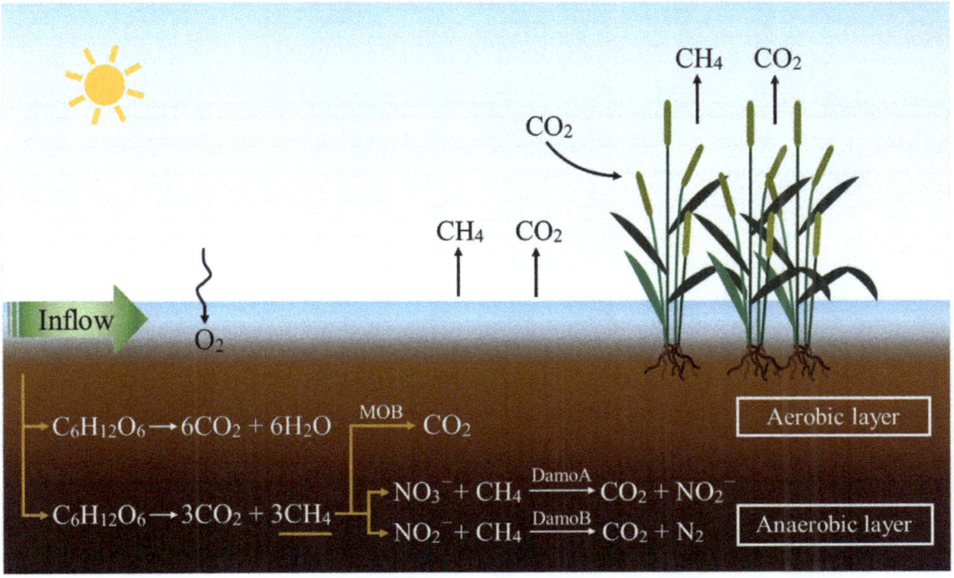

Figure 2. The generation and release of CO_2 and CH_4 in CWs.

3.2. Publication Number and Citation Frequency per Year

The number of publications and citation frequency in the field of GHGs in CWs per year is shown in Figure 3. The first study in this field was published in 1993. Overall, the number of published papers in the relevant field shows an increasing trend. The number of publications was limited from 1993 to 2000. It increased in 2001 and exhibited rapid growth after 2013, peaking at 121 publications in 2022. The citation frequency of literature also showed an increasing trend, especially after 2013, peaking at 4632 in 2022. The continuous rise in the publication number and the frequency of citations per year indicates a growing interest in this field, which is poised to propel the advancement of the field. In 2023, the number of published papers and the frequency of citations of literature both decreased, but the decline was not substantial. This could simply be a normal data fluctuation, not necessarily a research bottleneck.

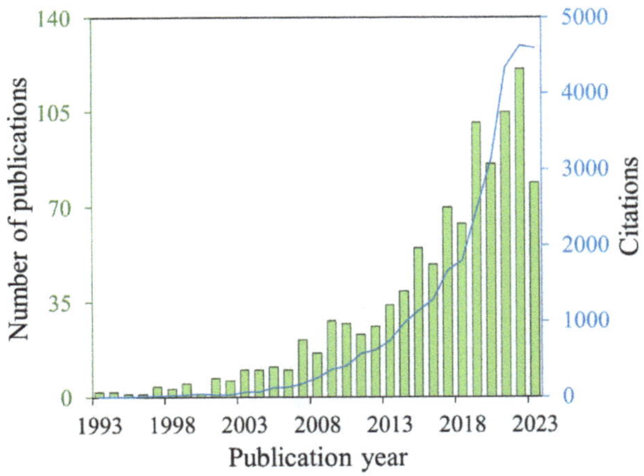

Figure 3. The publication number and citation frequency of research on GHG emissions in CWs per year.

3.3. Country and Author Analysis

From 1993 to 2023, the network of collaborating nations and territories comprised a total of 85 nodes (Figure 4). The cooperation network reflects extensive international and regional cooperation. Table 1 presents the 16 countries and territories that contributed the most to the total. The leading top 10 countries in terms of publication number are China, the United States, Canada, Spain, Germany, India, England, Australia, Estonia, and Italy. The color transition from the center to the periphery of the nodal circle illustrates the annual increase of pertinent research across various countries. Research in countries such as the United States, Germany, England, Estonia, and Japan commenced earlier. In contrast, research in China, Canada, Spain, India, and Australia is more recent. Although research on GHG emissions from CWs started later, China has the largest number of papers, with 453. It suggests a strong recent interest in this field among Chinese researchers. This can be attributed to the rapid economic growth in China, where the accelerated pace of industrialization and urbanization brings environmental challenges [9]. CWs offer an optimal alternative solution due to their effective purification effect, low cost, and low energy consumption. The demand promotes the related research of CWs in China [50]. In addition, from the centrality perspective, the United States (0.34) and China (0.28) have a significant international influence. They are the leading countries in this field.

Figure 4. A representation of national and regional cooperation networks.

Table 1. The top 16 countries and territories ranked by frequency.

Country and Region	Frequency	Centrality	Country and Region	Frequency	Centrality
China	453	0.28	Estonia	31	0.01
USA	172	0.34	Italy	29	0.17
Canada	57	0.02	France	27	0.11
Spain	50	0.11	Denmark	26	0.02
Germany	47	0.08	Japan	25	0
India	41	0.02	Netherlands	24	0.04
England	36	0.04	Sweden	21	0.08
Australia	34	0.05	Brazil	18	0.05

The ranking of the top 10 authors with the most record counts in the domain of GHG emissions from CWs is listed in Table 2. The publications authored by the 10 individuals account for 20.607% of the total 1019 publications. Zhang J ranks first, accounting for 2.846% of the full count of 1019, and holds the highest number of publications (29). Mandel Ü and Chang J follow, with 28 and 25 papers respectively, accounting for 2.784% and 2.453%, respectively. In addition, the authors ranking from fourth to tenth all have at least 15 papers, including Ge Y, Wu HM, Chen W, Luo HB, Zhang K, He SB, and Hu Z. The main authors of publications in a field have a relatively accurate grasp of the development context, research hotspots, and emerging trends in this domain. By continuously tracking the latest research results of the main authors and their teams, the mainstream research direction can be achieved in real time. Author cooperation was analyzed by CiteSpace 6.3.1 (Figure 5). Closely related author groups have been formed among the authors, such as Wu Haiming, Zhang Jian, Hu Zhen, Chen Yi, Guo Wenshan, He Qiang. Among these groups, Wu Haiming, Zhang Jian, and Hu Zhen have greater influence. As shown in Figure 5, there are five prominent author groups, with a higher proportion of Chinese authors. This could be attributed to the heightened interest of Chinese researchers in this particular field, as mentioned in the previous paragraph. Cooperation between researchers from different geographic areas remains to be seen. Furthermore, cooperation primarily takes place within teams. Limited cooperation occurs among different groups of authors, indicating a necessity to enhance inter-team cooperation.

Table 2. The ten authors with the most publications in the field of GHG emissions in CWs.

Authors	Record Count	% of 1019	Authors	Record Count	% of 1019
Zhang J	29	2.846%	Chen W	17	7.164%
Mander Ü	28	2.748%	Luo HB	17	7.066%
Chang J	25	2.453%	Zhang K	17	6.084%
Ge Y	24	2.355%	He SB	15	4.907%
Wu HM	23	2.257%	Hu Z	15	2.846%

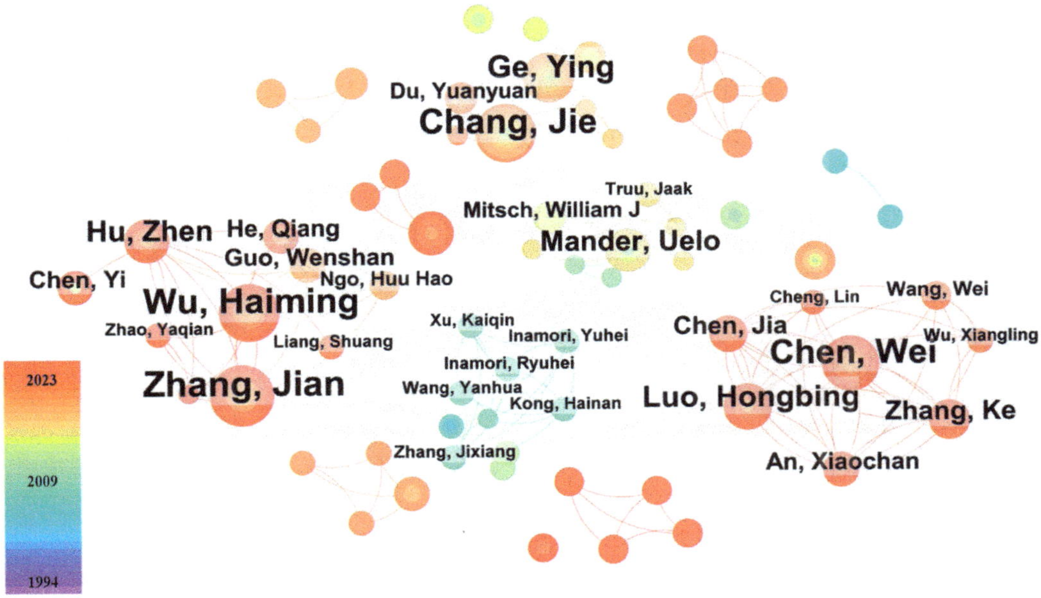

Figure 5. Author cooperation network in the field of GHG emissions in CWs.

3.4. Keyword Analysis

The analysis of keyword co-occurrence frequency is visualized in Figure 6. The details of the top 24 keywords (Table 3) show that "constructed wetlands", "constructed wetland", "waste water treatment", "removal", and "performance" exhibited higher frequencies compared to other terms, with respective counts of 379, 253, 199, 186, and 182. This is because one of the important purposes of the construction of CWs is to treat wastewater. Specifically, this involves the effective removal of nutrients (nitrogen, phosphorus, etc.) in wastewater through the processes of plants and microorganisms.

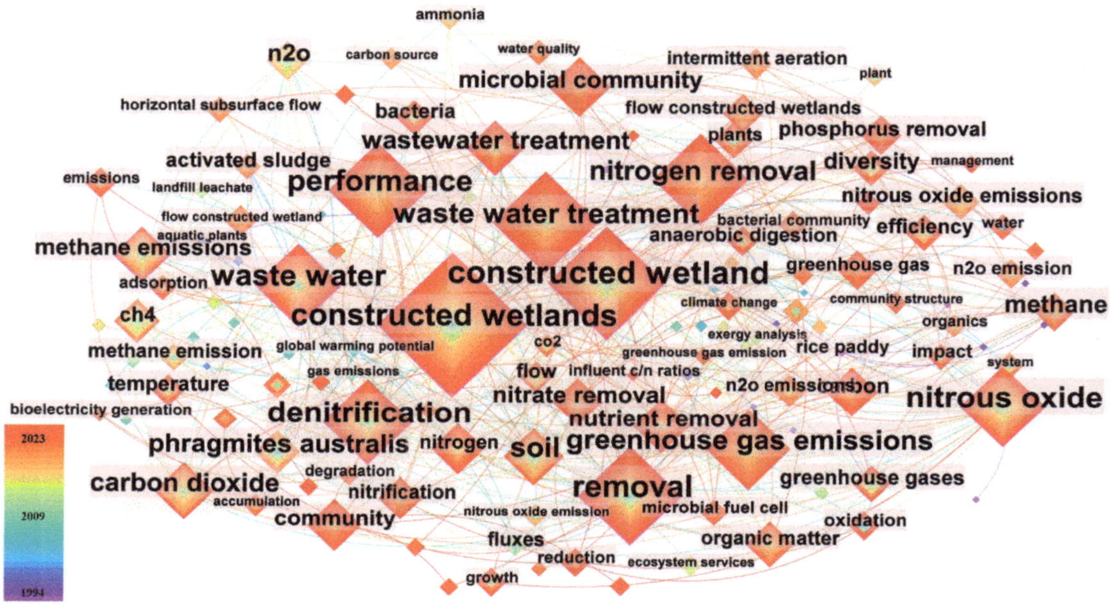

Figure 6. The co-occurrence network analysis of keywords on GHG emissions from CWs.

Table 3. Top 24 keywords with a high frequency in the domain of GHG emissions in CWs.

No.	Keywords	Frequency	Centrality	No.	Keywords	Frequency	Centrality
1	constructed wetlands	379	0.37	13	carbon dioxide	66	0.05
2	constructed wetland	253	0.32	14	nutrient removal	62	0.04
3	waste water treatment	199	0.11	15	wastewater treatment	61	0.1
4	removal	186	0.23	16	methane emissions	60	0.1
5	performance	182	0.14	17	community	57	0.06
6	nitrogen removal	158	0.12	18	nitrogen	54	0
7	denitrification	150	0.15	19	carbon	49	0.03
8	nitrous oxide	145	0.18	20	nitrate removal	47	0.06
9	waste water	138	0.21	21	organic matter	40	0.02
10	greenhouse gas emissions	132	0.08	22	CH_4	39	0.01
11	soil	70	0.08	23	phragmites australis	39	0.08
12	microbial community	68	0.04	24	nitrification	38	0.01

In the keyword clustering analysis (Figure 7), the Q value and S value of the graph parameters are 0.3645 and 0.7655, indicating good rationality and credibility. These studies are extensively distributed across 8 categories, and keyword clusters involving #0 constructed wetland, #5 constructed wetlands, and #6 greenhouse gas indicates that the graph is consistent with the research theme. Phosphorus removal is shown as #1 phosphorus removal in the cluster diagram. The timeline view of the keyword clusters (Figure 8) shows that the smaller cluster, #7 rice, appeared earlier, around 1995. The content is only about rice paddy, coarse fibers, and fields. Since then, there have been less relevant studies.

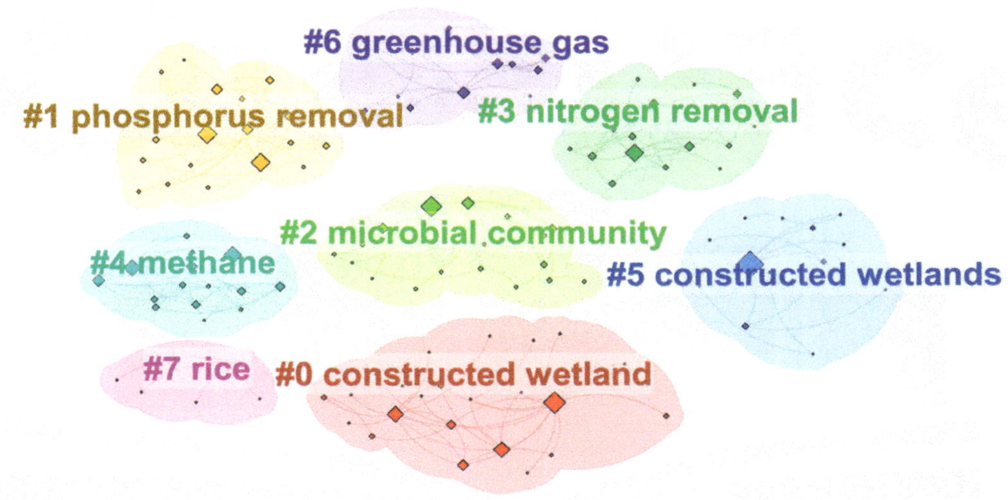

Figure 7. A visualization of the keyword cluster analysis on GHG emissions in CWs.

Figure 8. Timeline view of the keyword clusters on GHG emissions in CWs.

Other keywords that appear more frequently include "nitrogen removal", "denitrification", "nitrous oxide", "carbon dioxide", "nutrient removal", and "methane emissions". These mainly involve the migration and transformation of nitrogen and carbon and are closely related to the production and removal of GHGs, such as N_2O, CO_2, and CH_4. It is also expressed in clustering as #3 nitrogen removal and #4 methane. The role of microorganisms in this process is critically significant, as demonstrated by the #2 microbial community cluster. In addition, as indicated in Table 4, the emerging burst keywords following 2020 are "microbial fuel cell" (2021–2023) and "microbial community" (2021–2023), with strengths of 6.12 and 5.67, respectively. This highlights the current hotspots in research related to GHG emissions from CWs.

Table 4. Top 25 keywords with the strongest citation bursts. The red stripe represents the time period when the keyword burst, while the light blue stripe represents the time period when the keyword did not appear.

No	Keywords	Year	Strength	Begin	End	1994–2023
1	constructed wetlands	1998	5.03	1998	2003	
2	removal	2003	9.38	2003	2011	
3	oxidation	2003	4.92	2003	2007	
4	nitrous oxide	2002	5.72	2005	2011	
5	soil	1999	5.4	2005	2011	
6	phragmites australis	2000	7.74	2007	2016	
7	CH_4	2009	11.08	2009	2016	
8	N_2O	2000	10.39	2009	2017	
9	greenhouse gases	1995	5.15	2011	2015	
10	diversity	2011	4.77	2011	2016	
11	methane emissions	1995	6.68	2012	2017	
12	constructed wetland	1997	6.47	2012	2013	
13	flow	2013	5.28	2013	2017	
14	nitrogen	2004	5.13	2015	2016	
15	flow constructed wetlands	2017	7.01	2017	2020	
16	nitrous oxide emissions	2014	5.93	2017	2018	
17	intermittent aeration	2018	7.85	2018	2020	
18	flow constructed wetland	2018	5.97	2018	2021	
19	organics	2018	5.27	2018	2019	
20	horizontal subsurface flow	2019	6.39	2019	2020	
21	N_2O emission	2013	5.78	2019	2020	
22	N_2O emissions	2013	5.47	2019	2020	
23	greenhouse gas	2020	7.69	2020	2023	
24	microbial fuel cell	2021	6.12	2021	2023	
25	microbial community	2017	5.67	2021	2023	

3.5. Hotspot Analysis

The role of microorganisms is to run through the wastewater treatment process in CWs. As can be seen from Section 3.1, microorganisms are particularly important in GHG emissions in CWs. From the study of microbial species and functions to the investigation of the mechanism of action and the gene level, research related to microorganisms has always been carried out, and new discoveries and applications continue to be made. According to the keyword analysis in Section 3.4, the latest research trend of GHG emission reduction in CWs can be obtained. In recent years, microbial fuel cells (MFC) and microbial communities have become research hotspots.

The basic physical processes in CWs and MFCs are highly complementary, and combining them to operate MFCs in CWs can effectively control GHG emissions. Ke Zhang et al. studied the position of plant roots in relation to the electrodes and concluded that operating MFCs effectively reduced CH_4 emission irrespective of whether the plant roots were situated at the cathode or anode [51]. In the context of CW-MFC operating under sequencing batch conditions, the rhizosphere situated at the cathode was found to be more effective in suppressing CH_4 emission, while the rhizosphere situated at the anode was more advantageous for the generation of electricity [52]. The external resistance exhibited no significant effect on the CH_4 emission of CW-MFCs [52]. The study showed that the role of MFCs in CH_4 emissions was due to the competition between methanogens and electrogens. This interaction altered the structure of the biochemical process and microbial community in CWs [53]. Proteobacteria, the primary electricigen in CW-MFCs, were boosted with rhizospheres situated at the cathode, and the CH_4 emission exhibited a negative correlation with the abundance of proteobacteria [53]. Although the CW-MFC technology has a high potential for the control of CH_4 emissions, the relationship between CH_4 and CO_2 emissions needs to be further addressed [54]. Researchers have conducted a quantitative comparison of pollutant removal efficiency and gas emissions between batch-fed wetland systems (BF CWs) and MFC CWs. The findings indicated that MFC CWs demonstrated considerably lower global warming potential than BF CWs [55]. In terms of cathode materials, carbon fiber felt (CFF) has the lowest emissions of CH_4 and N_2O, compared to carbon cloth (CC) and stainless-steel wire mesh (SSM) [56]. Moreover, by controlling variable factors such as the C/N ratio and the pH of the influent, it is suggested that CW-MFCs provide an environment-friendly method for the management of GHG emissions [56].

The CW-MFC system presents significant potential for advancement in the field of wastewater treatment. However, it also faces several limitations and challenges. The mechanisms by which microbial activity is influenced remain incompletely understood. Further investigation is needed on issues such as the role of plant rhizospheres in relation to electrodes and the selection of optimal electrode materials. Additional studies are essential to enhance system configuration, improve treatment efficiency, and mitigate GHG emissions. Beyond the scope of laboratory exploration, the technology needs to be considered for more practical applications.

Not only MFCs, but also factors such as substrate types, plants, and supplementary carbon sources in CWs have an impact on GHG emissions to a large extent through microorganisms. The structure of a microbial community offers valuable insights into the function of CWs [57], and the related analysis has received more attention. To investigate the effects of iron and manganese oxides on microbial communities, Cheng et al. extracted DNA samples from vertical subsurface-flow CWs (VSSCWs) and performed high-throughput sequencing. The addition of manganese oxides improved the overall relative abundance of *Actinobacteria*, *Chloroflexi*, and *Proteobacteria*, resulting in increased total nitrogen (TN) removal and reduced N_2O fluxes, in contrast to quartz sand and iron oxides [58]. The relative abundance of *Euryarchaeota* in Fe-VSSCWs and Mn-VSSCWs were 0.40% and 0.19%, respectively. Both of these were lower than previously observed in the control group (0.51%). This discrepancy may contribute to the reduced CH_4 fluxes [58]. Compared with clay ceramsite, the amendment of biochar distinctly mitigated N_2O and CH_4 fluxes from CWs by promoting a higher abundance of *mcrA* and *nosZ* genes and higher ratios of *pmoA*/*mcrA* and *nosZ*/(*nirK* + *nirS*) [59]. Xushun Gu et al. concluded that the presence of plants supported the abundance of ammonia oxidation bacteria, such as *Nitrosomonas* and *Nitrosospira*, as well as the *amoA* gene, when an additional carbon source was provided [60]. Wetlands with plants primarily functioned as carbon sinks, exhibiting a net carbon dioxide absorption flux of approximately 13,000 mg m^{-2} d^{-1}. They had the capacity to offset emissions of N_2O and CH_4, with maximum values recorded at a maximum of 12.24 mg m^{-2} d^{-1} and 2.52 mg m^{-2} d^{-1}, respectively [60]. For supplementary carbon sources, alkali-heated corncobs improved the abundance of heterotrophic denitrifying bacteria and increased nitrogen functional genes while GHG fluxes were lower compared to common corncobs [61].

With the development of omics technologies, our understanding of microbial community structure and gene function has deepened significantly. This helps to precisely regulate the environment of CWs, improve purification efficiency, and reduce the GHG emissions of CWs.

4. Research Limitations and Prospects

Since not all of the included articles focused entirely on GHG research in CWs, this study has some limitations, but it is enough to provide a relatively comprehensive insight into this field. Despite the deepening of relevant studies, there is still a lack of long-term and applied discussions on the practical role of CWs in promoting GHG emission reduction, and a unified understanding has not been formed. The biological action and reaction mechanism involved in the production and release of GHGs still need to be further explored. Future research may be considered from the following aspects:

(1) The impact of different environmental factors is complex, and comprehensive consideration is needed for factors that affect GHG emissions from CWs, such as operation mode, substrate configuration, plant selection, and carbon source supplementation;
(2) Further in-depth research is needed on the GHG conversion process involving microorganisms within CWs, such as the interaction between multiple N_2O production pathways and the mechanism of CH_4 anaerobic oxidation;
(3) Genetic technology can be strategically employed to enhance microorganisms that are beneficial for mitigating GHG emissions in CWs;
(4) Intelligent supervision systems, in conjunction with information technology, can be developed to precisely control operating conditions and monitor the effectiveness of CWs.

5. Conclusions

Based on CiteSpace 6.3.1 and the Web of Science Core Collection, this study provides a clear knowledge map and many conclusions can be drawn.

(1) Organic nitrogen is converted to ammonia nitrogen by biological ammoniation and produces N_2O through nitrifying-denitrifying microbial nitrogen removal. The resulting product N_2O is eventually converted into N_2, which is released from CWs. Anammox, a process that directly transforms ammonia nitrogen to nitrogen, represents an attractive route for nitrogen removal.
(2) Organics in the influent of CWs are oxidized and the final product contains CO_2. Anaerobic fermentation and CO_2 reduction produce CH_4. The CO_2 and CH_4 are emitted through plant aeration tissue transport, bubble diffusion, and other forms. After that, CO_2 is fixed by plant photosynthesis.
(3) In the past 30 years, the number of published papers and the citation frequency in the relevant fields show an increasing trend. China and the United States published more papers. The top ten authors contributed to 20.607% of the total 1019, and the cooperation between different author groups needs to be strengthened.
(4) The emerging burst keywords following 2020 are "microbial fuel cell" and "microbial community", which highlights the current hotspots in research related to GHG emissions from CWs. Beyond the scope of laboratory exploration, the CW-MFC needs to be considered for more practical applications. The deepened understanding of microbial communities helps to precisely regulate the environment of CWs and reduce the GHG emissions of CWs.
(5) Despite relevant studies, there is still a lack of long-term and applied discussion on the role of CWs in promoting GHG emission reduction. The relevant reaction conditions and mechanisms need to be explored, and the possible research directions in the future can be genetic regulation and information technology.

Author Contributions: Conceptualization, H.W. and Z.G.; Methodology, J.D. and Z.G.; Software, R.Q.; Validation, Y.K. and Z.G.; Formal analysis, H.X.; Investigation, Z.G.; Resources, Z.G.; Data curation, R.Q. and J.D.; Writing—original draft, R.Q.; Writing—review & editing, J.D.; Visualization, R.Q.; Supervision, H.X., H.W., Z.H. and Z.G.; Project administration, Z.H. and Z.G.; Funding acquisition, Z.G. All authors have read and agreed to the published version of the manuscript.

Funding: This work was supported by the National Natural Science Foundation of China (Nos. 52270158, 52200196, 51925803, and 51908326).

Data Availability Statement: The data presented in this study are available on request from the corresponding author. The data are not publicly available due to restrictions.

Conflicts of Interest: The authors declare no conflict of interest.

References

1. Coenen, R.; Sardemann, G. Kyoto: Protocol on climate protection. *Atw—Int. Z. Kernenerg.* **1998**, *43*, 243–248.
2. World Council of Churches. Minute on UNFCCC Conference of the Parties—COP 15 in Copenhagen. *Ecum. Rev.* **2010**, *62*, 229–231. [CrossRef]
3. Falkner, R. The Paris Agreement and the new logic of international climate politics. *Int. Aff.* **2016**, *92*, 1107–1125. [CrossRef]
4. Liu, Q. The 28th United Nations Climate Change Conference (COP28)—30 November–12 December 2023. *China CDC Wkly.* **2023**, *5*, 1093. [CrossRef]
5. Beck, S.; Mahony, M. The IPCC and the new map of science and politics. *WIREs Clim. Change* **2018**, *9*, e547. [CrossRef]
6. Mora, C.; Spirandelli, D.; Franklin, E.C.; Lynham, J.; Kantar, M.B.; Miles, W.; Smith, C.Z.; Freel, K.; Moy, J.; Louis, L.V.; et al. Broad threat to humanity from cumulative climate hazards intensified by greenhouse gas emissions. *Nat. Clim. Change* **2018**, *8*, 1062–1071. [CrossRef]
7. Tiwari, T.; Kaur, G.A.; Singh, P.K.; Balayan, S.; Mishra, A.; Tiwari, A. Emerging bio-capture strategies for greenhouse gas reduction: Navigating challenges towards carbon neutrality. *Sci. Total Environ.* **2024**, *929*, 172433. [CrossRef]
8. Stankovic, D. Constructed wetlands for wastewater treatment. *Građevinar* **2017**, *69*, 639–652. [CrossRef]
9. Zhang, D.Q.; Gersberg, R.M.; Keat, T.S. Constructed wetlands in China. *Ecol. Eng.* **2009**, *35*, 1367–1378. [CrossRef]
10. Tao, W.D.; Bays, J.S.; Meyer, D.; Smardon, R.C.; Levy, Z.F. Constructed Wetlands for Treatment of Combined Sewer Overflow in the US: A Review of Design Challenges and Application Status. *Water* **2014**, *6*, 3362–3385. [CrossRef]
11. Vymazal, J. Constructed wetlands for treatment of industrial wastewaters: A review. *Ecol. Eng.* **2014**, *73*, 724–751. [CrossRef]
12. Mander, Ü.; Lohmus, K.; Teiter, S.; Mauring, T.; Nurk, K.; Augustin, J. Gaseous fluxes in the nitrogen and carbon budgets of subsurface flow constructed wetlands. *Sci. Total Environ.* **2008**, *404*, 343–353. [CrossRef]
13. Mitsch, W.J.; Bernal, B.; Nahlik, A.M.; Mander, Ü.; Zhang, L.; Anderson, C.J.; Jorgensen, S.E.; Brix, H. Wetlands, carbon, and climate change. *Landsc. Ecol.* **2013**, *28*, 583–597. [CrossRef]
14. Walter, B.P.; Heimann, M.; Matthews, E. Modeling modern methane emissions from natural wetlands 1. Model description and results. *J. Geophys. Res. Atmos.* **2001**, *106*, 34189–34206. [CrossRef]
15. Sha, C.Y.; Wang, M.; Wang, Q.; Lu, J.J. Wetland methane and carbon dioxide emission and affecting factors. *Shengtaixue Zazhi* **2011**, *30*, 2072–2079.
16. Bahram, M.; Espenberg, M.; Pärn, J.; Lehtovirta-Morley, L.; Anslan, S.; Kasak, K.; Koljalg, U.; Liira, J.; Maddison, M.; Moora, M.; et al. Structure and function of the soil microbiome underlying N_2O emissions from global wetlands. *Nat. Commun.* **2022**, *13*, 10. [CrossRef]
17. Kayranli, B.; Scholz, M.; Mustafa, A.; Hedmark, Å. Carbon Storage and Fluxes within Freshwater Wetlands: A Critical Review. *Wetlands* **2010**, *30*, 111–124. [CrossRef]
18. Salimi, S.; Almuktar, S.; Scholz, M. Impact of climate change on wetland ecosystems: A critical review of experimental wetlands. *J. Environ. Manag.* **2021**, *286*, 15. [CrossRef]
19. Maltais-Landry, G.; Maranger, R.; Brisson, J.; Chazarenc, F. Greenhouse gas production and efficiency of planted and artificially aerated constructed wetlands. *Environ. Pollut.* **2009**, *157*, 748–754. [CrossRef]
20. Cao, Q.Q.; Wang, H.; Chen, X.C.; Wang, R.Q.; Liu, J. Composition and distribution of microbial communities in natural river wetlands and corresponding constructed wetlands. *Ecol. Eng.* **2017**, *98*, 40–48. [CrossRef]
21. Maucieri, C.; Barbera, A.C.; Vymazal, J.; Borin, M. A review on the main affecting factors of greenhouse gases emission in constructed wetlands. *Agric. For. Meteorol.* **2017**, *236*, 175–193. [CrossRef]
22. Xu, D.; Sun, H.M.; Wang, J.; Wang, N.; Zuo, Y.J.; Mosa, A.A.; Yin, X.Q. Global trends and current advances regarding greenhouse gases in constructed wetlands: A bibliometric-based quantitative review over the last 40 years. *Ecol. Eng.* **2023**, *193*, 107018. [CrossRef]

23. Leung, X.Y.; Sun, J.; Bai, B. Bibliometrics of social media research: A co-citation and co-word analysis. *Int. J. Hosp. Manag.* **2017**, *66*, 35–45. [CrossRef]
24. Chen, C.M. Searching for intellectual turning points: Progressive knowledge domain visualization. *Proc. Natl. Acad. Sci. USA* **2004**, *101*, 5303–5310. [CrossRef]
25. Chen, C.M. CiteSpace II: Detecting and visualizing emerging trends and transient patterns in scientific literature. *J. Am. Soc. Inf. Sci. Technol.* **2006**, *57*, 359–377. [CrossRef]
26. Newman, M.E.J. Modularity and community structure in networks. *Proc. Natl. Acad. Sci. USA* **2006**, *103*, 8577–8582. [CrossRef]
27. Shibata, N.; Kajikawa, Y.; Takeda, Y.; Matsushima, K. Detecting emerging research fronts based on topological measures in citation networks of scientific publications. *Technovation* **2008**, *28*, 758–775. [CrossRef]
28. Rousseeuw, P.J. Silhouettes—A graphical aid to the interpretation and validation of cluster-analysis. *J. Comput. Appl. Math.* **1987**, *20*, 53–65. [CrossRef]
29. Yang, J.X.; Zheng, G.D.; Yang, J.; Wan, X.M.; Song, B.; Cai, W.; Guo, J.M. Phytoaccumulation of heavy metals (Pb, Zn, and Cd) by 10 wetland plant species under different hydrological regimes. *Ecol. Eng.* **2017**, *107*, 56–64. [CrossRef]
30. Amoo, A.E.; Babalola, O.O. Ammonia-oxidizing microorganisms: Key players in the promotion of plant growth. *J. Soil Sci. Plant Nutr.* **2017**, *17*, 935–947. [CrossRef]
31. Liu, Z.J.; Xie, H.J.; Hu, Z.; Zhang, J.; Zhang, J.D.; Sun, H.M.; Lan, W. Role of Ammonia-Oxidizing Archaea in Ammonia Removal of Wetland Under Low-Temperature Condition. *Water Air Soil Pollut.* **2017**, *228*, 1–11. [CrossRef]
32. Purkhold, U.; Pommerening-Röser, A.; Juretschko, S.; Schmid, M.C.; Koops, H.P.; Wagner, M. Phylogeny of all recognized species of ammonia oxidizers based on comparative 16S rRNA and amoA sequence analysis: Implications for molecular diversity surveys. *Appl. Environ. Microbiol.* **2000**, *66*, 5368–5382. [CrossRef] [PubMed]
33. Ritchie, G.A.F.; Nicholas, D.J. Identification of sources of nitrous-oxide produced by oxidative and reductive processes in *Nitrosomonas europaea*. *Biochem. J.* **1972**, *126*, 1181–1191. [CrossRef]
34. Stuven, R.; Vollmer, M.; Bock, E. The impact of organic-matter on nitric-oxide formation by *Nitrosomonas europaea*. *Arch. Microbiol.* **1992**, *158*, 439–443. [CrossRef]
35. Poughon, L.; Dussap, C.G.; Gros, J.B. Energy model and metabolic flux analysis for autotrophic nitrifiers. *Biotechnol. Bioeng.* **2001**, *72*, 416–433. [CrossRef]
36. Wrage, N.; Velthof, G.L.; van Beusichem, M.L.; Oenema, O. Role of nitrifier denitrification in the production of nitrous oxide. *Soil Biol. Biochem.* **2001**, *33*, 1723–1732. [CrossRef]
37. Colliver, B.B.; Stephenson, T. Production of nitrogen oxide and dinitrogen oxide by autotrophic nitrifiers. *Biotechnol. Adv.* **2000**, *18*, 219–232. [CrossRef]
38. Bai, J.H.; Wang, X.; Jia, J.; Zhang, G.L.; Wang, Y.Y.; Zhang, S. Denitrification of soil nitrogen in coastal and inland salt marshes with different flooding frequencies. *Phys. Chem. Earth* **2017**, *97*, 31–36. [CrossRef]
39. Wang, Y.L.; Wang, D.B.; Yang, Q.; Zeng, G.M.; Li, X.M. Wastewater Opportunities for Denitrifying Anaerobic Methane Oxidation. *Trends Biotechnol.* **2017**, *35*, 799–802. [CrossRef]
40. Chen, H.; Jin, R.C. Summary of the preservation techniques and the evolution of the anammox bacteria characteristics during preservation. *Appl. Microbiol. Biotechnol.* **2017**, *101*, 4349–4362. [CrossRef]
41. Roulet, N.T.; Lafleur, P.M.; Richard, P.J.H.; Moore, T.R.; Humphreys, E.R.; Bubier, J. Contemporary carbon balance and late Holocene carbon accumulation in a northern peatland. *Glob. Change Biol.* **2007**, *13*, 397–411. [CrossRef]
42. Picek, T.; Cízková, H.; Dusek, J. Greenhouse gas emissions from a constructed wetland -: Plants as important sources of carbon. *Ecol. Eng.* **2007**, *31*, 98–106. [CrossRef]
43. Lee, Y.J.; Romanek, C.S.; Wiegel, J. *Clostridium aciditolerans* sp nov., an acid-tolerant spore-forming anaerobic bacterium from constructed wetland sediment. *Int. J. Syst. Evol. Microbiol.* **2007**, *57*, 311–315. [CrossRef]
44. Conrad, R.; Klose, M.; Claus, P. Pathway of CH_4 formation in anoxic rice field soil and rice roots determined by ^{13}C-stable isotope fractionation. *Chemosphere* **2002**, *47*, 797–806. [CrossRef]
45. Le Mer, J.; Roger, P. Production, oxidation, emission and consumption of methane by soils: A review. *Eur. J. Soil Biol.* **2001**, *37*, 25–50. [CrossRef]
46. Chen, S.L.; Chen, J.F.; Chang, S.; Yi, H.; Huang, D.W.; Xie, S.G.; Guo, Q.W. Aerobic and anaerobic methanotrophic communities in urban landscape wetland. *Appl. Microbiol. Biotechnol.* **2018**, *102*, 433–445. [CrossRef]
47. Haroon, M.F.; Hu, S.H.; Shi, Y.; Imelfort, M.; Keller, J.; Hugenholtz, P.; Yuan, Z.G.; Tyson, G.W. Anaerobic oxidation of methane coupled to nitrate reduction in a novel archaeal lineage. *Nature* **2013**, *500*, 567–570. [CrossRef]
48. Ettwig, K.F.; Butler, M.K.; Le Paslier, D.; Pelletier, E.; Mangenot, S.; Kuypers, M.M.M.; Schreiber, F.; Dutilh, B.E.; Zedelius, J.; de Beer, D.; et al. Nitrite-driven anaerobic methane oxidation by oxygenic bacteria. *Nature* **2010**, *464*, 543–548. [CrossRef]
49. Feng, L.K.; He, S.F.; Yu, H.; Zhang, J.; Guo, Z.Z.; Wei, L.L.; Wu, H.M. A novel plant-girdling study in constructed wetland microcosms: Insight into the role of plants in oxygen and greenhouse gas transport. *Chem. Eng. J.* **2022**, *431*, 133911. [CrossRef]

50. Zhang, H.; Tang, W.Z.; Wang, W.D.; Yin, W.; Liu, H.L.; Ma, X.M.; Zhou, Y.Q.; Lei, P.; Wei, D.Y.; Zhang, L.T.; et al. A review on China's constructed wetlands in recent three decades: Application and practice. *J. Environ. Sci.* **2021**, *104*, 53–68. [CrossRef]
51. Zhang, K.; Wu, X.L.; Wang, W.; Luo, H.B.; Chen, W.; Chen, J. Effects of the bioelectrochemical technique on methane emission and energy recovery in constructed wetlands (CWs) and related biological mechanisms. *Environ. Technol.* **2023**, *44*, 540–551. [CrossRef] [PubMed]
52. Zhang, K.; Wu, X.L.; Wang, W.; Chen, J.; Chen, J.; Luo, H.B. Roles of external circuit and rhizosphere location in CH_4 emission control in sequencing batch flow constructed wetland-microbial fuel cell. *J. Environ. Chem. Eng.* **2021**, *9*, 106583. [CrossRef]
53. Zhang, K.; Wu, X.L.; Wang, W.; Luo, H.B.; Chen, W.; Ma, D.D.; Mo, Y.; Chen, J.; Li, L. Effects of plant location on methane emission, bioelectricity generation, pollutant removal and related biological processes in microbial fuel cell constructed wetland. *J. Water Process Eng.* **2021**, *43*, 102283. [CrossRef]
54. Liu, S.T.; Xue, H.P.; Wang, Y.; Wang, Z.; Feng, X.J.; Pyo, S.H. Effects of bioelectricity generation processes on methane emission and bacterial community in wetland and carbon fate analysis. *Bioresour. Bioprocess.* **2022**, *9*, 69. [CrossRef]
55. Zhu, H.; Niu, T.T.; Shutes, B.; Wang, X.Y.; He, C.G.; Hou, S.N. Integration of MFC reduces CH4, N2O and NH3 emissions in batch-fed wetland systems. *Water Res.* **2022**, *226*, 119226. [CrossRef]
56. Liu, S.T.; Xue, H.P.; Wang, M.X.; Feng, X.J.; Lee, H.S. The role of microbial electrogenesis in regulating methane and nitrous oxide emissions from constructed wetland-microbial fuel cell. *Int. J. Hydrog. Energy* **2022**, *47*, 27279–27292. [CrossRef]
57. Baddar, Z.E.; Xu, X.Y. Evaluation of changes in the microbial community structure in the sediments of a constructed wetland over the years. *Arch. Microbiol.* **2022**, *204*, 552. [CrossRef]
58. Cheng, S.Y.; Qin, C.L.; Xie, H.J.; Wang, W.X.; Zhang, J.; Hu, Z.; Liang, S. Comprehensive evaluation of manganese oxides and iron oxides as metal substrate materials for constructed wetlands from the perspective of water quality and greenhouse effect. *Ecotoxicol. Environ. Saf.* **2021**, *221*, 112451. [CrossRef]
59. Ji, B.H.; Chen, J.Q.; Li, W.; Mei, J.; Yang, Y.; Chang, J.J. Greenhouse gas emissions from constructed wetlands are mitigated by biochar substrates and distinctly affected by tidal flow and intermittent aeration modes. *Environ. Pollut.* **2021**, *271*, 116328. [CrossRef]
60. Gu, X.S.; Chen, D.Y.; Wu, F.; Tang, L.; He, S.B.; Zhou, W.L. Function of aquatic plants on nitrogen removal and greenhouse gas emission in enhanced denitrification constructed wetlands: Iris pseudacorus for example. *J. Clean. Prod.* **2022**, *330*, 129842. [CrossRef]
61. Liang, Z.H.; Hao, Q.J.; Hu, M.L.; Zhang, G.S.; Chen, K.Q.; Ma, R.Z.; Luo, S.X.; Gou, Y.X.; He, Y.J.; Chen, F.H.; et al. Application of alkali-heated corncobs enhanced nitrogen removal and microbial diversity in constructed wetlands for treating low C/N ratio wastewater. *Environ. Sci. Pollut. Res.* **2023**, *30*, 117624–117636. [CrossRef] [PubMed]

MDPI AG
Grosspeteranlage 5
4052 Basel
Switzerland
Tel.: +41 61 683 77 34

Water Editorial Office
E-mail: water@mdpi.com
www.mdpi.com/journal/water

www.ingramcontent.com/pod-product-compliance
Lightning Source LLC
Chambersburg PA
CBHW061437210926
51850CB00008B/1536